Plant Breeding: Theory and Techniques

Plant Breeding: Theory and Techniques

Edited by **Edgar Crombie**

New York

Published by Syrawood Publishing House,
750 Third Avenue, 9th Floor,
New York, NY 10017, USA
www.syrawoodpublishinghouse.com

Plant Breeding: Theory and Techniques
Edited by Edgar Crombie

© 2016 Syrawood Publishing House

International Standard Book Number: 978-1-68286-055-7 (Hardback)

This book contains information obtained from authentic and highly regarded sources. Copyright for all individual chapters remain with the respective authors as indicated. All chapters are published with permission under the Creative Commons Attribution License or equivalent. A wide variety of references are listed. Permission and sources are indicated; for detailed attributions, please refer to the permissions page and list of contributors. Reasonable efforts have been made to publish reliable data and information, but the authors, editors and publisher cannot assume any responsibility for the validity of all materials or the consequences of their use.

The publisher's policy is to use permanent paper from mills that operate a sustainable forestry policy. Furthermore, the publisher ensures that the text paper and cover boards used have met acceptable environmental accreditation standards.

Trademark Notice: Registered trademark of products or corporate names are used only for explanation and identification without intent to infringe.

Printed in the United States of America.

Contents

	Preface	IX
Chapter 1	Decreased row spacing as an option for increasing maize (*Zea mays* L.) yield in Trans Nzoia district, Kenya Owino Charles Onyango	1
Chapter 2	Line x tester analysis across locations and years in Sudanese x exotic lines of forage sorghum Maarouf I. Mohammed	4
Chapter 3	The somatic embryogenesis and plant regeneration from immature embryo of sweet corn inbred line Pitipong Thobunluepop	13
Chapter 4	Optimization and development of regeneration and transformation protocol in Indian mustard using lectin gene from chickpea [*Cicer arietinum* (L.)] V. V. Singh, Vandana Verma, Aniruddh K. Pareek, Monika Mathur, Rajbir Yadav, Poonam Goyal, Ajay Kumar Thakur, Y. P. Singh, K. R. Koundal, K. C. Bansal, A. K. Mishra, Arvind Kumar and Sandeep Kumar	19
Chapter 5	Identification of PCR-based DNA markers flanking three low phytic acid mutant loci in barley R. E. Oliver, C. Yang, G. Hu, V. Raboy and M. Zhang	24
Chapter 6	Intercropping and its application to banana production in East Africa: A review George Ouma	31
Chapter 7	A study on stigma receptivity of cytoplasmic-nuclear male-sterile lines of pigeonpea, *Cajanus cajan* (L.) Millsp. R. H. Luo, V. A. Dalvi, Y. R. Li and K. B. Saxena	34
Chapter 8	Grain filling rate is limited by insufficient sugar supply in the large-grain wheat cultivar Guohua Mi, Fanjun Chen and Fusuo Zhang	38
Chapter 9	Physicochemical and functional characteristics of cassava starch in Ugandan varieties and their progenies Ephraim Nuwamanya, Yona Baguma, Naushad Emmambux, John Taylor and Rubaihayo Patrick	43
Chapter 10	Combining ability for maize grain yield in *striga* endemic and non-endemic environments of the southern guinea savanna of Nigeria G. Olaoye and O. B. Bello	54

Chapter 11 **Seed germination of java plum (*Syzigium cumnii*) in three provenances western Kenya** 62
J. L. Okuto and G. Ouma

Chapter 12 **Structural features of a cytoplasmic male sterility source from *Helianthus resinosus*, CMS RES1** 72
F. Ardila, M. M. Echeverría, R. Rios and R. H. Rodríguez

Chapter 13 **Behavioural pattern of upland rice agronomic parameters to variable water supply in Nigeria** 77
Christopher O. Akinbile

Chapter 14 **Assessment of dissimilar gamma irradiations on barley (*Hordeum vulgare* spp.)** 85
S. Sarduie-Nasab, G. R Sharifi-Sirchi and M. H. Torabi-Sirchi

Chapter 15 **Grain yield and yield components of maize (*Zea mays* L.) as affected by crude oil in soil** 90
O. M. Agbogidi

Chapter 16 **Breeding for improved organoleptic and nutritionally acceptable green maize varieties by crossing sweet corn (*Zea mays saccharata*): Changes in quantitative and qualitative characteristics in F_1 hybrids and F_2 populations** 94
G. Olaoye, O. B. Bello, A. K. Ajani and T. K. Ademuwagun

Chapter 17 **Variability in sucrose content at grand growth phase in tissues of *Saccharum officinarum* × *Saccharum spontaneum* inter-specific hybrid progeny** 102
Vandana Vinayak, A. K. Dhawan and V. K. Gupta

Chapter 18 **Evaluation of head yield and participatory selection of horticultural characters in cabbage (*Brassica oleraceae* var. *capitata*)** 110
O. T. Adeniji, I. Swai, M. O. Oluoch, R. Tanyongana and A. Aloyce

Chapter 19 **Effect of different sowing dates on the yield and yield Components of direct seeded fine rice (*Oryza sativa* L.)** 118
Nadeem Akbar, Asif Iqbal, Haroon Zaman Khan, Muhammad Kashif Hanif and Muhammad Usman Bashir

Chapter 20 **Combining the yield ability and secondary traits of selected cassava genotypes in the semi-arid areas of Eastern Kenya** 122
Joseph Kamau, Rob Melis, Mark Laing, John Derera, Paul Shanahan and Eliud Ngugi

Chapter 21 **Evaluation of intra and interspecific rice varieties adapted to valley fringe conditions in Burkina Faso** 133
M. Sié, S. A. Ogunbayo, D. Dakouo, I. Sanou, Y. Dembélé, B. N'dri, K. N. Dramé, K. A. Sanni, B. Toulou and R. K. Glele

Chapter 22 **Biochemical and molecular characterization of submergence tolerance in rice for crop improvement** 144
Bishun D. Prasad, Ganesh Thapa, Samindra Baishya and Sangita Sahni

Chapter 23 **Participatory selection and characterization of quality protein maize (QPM) varieties in Savanna agro-ecological region of DR-Congo** 155
K. Mbuya, K. K. Nkongolo, A. Kalonji-Mbuyi and R. Kizungu

Chapter 24	**Effects of Gibberellic acid and 2,4-dichlorophenoxyacetic acid spray on fruit yield and Quality of tomato (*Lycopersicon esculentum* Mill.)** Dandena Gelmesa, Bekele Abebie and Lemma Desalegn	163
Chapter 25	**Studies on effectiveness and efficiency of gamma rays, EMS and their combination in soybean [*Glycine max* (L.) Merrill.]** Mudasir Hafiz Khan and Sunil Dutt Tyagi	172
Chapter 26	**Indigenous pest and disease management practices in Traditional farming systems in north east India. A review** Gopal Kumar Niroula Chhetry and Lassaad Belbahri	176
Chapter 27	**Comparative effects of water deficit on *Medicago laciniata* and *Medicago truncatula* lines sampled from sympatric populations** Mounawer Badri, Soumaya Arraouadi, Thierry Huguet and Mohamed Elarbi Aouani	187

Permissions

List of Contributors

Preface

I am honored to present to you this unique book which encompasses the most up-to-date data in the field. I was extremely pleased to get this opportunity of editing the work of experts from across the globe. I have also written papers in this field and researched the various aspects revolving around the progress of the discipline. I have tried to unify my knowledge along with that of stalwarts from every corner of the world, to produce a text which not only benefits the readers but also facilitates the growth of the field.

Plant breeding has revolutionized agriculture through genetic modifications in plants to get desirable characteristics or species. It emphasizes on development of high-productivity crops either through simple propagation or by making changes at genetic level. This book is compiled in such a manner, that it will provide in-depth knowledge about the theory and practice of crop improvement, pollen behavior and food security. The various studies that are constantly contributing towards advancing technologies and evolution of this field are examined in detail. As this field is emerging at a rapid pace, the contents of this book will help the readers understand and analyze the modern concepts and applications of the subject.

Finally, I would like to thank all the contributing authors for their valuable time and contributions. This book would not have been possible without their efforts. I would also like to thank my friends and family for their constant support.

Editor

Decreased row spacing as an option for increasing maize (Zea mays L.) yield in Trans Nzoia district, Kenya

Owino Charles Onyango

Kenya Plant Health Inspectorate Service. P.O Box 249, Kitale, Kenya. E-mail: onyangochrls@yahoo.com.

Maize yield trend has been declining in recent past. Farmers in Trans Nzoia District rarely intercrop maize with other crops like beans. With good nutrition and favourable weather conditions, decreased maize row spacing can maximise maize production per unit land area by increasing plant population density, optimal light interception and nutrient uptake. The experiment was carried out during the long rain season (April - September) for two successive years starting 2006. There were significant treatment differences (at 5%) between row spacing, varieties and interaction between row spacing and varieties. The mean yield increased with decreasing row spacing. Decreased row spacing combined with improved maize varieties is a possibility of increasing maize yield in Trans Nzoia District.

Key words: Row spacing, maize varieties, yield, interaction.

INTRODUCTION

The significance of maize (Zea mays L.) in households in Trans Nzoia District cannot be overemphasised (Onyango et al., 2000). In fact, Nyamangara et al. (2003) reported that the smallholder cropping in much of southern and eastern Africa is based on maize, the staple food crop. Other than being staple food crop, maize is a cash crop as well as source of employment both at farm and industry levels thereby directly or indirectly affecting livelihoods of many people in the District. Unfortunately, maize yield trend has been declining in the recent past. The cost of production of this crop is often in excess of the accrued cash returns thereby discouraging its production. This has negatively impacted not only on the people of Trans Nzoia but also those beyond the District since Trans Nzoia is a net exporter of maize, stereotyped 'grain basket' of Kenya. The declining yield trend has been partly attributed to increased human population against non-expansible land as a natural resource. The increased population pressure on land has caused subdivision of large tracks of agricultural farmland into individual small parcels for human settlement thereby reducing land area under arable agriculture. To sustain this increased population, it is only wise to increase the productivity of the remaining farmland.The blanket traditional maize inter row spacing of 75 cm has been used indiscriminately since time immemorial, without taking into account the myriad morphological and genetic differences that exist between and among maize varieties. Moreover, farmers in Trans Nzoia District rarely intercrop maize with other crops like beans. It is hypothesised that with good nutrition and water supply, decreased maize row spacing can maximise yield per unit land area by increasing plant population density, optimum light interception and nutrient uptake. It is against this background of understanding that this experiment was conceived: to explore reduced row spacing as an economically viable, ecologically non-degrading and socially acceptable cultural practice that can enhance yield per unit land area in Trans Nzoia, thereby leading to increased food self-sufficiency, food security, employment, industrial raw material and possibly increased foreign exchange earnings.

MATERIALS AND METHODS

Plant material and experimental design

The experiment was conducted at KEPHIS-Kitale Regional Office farm. The treatments included three inter row spacing (75, 60 and 50 cm) and some five common late maturity maize varieties (H 614D, H 6213, H 9401, H 628 and H 629) at a standard intra-row spacing of 25 cm. This was a factorial experiment laid out in a complete randomised block design (CRBD) with row spacing being main factor and variety as sub-factor, giving a total of 15 treatments. The block was folded 3 times. The treatments were replicated three times.

Each plot measured 5 m × 3 m. Planting was done at the onset of rains each season. Two seeds were planted in each hole and later thinned to one plant per hill soon after emergence. All agronomic

Table 1. Analysis of variance for mean treatment effects (ANOVA).

Source	df	SS	MS	F value	P>F
Year	1	0.0001111	0.0001111	0.00	0.9959[ns]
Rep (year)	4	21.2859259	5.3214815	1.32	0.2852[ns]
Row spacing	2	86.1942963	43.0971481	10.6	0.0003**
Rep × row spacing (year)	8	64.8197037	8.1024630	2.00	0.0782[ns]
Year × row spacing	2	0.0000741	0.0000370	0.00	1.0000[ns]
Variety	4	56.6240247	14.1560062	3.50	0.0177*
Rep × variety (year)	16	64.1758025	4.0109877	0.99	0.4883[ns]
Year × variety	4	0.0032840	0.0008210	0.00	1.0000[ns]
Row spacing × variety	8	105.565822	13.1957284	3.26	0.0079*
Year × row spacing × variety	8	0.0030123	0.0003765	0.00	1.0000[ns]

KEY: Ns, not significant at 5%; *, Significant at 5%; **, Significant at 1% (highly significant); CV = 19.0%.

Table 2. Probabilities of row spacing differences.

		Probabilities		
Row spacing	Yield (KG)	50 cm	60 cm	75 cm
50 cm	11.66[a]	.	0.2283	0.0089
60 cm	10.76[a]		.	0.0659
75 cm	9.29[b]			.

practices were uniformly applied across the treatments. At physiological maturity, plots were harvested separately and yield subdivided into clean and rotten/sprouted cobs. The yield was shelled, dried to 13.5% kernel moisture content and weighed. The experiment was conducted for 2 consecutive years beginning 2006. Data was statistically analysed using SAS computer package (SAS, 1998).

RESULTS AND DISCUSSION

There were significant treatment differences between row spacing, variety and interaction between row spacing and variety (Table 1).

The data presented is for two years combined, each time the treatments are being replicated thrice.

Effect due to row spacing

The mean yield increased with decreasing row spacing. However, there were no significant differences in yield between 50 and 60 cm row spacing (Table 2). Decreased row spacing implies high plant density, which is concomitantly equal to high yield with every successful ear formation per plant.

It also improves water use efficiency since evaporation losses are reduced as ground cover increases. Moreover, the dense crop canopy smothers weeds thereby reducing resource competition. This finding is in contrast to research findings in Argentina by Maddonni et al. (2006) which shows that maize grain yield was stable in response to changes in plant spatial arrangement at all plant population densities. Tollenaar et al. (2006) in their research finding argued that a moderate increase in plant-spacing variability does not influence maize grain yield at the canopy level because reductions in grain yield of plants that experience enhanced crowding stress is compensated, in part, by increased yield of plants that experience reduced crowding stress.

However, it is worth mentioning that decreasing row spacing has socio economic implications: high plant population densities means upward adjustment of the amount of agro inputs used (seed rate and fertilizer). Manual weeding, harvesting and other agronomic maintenance operations would take more labour and time, as it is difficult working through the dense crop stand.

Effect due to variety

New varieties performed better than the old varieties, with H 6213 being the best and H 614D giving the least yield (Table 3). This is in agreement with findings by Owino (unpublished data). The findings reaffirm breeders' commitment to produce new and improved high yielding varieties. However, there were no significant yield differences among H 6213, H 9401 and H 628 on one hand and between H 614D and H 629 on the other hand. The challenge is to convince farmers to adopt these new varieties and drop the old ones.

Conclusion and Recommendation

The observed increased maize grain yield under decreased row spacing may be attributed to improved intercepted photosynthetic active radiation (IPAR), radiation use efficiency (RUE) and azimuth leaf distribution of a genotype so long as critical leaf area index (LAI) is not exceeded. "Plastic genotypes" re-orientate their leaves in horizontal plane to fill empty space inorder to maximise photo-interception as opposed to "rigid genotypes" that present random leaf azimuth independently of spatial

Table 3. Probabilities of variety differences.

Variety	Yield	Probabilities				
		H 6213	H 9401	H 628	H 629	H 614D
H 6213	11.34^a	.	0.7157	0.7016	0.0911	0.0059
H 9401	11.09^a		.	0.9848	0.1727	0.0128
H 628	11.07^a			.	0.1783	0.0133
H 629	10.13^{ab}				.	0.1881
H 614D	9.21^b					.

arrangement. Decreasing row spacing seems to be an alternative that can be used to intensify crop production per unit land area. However, varieties are likely to perform differently under different planting densities owing to their different genetic and phenotypic characteristics. More work should be done to develop and identify varieties that are best suited to closer row spacing with high radiation use efficiency. The experiment was done in one site and needs to be repeated in several sites to confirm the so far observed trend.

ACKNOWLEDGEMENT

The author is indebted to Kenya Plant Health Inspectorate Service (KEPHIS) with whose resources the study was conducted.

REFERENCES

Maddonni GA, Alfredo GC, Otegui ME (2006). Row width and maize grain yield. Agron. J. 98: 1532-1543.
Nyamangara J, Bergstron LF, Piha MI, Giller KE (2003). Fertilizer Use Efficiency and Nitrate Leaching in a tropical Sandy Soil. J. Environ.Qual., 32: 599-606.
Onyango RMA, Mwangi TJ, Wanyonyi M, Barkuto JK, Lunzalu EN (2000). Verifying the Potential Use of Inorganic and organic Fertilizers and their Combinations in Small Holder Maize Production Farms in Trans Nzoia District, KARI-NARC-Kitale.
Owino CO (Unpublished). Fertilizer Options for Sustainable Maize Production in Trans Nzoia District-Kenya
SAS Institute (1998). SAS/STAT Users guide Version 7. ed SAS Inst.Cary, NC USA.
Tollenaar M, Deen W, Echarte L, Weidong L (2006). Effect of Crowding Stress on Dry Matter Accumulation and Harvest Index in Maize. Agron. J. 98: 930-937.

Line x tester analysis across locations and years in Sudanese x exotic lines of forage sorghum

Maarouf I. Mohammed

Shambat Research Station, ARC, P.O. Box 30, Khartoum North. Sudan. E-mail: ibrahimarof@yahoo.com.

Combining ability in forage sorghum [*Sorghum bicolor* (L.) Moench] was not as much investigated as in grain sorghum. In the present study, 4 local stocks (Testers) and 7 exotic stocks in A3 cytoplasm were crossed in line x tester fashion to investigate combining ability in forage sorghum. The hybrids and their parents were evaluated across two years (2002 - 2003) and at two locations in Khartoum State, Sudan. Forage yield, days to flower, plant height, leaf to stem ratio and stem diameter were studied. Significant general (GCA) and specific combining ability (SCA) effects in desirable direction were detected for most traits. The best general combiners for forage yield and earliness in flowering were identified. Both additive and non-additive gene actions were important in the expression of all characters with the preponderance of additive actions for days to flower, forage yield, stem diameter, leaf to stem ratio and non-additive actions for plant height. GCA effects were more stable over years than specific ones. Selection in early generations was suggested for characters predominately controlled by additive genes. Heterosis breeding was recommended for forage yield improvement. For leaf to stem ratio, selection must be based on more genetically diverse materials.

Key words: A3 cytoplasm, additive, ankolib, combining ability, islang, hybrid sorghum, shambat.

INTRODUCTION

Forage sorghum [*Sorghum bicolor* (L). Moench] has recently witnessed an increasing importance in the semi arid tropics and drier parts of the world where livestock constitutes a major component of the production system. Compared to other cereals, especially maize, sorghum is more droughts tolerant, less input demanding and can thrive better under harsh conditions. Most of sorghum improvement programs are grain oriented. Improvement for non-grain attributes has been limited. Kelley et al. (1991) questioned the current strategy of strictly adopting grain-yield criteria in evaluating sorghum genotypes, arguing that fodder's contribution to the total value of sorghum production has increased considerably. They reported that the grain/straw price ratio of sorghum has dropped from 6:1 in 1970 to 3:1 in 1990 and is likely to decline further. In the Sudan, where the second largest animal wealth in Africa exists, forage sorghum constitutes the bulk of the animal feed in the country (Mohammed, 2007). Very little or no attempts have been made to develop improved forage types. The first fully devoted forage improvement program in the country was initiated in 2000 (Mohammed et al., 2008). One of the program objectives was to develop locally adapted forage sorghum hybrids. Knowing general (GCA) and specific (SCA) combing ability effects of genetic materials is of practical value in breeding programs. Both components play an important role in selecting superior parents for hybrid combinations (Duvick, 1999) and represent a powerful method to measure the nature of gene action involved in quantitative traits (Baker, 1978). GCA effects represent the fixable component of genetic variance, and are important to develop superior genotypes. SCA represents the non-fixable component of genetic variation, it is important to provide information on hybrid performance. Most of our present information about combining ability in sorghum was based on studies carried under temperate environments with materials limited to photoperiod conversion program (Maunder, 1992). The objectives of this study were to investigate combining ability for some agronomic traits in introduced and local forage sorghum genetic stocks using line x tester analysis to identify parents with desirable GCA effects and cross combinations with desirable SCA effects and to study the nature of gene action involved in

Table 1. Genetic stock designation, recurrent parent and cytoplasm source of the seven female parents used as lines in the study.

Genetic stock	Recurrent parent	Cytoplasm source	Pericarp color	Mid-rib color
A3N166	Blue Ribbon	A3Tx 398	brown	green
A3N168	Hastings	A3Tx 398	brown	green
A3N169	E-35-1	A3Tx 430	white	green
A3N159	N 100	A3Tx 398	brown	green
A3N173	N 109	A3Tx 398	white	green
A3N154	Sugar Drip	A3Tx 398	brown	green
A3N151	Dale	A3Tx 398	brown	green

yield and related traits in forage sorghum.

MATERIALS AND METHODS

Plant materials

Seven forage sorghum genetic stocks in A3 cytoplasm chosen from the materials received from J. F. Pedersen, USDA-ARS, USA were used as females (Lines) in this study. They include: Blue Ribbon, Hastings, Sugar Drip, Dale, N100, E-35-1, and N109. Table 1 reflects the genetic stock designation, recurrent parent and cytoplasm source of the seven selected females. The males (Testers) comprised four local genetic stocks, two of which, namely: S.70 and S. 186 represent the two major types of the traditional cultivar 'Abu Sab'in', known as Alyab and Rubatab, respectively. The other two were: 'Garawi', a cultivated forage type of Sudan Grass (*Sorghum sudanense* (Piper) Stapf) and 'Ankolib', a local sweet sorghum cultivar. They are heterogeneous land race cultivars with broad genetic base, desirable for providing information about the general combining ability of a line. Abu Sab'in selections, on the other hand, are expected to show good performance in specific hybrid combinations with the selected lines.

The experiment

The seven lines and the four testers were grown together with their 28 hybrids for two years (2002 and 2003) and at two locations in Khartoum State, namely: Shambat (lat.15° 39' N; long. 32° 31' E) and Islang (lat.15° 53' N; long. 32° 32' E). The soil at Shambat site is heavy clay with pH 8.5. The physical properties of Islang soil varies from silty clay to silty loam. The growing season of the year 2002 compared to that of 2003 was characterized by increased maximum temperature, reduced total rain fall and lower relative humidity. In the year 2002, sowing date was on the 12th and 25th of July at Shambat and Islang, respectively, while in the year 2003, sowing date was on the 24th of June and 11th of July at Shambat and Islang, respectively. Apart from that, other methods and materials were similar for different years and locations. The treatments were arranged in a randomized complete block design with three replicates. The plot size was 7.5 x 0.7 m ridge. Three to four seeds were sown in holes spaced at 10 cm along the ridge. The plants were later thinned to one plant per hole. The experiment was watered every 7 - 10 days. Harvesting was carried out 15 days after each entry had completed 50% flowering, which simulates the common farmer practice.

Green matter yield (GMY) was estimated from 6.5 m harvested from each plot leaving 0.5 m from each side. The dry matter yield (DMY) was estimated from a random sample of 0.5 kg taken from the harvested plot after determining GMY and oven-dried at 75 °C for 48 h. Yield related traits include: Days to flower, plant height, stem diameter and leaf to stem ratio.

Statistical analysis

Single analysis of variance was performed for all characters prior to combine analysis. Line x tester analysis was performed based on data combined over years and locations. Source of variation due to entry and its interaction with year and location were subdivided into variations due to hybrids and parents. Similarly, the hybrid source of variation was partitioned into variations due to lines, testers and line x testers. Estimates of general (GCA) and specific (SCA) combining ability based on data combined over years and locations were worked out following the procedure of Biel and Atkins (1967) which is comparable to the analysis of a two-way classification model with interaction component being a measure of the SCA effects. Estimate of GCA of a tester (male) was obtained in terms of its performance in F1 hybrid combinations with all possible lines (females). Likewise, GCA of a line was determined in terms of its performance in F1 hybrid combinations with all possible testers. The lines and testers were considered as fixed effects. Years and location were considered as random effects. GCA and SCA effects were determined for each trait as follows:

GCA lines (L) = $\bar{X}_{j\cdot} - \bar{Y}$
GCA tester (T) = $\bar{X}_{i\cdot} - \bar{Y}$
SCA (L x T) = $\bar{X}_{ij} - \bar{X}_{j\cdot} - \bar{X}_{i\cdot} - \bar{Y}$,

Where:

$\bar{X}_{j\cdot}$ = the mean of hybrid with a given line (female) averaged over all replications, years, locations and testers (males),

$\bar{X}_{i\cdot}$ = the mean of hybrid with a given tester (male) averaged over all replications, years, locations and lines (females),

$\bar{X}_{ij}$ = the mean of a given hybrid (L x T) averaged over replications, years and locations,

$\bar{Y}$ = the experimental mean.

Standard errors (SE) for general and specific combining ability were calculated following Groz et al. (1987) as follows:

$SE_{Lines} = (M_{fyl}/rmyl)^{1/2}$, $SE_{Testers} = (M_{myl}/rfyl)^{1/2}$, and $SE_{Line \times Tester} = (M_{fmyl}/ryl)^{1/2}$,

Where

M_{fyl} and M_{myl} are the respective mean squares of line x year x location and tester x year x location divided by number of observations (replicates, years, locations, males or females). M_{fmyl}

Table 2. Mean squares from combined ANOVA for green (GMY), dry (DMY) matter yield and yield related-traits of 28 forage sorghum hybrids and their parents tested over 2 years (2002-03) and 2 locations (Shambat, Islang).

Source of variation	d.f.	Mean squares					
		GMY (t/ha)	DMY (t/ha)	Days to flower	Plant height (cm)	Stem diameter (cm)	Leaf/stem ratio (%)
Year (Yr)	1	436.70 **	1.668 NS	2239.59 **	1868.29 **	0.799 **	204.719 **
Location (Lo)	1	1135.55 **	16.193 **	208.154 **	19192.88 **	1.801 **	1569.0 **
Yr x Lo	1	9.798 NS	0.708 NS	49.51 **	4598.93 **	0.006 NS	20.459 NS
Rep (Lo x Yr)	8	76.629 *	2.194 **	65.418 **	9975.56 **	0.027 NS	121.118 **
Entry (E)	38	854.685 **	15.247 **	728.104 **	8885.22 **	0.412 **	183.739 **
Parent (P)	10	618.172 **	10.445 **	838.553 **	13064.32 **	0.488 **	384.088 **
Hybrid (H)	27	364.729 **	8.188 **	701.545 **	892.809 **	0.282 **	63.359 **
P vs H	1	16448.65 **	253.87 **	340.69 **	182889.5 **	3.157 **	1430.49 **
Yr x E	38	71.512 **	1.208 **	28.654 **	752.389 **	0.060 **	19.696 *
Yr x H	27	73.063 **	1.313 *	25.930 **	944.406 **	0.057 **	22.027 *
Yr x P	10	56.451 **	0.979 **	36.296 **	255.725 NS	0.027 **	15.291 *
Lo x E	38	2.059 NS	0.047 NS	0.958 NS	57.063 NS	0.002 NS	2.936 NS
Lo x H	27	1.588 NS	0.050 NS	0.477 NS	36.683 NS	0.002 NS	1.824 NS
Lo x P	10	3.034 NS	0.045 NS	2.061 NS	105.924 NS	0.002 NS	3.382 NS
Yr x Lo x E	38	3.320 NS	0.066 NS	1.154 NS	54.654 NS	0.006 NS	4.501 NS
Yr x Lo x H	27	1.997 NS	0.049 NS	0.460 NS	24.475 NS	0.003 NS	0.761 NS
Yr x Lo x P	10	6.842 NS	0.092 NS	2.212 NS	111.420 NS	0.012 NS	11.161 NS
Error	304	31.530	0.671	6.486	253.073	0.016	12.122

*, **: Significant at 0.05 and 0.01 probability level respectively. NS: Non-significant at 0.05 probability level.

is the mean square for (line x tester) x year x location divided by number of observations (replicates, years, locations).

The critical difference (C.D.) was calculated as follows: C.D. = SE x t (tabulated). If the absolute effect of GCA or SCA is greater than the C.D., it is considered significantly different from zero. Data analysis was performed using the statistical package of GenStat (2006).

RESULTS

Table 2 reveals that the entries and their sub-sources of variation (parents, hybrids, and parents vs hybrids) differed significantly ($p < 0.01$) for all characters. Their interactions with years, unlike those with locations, were significant for most characters. Table 3 shows that differences among lines, testers and line x tester were significant ($p < 0.01$) for all characters. The interaction of lines with years was significant ($p < 0.01$) for days to flower, plant height and stem diameter and that of testers was significant ($p < 0.05$) for DMY and plant height. The line x tester interaction with years, unlike that with locations, was significant for all characters.

General and specific effects

Tables 4 and 5 show that significant GCA effects were expressed by some lines and testers in nearly all characters. The exceptions being plant height for both lines and testers and leaf to stem ratio for lines (Table 5). More significant cases were displayed by testers compared to lines, with significant cases being more frequent for number of days to flower. Table 4 shows that the highest significant ($p < 0.01$) positive GCA effects for yield was expressed by E-35-1 and S.70 among lines and testers, respectively. Yield ranking indicated that both entries were among the top yielders. Positive, but insignificant, GCA effects were shown by the line Dale and the tester Ankolib. For days to flower, where negative effects are desirable, Blue Ribbon from lines and S.186 from testers showed the highest significant ($p < 0.01$) negative GCA effects, followed by N 109 and Garawi from lines and testers respectively (Table 5). Positive significant ($p < 0.01$) GCA effects were expressed by the line E-35-1 and the tester Ankolib. For leaf to stem ratio, Garawi, among testers, gave the highest significant ($p < 0.01$) positive GCA effect followed by Ankolib. No significant GCA effects were displayed by lines for leaf to stem ratio. For Plant height, no significant GCA effects were displayed by both lines and testers.

Table 6 shows that significant ($p < 0.01$) positive SCA effect for GMY was shown by 6 hybrids, of which Sugar Drip x Ankolib, and Blue Ribbon x Ankolib scored the highest SCA estimates. Yield ranking indicated that hybrids with significant positive SCA effects were also among the best in per se performance. However, the hybrid E-35-1 x S.70, which was the top yielder in the whole material, showed insignificant negative SCA effects.

Table 3. Mean squares from line x tester analysis based on data combined over 2 years and 2 locations for green (GMY), dry (DMY) matter yield and yield-related traits of 28 forage sorghum hybrids.

Source of variation	d.f.	Mean squares					
		GMY (t/ha)	DMY (t/ha)	Days to flower	Plant height (cm)	Stem diameter (cm)	Leaf/stem ratio (%)
Line (L)	6	796.872 **	20.309 **	1401.07 **	1094.33 **	0.584 **	42.740 **
Tester (T)	3	746.501 **	18.101 **	2795.20 **	1150.74 **	0.800 **	288.405 **
L x T	18	157.164 **	2.495 **	112.302 **	783.923 **	0.092 **	32.893 **
Yr x L	6	63.994 NS	0.758 NS	28.563 **	1126.39 **	0.097 **	3.818 NS
Yr x T	3	74.368 NS	2.104 *	14.283 NS	895.652 *	0.040 NS	13.478 NS
Lo x L	6	1.006 NS	0.054 NS	0.323 NS	58.491 NS	0.002 NS	2.246 NS
Lo x T	3	1.208 NS	0.011 NS	0.243 NS	42.494 NS	0.001 NS	1.211 NS
Yr x Lo x L	6	4.669 NS	0.072 NS	0.693 NS	43.438 NS	0.004 NS	1.212 NS
Yr x Lo x T	3	0.903 NS	0.024 NS	2.262 NS	12.644 NS	0.003 NS	0.825 NS
Yr x L x T	18	75.708 **	1.372 *	26.458 **	895.494 **	0.046 **	29.557 **
Lo x L x T	18	1.848 NS	0.054 NS	0.968 NS	27.930 NS	0.002 NS	1.760 NS
Yr x Lo x L x T	18	1.268 NS	0.045 NS	0.684 NS	19.778 NS	0.002 NS	0.599 NS
Error	216	35.207	0.761	7.997	291.41	0.018	14.060

*, **: Significant at 0.05 and 0.01 probability level respectively. NS. : Non-significant at 0.05 probability level.

Table 4. Estimates of general combining ability (GCA) in forage sorghum for green (GMY) and dry (DMY) matter yield based on data combined over years and locations.

Parents	GMY			DMY		
	GCA_{Lines}	(t/ha)	Rank	GCA_{Lines}	(t/ha)	Rank
Lines (Females)						
E-35-1	2.953**	30.89	3	0.474**	4.28	3
Hastings	- 0.234	22.41	6	- 0.116	2.74	8
Blue Ribbon	- 0.294	21.73	8	- 0.115	2.49	10
N 109	- 0.816	16.51	11	- 0.048	2.06	11
Dale	0.047	20.12	10	0.036	2.78	7
N 100	- 1.117*	22.08	7	- 0.136*	3.00	5
Sugar Drip	- 0.538	20.47	9	- 0.095	2.61	9
S.E. $_{GCA\ Lines}$	0.312			0.039		
Testers (Males)	$GCA_{Testers}$			$GCA_{Testers}$		
S.70	1.172**	38.59	1	0.204**	4.91	1
S.186	- 0.603**	37.02	2	- 0.094*	4.48	2
Garawi	- 1.013**	28.27	4	- 0.143**	3.53	4
Ankolib	0.445	23.49	5	0.033	2.88	6
S.E. $_{GCA\ Testers}$	0.104			0.017		

*, **: Significantly different from zero at 0.05 and 0.01 probability level respectively.

Table 7 shows that 6 hybrids expressed significant (p < 0.01) negative SCA effects for days to flower 3 of which involved the line Hastings. For leaf to stem ratio, significant positive specific effects were displayed by 2 hybrids, namely, N100 x Ankolib and E-35-1 x Ankolib.

GCA and SCA variance estimates

Table 8 shows that, for all characters, variances of the main effects were significant (p < 0.01) and mostly higher in magnitude than the interaction effects. However, inte-

Table 5. Estimates of general combining ability (GCA) in forage sorghum for yield-related traits based on data combined over years and locations.

Parents	Days to flower	Plant height (cm)	Stem diameter (cm)	Leaf/stem ratio (%)
GCA Lines (Females)				
E-35-1	3.793**	1.943	0.071**	- 0.099
Hastings	0.392	- 0.173	0.020	- 0.373
Blue Ribbon	- 1.570**	0.853	- 0.014	- 0.415
N 109	- 1.252**	- 0.858	- 0.029	0.079
Dale	- 0.054	1.259	- 0.017	0.212
N 100	- 0.547*	- 1.941	- 0.036*	0.446
Sugar Drip	- 0.763**	- 1.084	0.005	0.150
S.E. GCA Lines	0.120	0.951	0.009	0.159
GCA Testers (Males)				
S.70	- 0.075	1.275	0.014	- 0.470**
S.186	- 1.521**	0.173	- 0.016	- 0.593**
Garawi	- 1.138**	- 0.161	- 0.035*	0.586*
Ankolib	2.734**	- 1.286	0.038*	0.478*
S.E. GCA Testers	0.164	0.388	0.006	0.099

*, **: Significantly different from zero at 0.05 and 0.01 probability level respectively.

Table 6. Estimates of specific combining ability (SCA) in forage sorghum for green (GMY) and dry (DMY) matter yield based on data combined over years and locations.

Hybrid			GMY			DMY	
		SCA	(t/ha)	Rank	SCA	(t/ha)	Rank
E-35-1	X S.70	- 0.354	50.1	1	- 0.084	6.68	1
E-35-1	X S.186	1.009*	48.9	2	0.042	6.17	4
E-35-1	X Garawi	1.242*	48.3	3	0.216*	6.54	2
E-35-1	X Ankolib	- 1.898**	43.3	7	- 0.174	5.88	5
Hastings	X S.70	1.016*	44.6	6	0.090	5.44	7
Hastings	X S.186	- 0.660	34.3	22	- 0.095	3.99	26
Hastings	X Garawi	0.896	39.2	11	0.034	4.23	21
Hastings	X Ankolib	- 1.252*	35.7	19	- 0.028	4.57	17
B.Ribbon	X S.70	- 0.871	38.8	12	- 0.129	4.78	13
B.Ribbon	X S.186	- 0.229	35.4	20	- 0.039	4.16	23
B.Ribbon	X Garawi	- 0.253	34.1	23	0.024	4.20	22
B.Ribbon	X Ankolib	1.352**	43.3	7	0.144	5.09	11
N 109	X S.70	- 0.720	37.7	15	- 0.083	5.12	10
N 109	X S.186	0.034	34.6	21	0.033	4.58	16
N 109	X Garawi	0.036	33.4	25	- 0.106	4.01	25
N 109	X Ankolib	0.651	39.6	10	0.156	5.33	8
Dale	X S.70	1.054*	45.6	4	0.219*	6.28	3
Dale	X S.186	0.491	38.6	13	0.075	4.96	12
Dale	X Garawi	- 0.761	33.6	24	- 0.085	4.33	20
Dale	X Ankolib	- 0.783	37.9	14	- 0.208*	4.49	18
N 100	X S.70	0.892	41.6	9	0.156	5.57	6
N 100	X S.186	- 0.866	31.0	27	- 0.113	3.88	27
N 100	X Garawi	0.280	33.2	26	0.027	4.15	24
N 100	X Ankolib	- 0.306	35.8	18	- 0.070	4.39	19
Sugar Drip	X S.70	- 1.016*	37.6	16	- 0.167	4.75	14
Sugar Drip	X S.186	0.221	36.0	17	0.096	4.63	15
Sugar Drip	X Garawi	- 1.441**	29.8	28	- 0.109	3.86	28
Sugar Drip	X Ankolib	2.236**	45.2	5	0.180	5.26	9
S.E.		0.325			0.061		

*, **: Significantly different from zero at 0.05 and 0.01 probability level respectively.

Table 7. Estimates of specific (SCA) combining ability in forage sorghum for yield-related traits based on data combined over years and locations.

Hybrid		Days to flower	Plant height (cm)	Stem diameter (cm)	Leaf/stem ratio (%)
E-35-1	X S.70	0.484	- 3.440	0.034	- 0.259
E-35-1	X S.186	0.497	4.059*	0.029	0.119
E-35-1	X Garawi	0.244	2.332	- 0.004	- 0.544
E-35-1	X Ankolib	- 1.225**	- 2.952	- 0.059**	0.684*
Hastings	X S.70	- 0.922*	2.287	- 0.001	- 0.109
Hastings	X S.186	- 0.922*	2.046	- 0.007	0.296
Hastings	X Garawi	- 1.059**	- 0.308	0.002	- 0.110
Hastings	X Ankolib	2.902**	- 4.025*	0.005	- 0.076
B.Ribbon	X S.70	- 0.183	- 1.383	- 0.013	0.398
B.Ribbon	X S.186	0.321	- 0.314	- 0.013	0.048
B.Ribbon	X Garawi	0.587	- 0.148	0.003	0.365
B.Ribbon	X Ankolib	- 0.725*	1.845	0.024	- 0.811*
N 109	X S.70	- 0.124	0.791	- 0.007	- 0.202
N 109	X S.186	0.239	- 1.867	- 0.008	0.056
N 109	X Garawi	0.332	0.070	0.011	0.314
N 109	X Ankolib	- 0.447	1.006	0.004	- 0.168
Dale	X S.70	0.931*	0.945	- 0.027	0.137
Dale	X S.186	- 0.019	1.543	0.014	- 0.872*
Dale	X Garawi	0.268	- 1.460	0.019	0.472
Dale	X Ankolib	- 1.181**	- 1.028	- 0.005	0.263
N 100	X S.70	- 0.463	2.911	0.031	- 0.362
N 100	X S.186	- 0.259	- 4.410*	- 0.008	- 0.125
N 100	X Garawi	- 0.323	1.123	0.006	- 0.455
N 100	X Ankolib	1.045**	0.376	- 0.029	0.942**
Sugar Drip	X S.70	0.276	- 2.112	- 0.017	0.397
Sugar Drip	X S.186	0.143	- 1.057	- 0.007	0.479
Sugar Drip	X Garawi	- 0.050	- 1.610	- 0.037	- 0.042
Sugar Drip	X Ankolib	- 0.369	4.779*	0.061**	- 0.834*
S.E.		0.239	1.284	0.013	0.223

*, **: Significantly different from zero at 0.05 and 0.01 probability level respectively.

raction of SCA variance ($\sigma^2 SCA_{LxT \times Y}$) with years was exceptionally sizable, especially for plant height. The variances of GCA for lines ($\sigma^2 GCA_{Lines}$) were higher than those for testers ($\sigma^2 GCA_{Tester}$) for most characters. The exception being days to flower and leaf to stem ratio. The interaction effects of $\sigma^2 GCA_{Lines}$ with years were higher than those of $\sigma^2 GCA_{Tester}$ for plant height, days to flower and stem diameter. Higher order interactions of SCA variance ($\sigma^2 SCA_{LxT \times Y \times LO}$) were considerably low for all characters. The variance ratio of general to specific effects ($\sigma^2 GCA / \sigma^2 SCA$) is above unity for all characters, except for plant height. The SCA variance for plant height was about more than three times greater than the sum of its GCA variance for line and tester. Number of days to flower showed the highest $\sigma^2 GCA / \sigma^2 SCA$ ratio compared to other characters. Table 9 shows that the contribution of lines was greater than that of testers for GMY, DMY, plant height and stem diameter. On the other hand, the contribution of testers was greater than that of lines for leaf to stem ratio. Equal contributions to the total variance were noticed for number of days to flower. The contribution of either lines or testers was greater than that of lines x testers for all characters with the exception of plant height

DISCUSSIONS

The data presented in Table 2 point to the high degree of genetic variability existing among parents and hybrids for all characters studied. The variability among hybrids was less than that among parents for all characters. The contrast of parents' vs hybrids was sizable and highly significant for all characters, pointing to the potential of heterotic effects among hybrids. Both hybrids and parents performed consistently over locations, but not

Table 8. Variance components and ratio for general (σ^2 GCA) and specific (σ^2 SCA) genetic effects, their interactions over years (Yr) and locations (LO) for green (GMY), dry (DMY) matter yield and yield-related traits based on data from 28 forage sorghum hybrids.

Variance components[#]	GMY (t/ha)	DMY (t/ha)	Days to flower	Plant height (cm)	Stem diameter (cm)	Leaf/stem ratio (%)
σ^2 GCA$_{Lines\,(L)}$	13.33**	0.371**	26.85**	6.467**	0.010**	0.205**
σ^2 GCA$_{Tester\,(T)}$	7.016**	0.186**	31.94**	4.367**	0.008**	3.042**
σ^2 SCA$_{LxT}$	10.16**	0.145**	8.692**	41.04**	0.006**	1.569**
σ^2 GCA$_{L \times YR}$	- 0.488	- 0.026	0.088**	9.621**	0.002**	- 1.072
σ^2 GCA$_{T \times YR}$	- 0.032	0.017*	- 0.290	0.004*	0.0001	- 0.383
σ^2 GCA$_{L \times LO}$	- 0.035	0.000	- 0.027	1.273	0.000	0.028
σ^2 GCA$_{T \times LO}$	- 0.015	- 0.001	- 0.017	0.347	0.000	- 0.013
σ^2 GCA$_{L \times YR \times LO}$	0.283	- 0.002	0.001	1.972	0.0002	0.051
σ^2 GCA$_{T \times YR \times LO}$	- 0.017	- 0.001	0.075	- 0.340	0.000	0.011
σ^2 SCA$_{LxT \times YR}$	12.41**	0.221*	4.296**	145.95**	0.007**	4.826**
σ^2 SCA$_{LxT \times LO}$	0.097	- 0.002	0.047	1.359	0.000	0.194
σ^2 SCA$_{LxT\,YR \times LO}$	- 11.31	- 0.239	- 2.438	- 90.54	- 0.005	- 4.487
Error mean square	35.207	0.761	7.997	291.41	0.018	14.06
σ^2 GCA / σ^2 SCA ratio	2.003	3.841	6.764	0.264	3.0	2.069

*, **: Significant at 0.05 and 0.01 probability level respectively.
: Negative component interpreted as zero.

Table 9. Contribution of lines, testers, and lines x testers to the total variance for six characters in forage sorghum based on data combined over years and locations.

Character	Contribution (%)		
	Lines	Testers	Lines x Testers
Green matter yield (t/ha)	48.6	22.7	28.7
Dry matter yield (t/ha)	55.1	24.6	20.3
Days to flower	44.4	44.3	10.7
Plant height (cm)	26.1	14.3	58.5
Stem diameter (cm)	46.0	31.5	21.8
Leaf/stem ratio (%)	15.0	50.6	34.6

over years with hybrids being more consistent in their performance over environments than parents. The testers were more variable than lines for most characters (Table 3). This is expected since they represent diverse groups of forage sorghum (grass, sweet and grain forage sorghums), whereas the lines represent one group (sweet sorghum). The interaction of lines with testers was highly significant for most characters indicating the presence of specific effects.

Given that testers are more genetically diverse, more significant GCA cases were noted among them than lines (Tables 4 and 5). The insignificant GCA effects noted for plant height (Table 5) might be attributed to the high interaction of lines and testers with years observed for this character (Tables 3). The mean squares of lines for leaf to stem ratio though significant, was relatively low, which might explain the absence of significant GCA effects for this character among lines.

E-35-1 from lines and S.70 from testers appeared to be the best general combiners for forage yield and may be expected to do well in hybrid combinations with other parents. The line E-35-1 was involved in 3 out of the 4 top yielding hybrids. Unfortunately, it turned to be the poorest combiner for earliness (Table 5). Earliness was one of the most desirable characters under the local system of forage production. Furthermore, E-35-1 and S. 70 were poor general combiners for leaf to stem ratio, especially the latter. Leaf to stem ratio was considered by many workers (e.g. Chacon and Stobbs, 1976; Chacon et al., 1978; Forbes and Colman, 1993) as being essential in determining forage quality, diet selection and forage intake. The line Dale although ranking third in general effects for yield, could be regarded as the best choice as it possesses acceptable GCA effect for yield while maintaining desirable general effects for other traits. The best general combiners for earliness were Blue Ribbon, N109 from lines and S.186 from testers. Most of the top yielding hybrids showed significant SCA values for forage yield, indicating the involvement of specific effect in the expression of yield of these hybrids. However, ranking of hybrids' yields along with their respective SCA effects (Table 6) showed that the highest mean values for a trait did not necessarily imply significant SCA effects or vice versa. Such patterns of combining ability effects were encountered by Ross et al. (1983) and Satyanarayana (1998).

Both additive and non-additive gene actions are expected to be important in the expression of the studied characters, with the preponderance of additive gene actions for days to flower, forage yield, stem diameter, leaf to stem ratio and non-additive actions for plant height (Table 8). The magnitude of GCA/SCA variance ratio for number of days to flower was specifically sizable, indicating the predominance of additive gene action; however, the specific effects were also highly significant, suggesting the involvement of non-additive effects in controlling this character. For forage yield, these results were in agreement with those reported by Blum (1968), Gupta et al. (1976) and Dangi et al. (1980); and disagree with the results obtained by Gupta and Paliwal (1976) and Sanghi and Monpara (1981). The data presented by Blum (1968) showed that GCA variance was 20.5 times greater than SCA variance for forage yield. Gupta et al. (1976) reported up to 12 GCA/SCA variance ratios for the same character. In this study the magnitude of GCA/SCA variance ratio was much lower (< 4) indicating the relative importance of non-additive gene action in controlling forage yield. For days to flower, our results agree with those of Liang (1967), Bijapur (1980) and Meng et al. (1998), but disagree with those of Kukadia and Singhania (1980) and Sanghi and Monpara (1981). For stem diameter, our results agree with those of Kirby and Atkins (1968) but for leaf to stem ratio, they disagree with those of Kukadia and Singhania (1980). On the other hand non-additive gene actions were more important than additive ones in controlling plant height (Table 8). This was in accordance with Sanghi and Monpara (1981) but was not in agreement with those of many workers (e.g. Kirby and Atkins, 1968; Shankaregowda et al., 1972; Singhania, 1980 and Meng et al., 1998).

The low interaction of GCA variance with years as compared to those of SCA variance indicate that general effects were more stable over years compared to specific effects (Table 8). Kambal and Webster (1965) studying general and specific effects in grain sorghum reported similar results.

Being predominately controlled by additive genes, days to flower could be improved by selection in early generations. With respect to forage yield, stem diameter and leaf to stem ratio which were under control of both additive and non-additive effects, reciprocal recurrent selection is usually suggested as it permits simultaneous exploitation of both general and specific effects. This breeding technique has been recently adapted for crops like sorghum (Brengman, 1995) where mass genetic recombination is facilitated by the use of the dominant fertility restoration gene 'Rf$_1$' in A$_1$ cytoplasm. However, such system will not work under the A3 cytoplasm due to the lack of genes that restore fertility. Nonetheless, the chance to capitalize on heterotic effects still exists for forage yield since appreciable non-additive effects were indicated by the highly significant mean squares observed for SCA and contrast of parents vs hybrids. Heterosis breeding is, therefore, suggested for improving forage yield. The results obtained for contributions of lines, testers and their interaction to the total variance (Table 9) substantiate the previous findings that general effects were more important than specific ones in the expression of these characters.

Conclusion

The line Dale seemed to receive the top priority as it desirable GCA effects for many characters. E-35-1 from lines and S.70 from testers could make a good couple to improve yield under production systems where lateness in flowering is not a major problem. Blue Ribbon and N109 were promising general combiner for earliness. Selection in early generations might be effective in improving characters predominately controlled by additive genes like days to flower. Heterosis breeding was recommended for forage yield improvement. For improvement of leaf to stem ratio, selection program based on more genetically diverse material with increased number of lines was suggested.

AKNOLOWEDGMENTS

Thanks are due to J F Pedersen, Research geneticist-University of Nebraska, USDA-ARS and the Agricultural Research Division, Institute of Agriculture and Natural Resources, University of Nebraska, for providing the female lines in A3 cytoplasm

REFERENCES

Baker RJ (1978). Issues in diallel analysis. Crop Sci. 18: 533-536.

Biel GM, Atkins RE (1967). Estimates of general and specific combining ability in F1 hybrids for grain yield and its components in grain sorghum, Sorghum vulgare Pers. Crop Sci. 7: 225-228.

Bijapur UK (1980). Evaluation and stability parameters of experimental rabi sorghum (Sorghum bicolor L. Moench) hybrids under three environments. Plant Breeding Abstract. 52(2): 1320.

Blum A (1968). Estimates of general and specific combining ability for forage yield in F1 hybrids of forage sorghum. Crop Sci. 8: 392-393.

Brengman RL (1995). The Rf1 gene in grain sorghum and its potential use in alternative breeding methods. Queensland Department of Primary Industries. Information Series No QI95003 p. 16.

Chacon EA, Stobbs TH (1976). Influence of progressive defoliation of a grass sward in the eating behaviour of cattle. Aust. J. Agric. Res. 27:

709-727.

Chacon EA, Stobbs TH, Dale MB (1978). Influence of sward characteristics on grazing behaviour and growth of Hereford steers grazing tropical grass pasture. Aust. J. Agric. Res. 29: 89-102.

Dangi OP, Ram H, Lodhi GP (1980). Line x tester analysis for combining ability in forage Sorghum. Sorghum Newsl. 23: 8-9.

Duvick DN (1999). Commercial strategies for exploitation of heterosis: The Genetics and Exploitation of Heterosis in Crops. Wisconsin, USA. p. 19-29.

Forbes TDA, Coleman SW (1993). Forage intake and ingestive behavior of cattle grazing old world bluestems. Agron. J. 85: 808-816.

GenStat for Windows (2006). 9th edition, Version-9.1.0.174, VSN International. UK.

Gupta SC, Paliwal RL (1976). Diallel analysis of forage yield and quality characters in sorghum. Egypt. J. Genetic. Cytol. 5: 281-287.

Gupta SC, Paliwal RL, Nanda JS (1976). Combining ability for forage yield and quality characters in sorghum. (*Sorghum bicolor* L. Moench). Egypt. J. Genetic. Cytol. 5: 89-97.

Groz HJ, Haskins FA, Pedersen JF, Ross WM (1987). Combining ability effects for mineral elements in forage sorghum hybrids. Crop Sci. 27: 216-219.

Kambal AE, Webster OJ (1965). Estimates of general and specific combining ability in grain sorghum. *Sorghum vulgare* Pers. Crop Sci. 5: 521-523.

Kelley TG, Rao PP, Walker TS (1991). The relative value of cereal-straw fodder in the semi-arid tropics of India: Implications for cereal breeding programs at ICRISAT. Progress Report, ICRISAT, India, 105: 33.

Kirby JS, Atkins RE (1968). Heterotic response for vegetative and mature plant characters in grain sorghum, *Sorghum bicolor* (L.) Moench. Crop Sci. 8: 335-339.

Kukadia MU, Singhania DL (1980). Diallel analysis of certain quantitative traits in forage sorghum. Indian J. Agric. Sci. 50: 294-297.

Liang GHL (1967). Diallel analysis of agronomic characters in grain sorghum. *Sorghum vulgare* Pers. Can. J. Genet. Cytol. 9: 269-276.

Maunder AB (1992). Identification of useful germplasm for practical plant breeding programs: Plant Breeding in the 1990s. Wallingford, UK. Pp. 147-149.

Meng CG, An XM, Zhang FY, Zheng JB, Wang LX, Li PL (1998). Analysis of combining ability of newly developed sorghum male-sterile lines. Acta Agriculturae Boreali-Sinica (Summary: En) 13: 81-85.

Mohammed Maarouf I (2007). Potential of locally developed forage sorghum hybrids in the Sudan. Sci Res. Essay. 2: 330-337.

Mohammed MI, Gamal EK, Ghada HA, Mohammed IE (2008). Improvement of the traditional forage sorghum cultivar 'Abu Sab'in'. Sudan J. Agric. Res., 11: 25-33.

Ross WM, Groz HJ, Haskins FA, Hookstra GH, Rutto JK, Ritter R (1983). Combining ability effects for forage residue traits in grain sorghum hybrids. Crop Sci. 23: 97-101.

Sanghi AK, Monpara BA (1981). Diallel analysis in forage yield and its components in sorghum. Madras Agric. J. 68: 296-300.

Satyanarayana PV (1998). Studies on combining ability and heterosis in rice. International Rice Research Notes 23(3): 10.

Shankaregowda BT, Madhav R, Mensinkal SW (1972). Heterosis and line x tester analysis of combining ability in selected lines of sorghum (*Sorghum vulgare* Pers.). II. Combining ability. Mysore J. Agric. Sci. 6: 242-253.

Singhania DL (1980). Heterosis and combining ability studies in grain sorghum. Indian J. Gen. Plant Breed. 40: 463-471.

The somatic embryogenesis and plant regeneration from immature embryo of sweet corn inbred line

Pitipong Thobunluepop

Faculty of Technology, Department of Agricultural Technology, Mahasarakham University, Talard Sub-District, Muang District, Maha Sarakham 44000, Thailand, 44000. E-mail: pitipnogtho@yahoo.com.

Synthetic seed consisting of somatic embryos enclosed in protective coating are a suitable tool for clonal mass propagation of elite plant varieties. The *in vitro* study was aimed to evaluate the optimizing medium for sweet corn somatic embryogenesis, synthetic seed production which leads to increasing germination and seed viability percentage. The *in vitro* study was aimed to evaluate the optimizing medium for sweet corn somatic embryogenesis, synthetic seed production which leads to increasing germination and seed viability percentage. Sweet corn (*Zea mays* var. *saccharata*) variety FAH01 embryogenic callus were derived from culturing immature zygotic embryos at 11 days after pollination on N6 medium that contained 2, 4-D 2 mgl^{-1} and sucrose 60 gl^{-1}. Somatic embryos was developed after transferred embryogenic callus to N6 medium contained with 2 mgl^{-1} 2, 4-D and 30 gl^{-1} sucrose. Sweet corn synthetic seed was produced by somatic embryos encapsulated into a protective calcium-alginate matrix with provides mechanical support, protection and is coated with a wax film to prevent desiccation. Synthetic seed were produced, it was found that when synthetic seed were treated with 60 gl^{-1} sucrose and stored at 15 ± 2 degree Celsius for 2 weeks, the percentage of germination of synthetic seeds were 42%, percentage of normal seedling was 91% and abnormal seedling was 8%, and they germinated for 8 - 9 days and could produce normal plantlet. When the synthetic seed were dehydrated by silica gel until remained 60% of their moisture content and then stored for 2 weeks, they could germinated at level 23%, which 83% of normal seedling and 17% of abnormal seedling. The survival ratio in sweet corn synthetic seed in this investigation indicated that there is still some more research required to increase number of the survival seeds and the optimum storage technique to prolong their viability is also needed

Key words: Somatic embryo, synthetic seed, sweet corn, artificial seed, tissue culture.

INTRODUCTION

Sweet corn is a herbaceaus monocot with an annual cycle. Its embryo lies embedded in the endosperm at one side. Abbe and Stein (1954) have described eight stages of zygotic embryo development and this order of events has been confirmed and more detail by Van Lammeren (1986). Sweet corn has contributed extensively in the area of genetics, biochemistry, physiology and specially molecular biology. However, the development of somatic embryogenesis system capable of plantlet regeneration is still in its beginning. Somatic embryogenesis has been demonstrated in immature embryos culture. Sweet corn somatic embryogenesis has been reported by a number of workers e.g. Armstrong and Green (1985), Vasil (1986), Fransz and Schel (1991a), Emans and Kieft (1991). Plant regeneration via somatic embryogenesis starts with one or only a few cells, this type of regeneration is important for plant production and plant biotechnology such as somaclonal propagation, multiplication and especially genetic transformation. This collection of techniques can be directed toward production of identical plants or to induce variability (Gordon-Kamm et al., 1990). The conventional techniques of crop improvement in agricultural system involve a search for stain of plant. Redenbaugh (1991) suggested that it is possible to produce asexual embryos in vitro. Synthetic seeds technology is one of the important applications of somatic embryogenesis. In these first synthetic seed system, somatic embryo is encapsulation in protective

Table 1. Effect of medium culture compositions on sweet corn callus initiation.

Medium	Sucrose (gl^{-1}.)	2, 4–D (mgl^{-1}.)	Callus generated (Percentage) Time (Weeks)		
			2	6	10
N6	30	2	54c	75c	89c
		3	57ab	80bc	92b
		4	54c	78c	90c
	60	2	73a	89a	96a
		3	69b	84bc	92b
		4	69b	86b	94b

*: The different letters indicate the statistically significant difference by LSD at 5% level.

encapsulation in protective alginate matrix which provides mechanical support, protection and was coated with a wax film to prevent dessication Radenbaugh (1986). However, the stage life and vigor of synthetic seed is limited. The main objective of the study was to find the optimizing medium for sweet corn somatic embryogenesis, synthetic seed production which lead to increasing germination rate and seed viability.

MATERIALS AND METHODS

The experiment was conducted at Department of Agricultural Technology, Department of Technology, Maha Sarakham University, and Department of Agronomy, Faculty of Agriculture, Chiang Mai University, Thailand in 2007 - 2009. Sweet corn var. FAH01, fresh immature zygotic embryos; 11 days after pollination were collected and were sterilized with 10% clorox solution for 5 min, followed to through rinsed with sterile water and were incubated on various culture mediums. The experiment was conducted in factorial in complete randomized design with 4 replications. Sweet corn callus was induced from sterilized zygotic embryos which were cultured on agar-solidified N6 medium, containing with different levels of sucrose (30 and 60 gl^{-1}) and 2, 4-D (2, 3 and 4 mgl^{-1}). In all experiments culture temperature was 25 ± 2°C and was incubated in the dark. Fast-growing friable type II embryogenic callus was selected and maintained by sub-cultured once every 2 weeks intervals depending on growth rate. Callus maintenance was incubated under temperature 25 ± 2°C in the dark. Somatic embryo was developed by transferring callus aggregates on regenerate-medium N6 contained 1 mgl^{-1} NAA and incubated under temperature 25 ± 2°C in the dark for 2 weeks. Then, embryogenic callus were transferred to plant growth regulator-free MS medium and cultured under 25 ± 2°C in the light condition. Sweet corn synthetic seeds were produced by inserting single somatic embryo into visco liquid beads of 3% (w/v) sodium alginate solution, single embryo was encapsulated by putting into 100 mM calcium nitrate solution for 20 min to develop calcium alginate gel complex and then, rinsed in sterilized water. Sweet corn synthetic seed germination, normal seedling, abnormal seedling were recorded. The analysis of variance was performed for data analysis and differentiated with last significant different (LSD) test at p < 0.05 using the software SX release 8.0 (Analytical software, Tallahassee, USA).

RESULTS AND DISCUSSION

The effect of sucrose and 2, 4 - D supplemented in N6 medium was significantly affected on sweet corn callus initiation. 11 days after pollinated immature zygotic embryos were placed on N6 medium supplemented with 2 mgl^{-1} of 2, 4 - D and 60 gl^{-1} of sucrose and resulted in the highest of callus induction rate. The callus initiation percentages were 73, 89 and 96 after cultured 2, 6 and 10 weeks respectively. On the other hand, the application of 30 g L^{-1} of sucrose and 2 mgl^{-1} of 2, 4 - D resulted in the lowest of callus induction (Table 1). This result congruence with Bates (1993) who reported the increasing of sucrose concentration increased the induction and development of callus.

The embryogenic callus was obvious that entirely developed from scutellum node, non-embryogenic callus was developed from shoot apex and the radicle of immature zygotic embryo was not developed to callus at all (Figure 1A). Two type of embryogenic calli were distinguished in 2 types; type I callus was compacted, type II callus so called friable callus was differentiated lesser than regenerable callus (Figure 1B). Type I callus might consist of aggregates of undifferentiated calli, interspersed with vascular cells, and groups of small isodiametric meristemic cells on the outside of the aggregated as similar results reported by Emons and Kieft (1991) (Figure 1C), which are tightly packed, thin-walled, richly cytoplasmic, many small vascular (Fransz and Schel 1991a, 1994). These groups of cell are comparable to the proembryogenic masses. After sub-culturing, these embryogeic calluses remained similar. In the fact, they were the dividing cells giving rise to new embryogenic cells. (Figure 1D). Single globules are unable to regenerate into plantlets; they form only roots (Figure 1E), which was correlated with Carvolho et al. (1997) results.

The embryogenic calli was selected and incubated on N6 medium contained with 2 mgl^{-1} of 2, 4-D and 60 gl^{-1} of

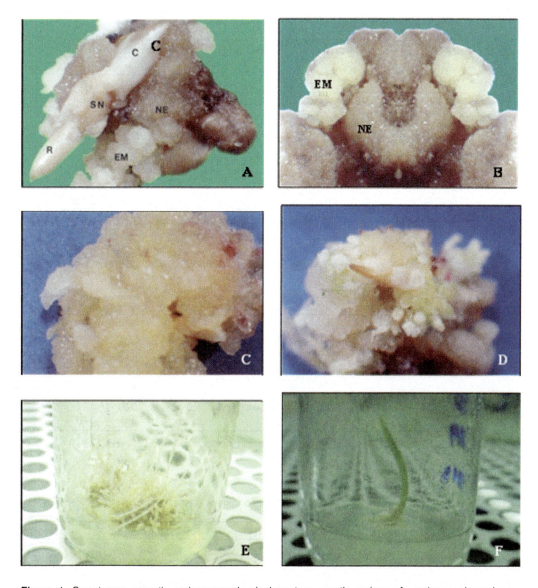

Figure 1. Sweet corn somatic embryogenesis via immature zygotic embryo, A: embryogenic and non-embryogenic callus development position, C: colepotile, R: radicle, SN: scutellum node, EM: embryogenic callus, NE: non-embryogenic callus, B: compact embryogenic callus and non-embryogenic callus, C: softy yellow-white friable non-embryogenic callus, D: friable embyogenic callus, E: rhizogenic callus, and F: plantlet regeneration via somatic embryogenesis

of sucrose, callus growth was monitored, callus size and fresh weight increased significantly through this culturing. It was shown that, callus fresh increased 0.0990 - 0.1103 g and callus size 2.16 - 4.88 mm. after 4 and 10 weeks of culturing (Table 2). Fahye et al. (1986) reported that sucrose was provided as carbon source for callus cell using for the increasing of cell metabolism, which was finally resulted the induction of callus size and fresh weight.

After the embryogenic callus were cultured on N6 medium contained with 30 gl^{-1} of sucrose and 2 and 3 mgl^{-1} of 2,4-D, then, the embryogenic callus were transferred to plant growth regulator-free MS medium containing 60 gl^{-1} of sucrose. The somatic embryos were allowed to regenerated and mature, the shoot meristem from these single somatic embryos was highest developed for 65 and 62% (Table 3). The somatic embryo continued to develop to plantlet and ready to be transplanted within 4 weeks (Figure 1F). While, the culture medium containing the higher levels of 2, 4 - D resulted the decreasing of percentage of sweet corn somatic embryos development. Based on this observation, it refers to the developmental sequence as sweet corn somatic embryogenesis. Green and Philips (1975) and Green et al. (1974) reported the optimum concentration of plant growth hormone especially on auxin and cytokinin was play a important role of sweet corn somatic embryogenesis, however, the over requirement levels

Table 2. Effect of Medium culture composition on sweet corn callus size.

Medium	Sucrose (gl^{-1})	2,4–D (mgl^{-1})	Callus size (mm) Time (weeks)		
			2	6	10
N6	30	2	2.16	2.94b	3.99c
		3	2.19	3.04ab	4.08b
		4	2.11	2.91b	3.93c
	60	2	2.16	3.10a	4.88a
		3	2.19	3.20a	4.40a
		4	2.11	3.12a	4.23ab

*The different letters indicate the statistically significant difference by LSD at 5% level.

Table 3. Effect of medium culture compositions on percentage of sweet corn somatic embryos development.

Medium	Sucrose (gl^{-1})	2,4 - D (mgl^{-1})	Somatic embryos (percentage)
N6	30	2	65a
		3	62a
		4	54b
	60	2	54b
		3	52b
		4	43c

*: The different letters indicate the statistically significant difference by LSD at 5% level.

Table 4. The effect of sucrose concentration and storage temperature conditions on sweet corn synthetic seed viability.

Temperature (°C)	Sucrose (gl^{-1})	Germination (%)	Normal seedling (%)	Abnormal seedling (%)
15 ± 2	0	43^c	90^c	10^a
	30	50^b	92^b	8^b
	60	45^c	92^b	8^b
25 ± 2	0	49^{bc}	92^b	8^b
	30	56^a	95^a	6^c
	60	51^b	95^a	5^c

*: The different letters indicate the statistically significant difference by LSD at 5% level.

requirement levels of their compounds could also reducing the somatic embryo induction. Its might be the toxicity of their compounds (Table 4).

Single somatic embryos were encapsulated with protective seed coat calcium-alginate matrix, so-called synthetic seed. Seed qualities were observed. It was found that, somatic embryos were treated with 30 and 60 gl^{-1} sucrose significantly affected on germination percentage and date of germination, while synthetic seeds germinated for 57 and 46% respectively and their germinated after 8 and 9 days after planted (Figure 2). For seedling characteristics, the sweet corn synthetic seed could produce 89 - 94% of normal seedlings and 6 - 11% of abnormal seedlings (Figure 3). After stored at 25 ± 2 °C for 2 weeks, the high level of sucrose concentration significantly produced the highest level of germination and normal seedling percentage. The percentage of germination was 56%, normal and abnormal seedling was 91 and 8%, respectively, they germinated after 6 days to produce normal plantlet. On the other hand, the germination and normal seedling percentage was decreased significantly when reducing the concentration

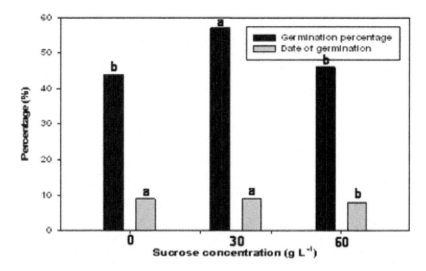

Figure 2. The effect of sucrose concentration on sweet corn synthetic seed germination.

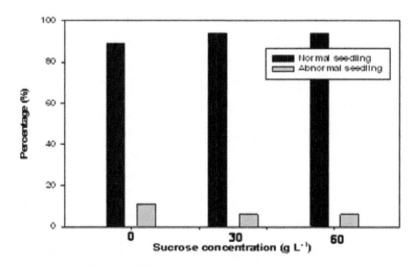

Figure 3. The effect of sucrose concentration on sweet corn seedling characteristics after germinated from sweet corn synthetic seeds

of sucrose (Table 5). Lecouteux et al. (1994) reported that pretreatment of sucrose was an important factor for the preservation of somatic embryos before the dehydration. Sucrose was provided as energy source to supported cell metabolism. Moreover, it could be incorporated into the cell wall to improve cell wall structure to tolerance to the dehydration. After synthetic seeds were dehydrated by silica gel method until remained 60% less of their moisture content and then stored 2 weeks, the synthetic seeds were shown low germinated was 23%, which was resulted the normal and abnormal seedling for 83 and 17% respectively. However, the prolongation of sweet corn synthetic seed viability could do under the high level of moisture content, which was resulted the highest percentage of germination and normal seedling (Table 5). Machii (1993) reported the best storage condition of somatic embryo especially under synthetic seeds form was cold temperature and high moisture content.

Sweet corn somatic embryogenesis by using N6 medium contained with sucrose and 2, 4-D could be produce embryogenic calli from immature zygotic embryos. This friable type II callus can be maintained in culture and regenerated (Takahata et al., 1993). The callus consists of aggregated, globular stage somatic embryos attached to a mass of vaculoe cells; regeneration can take place in small cell clusters that first from globules and then develop into callus aggregate. Single globules are unable to regenerate into plantlets; they form only roots (Pareddy and Petolino, 1990). The somatic embryos developed depending on abundant sucrose is necessary. Without this phase, calli were regenerate via organogenesis (Duncan et al., 1985). The survival rate of sweet corn synthetic seeds in this

Table 5. The effect of dehydrated levels and storage time on sweet corn synthetic seed viability.

Storage time (Weeks)	Moisture lose (%)	Germination (%)	Normal seedling (%)	Abnormal seedling (%)
1	20	53^a	92^a	8^c
	40	45^b	90^a	10^b
	60	30^c	86^b	14^b
2	20	46^b	89^b	11^b
	40	38^c	87^b	13^b
	60	23^d	83^c	17^a

*: The different letters indicate the statistically significant difference by LSD at 5% level.

investigation indicated that there are still some more research required to increase the number of the survival seeds and the optimum storage techniques to prolong their viability is also need.

ACKNOWLEDGMENTS

This research is an activity of the programme Subject-related partnership between the University of Göttingen (Germany) and Faculty of Agriculture, Department of Agronomy, Chiang Mai University (Thailand) in the area of Academic Co-operation in Teaching and Research, and acknowledged to the Department of Agricultural Technology, Faculty of Technology, Maha Sarakham University, Thailand.

REFERENCES

Armstrong CL, Green CE (1985). Establishment and maintenance of friable, embryogenic maize callus and the involvement of L-proline. Planta, 164: 207-214.

Bates KJ (1993). Callus induction from various genotypes. Maize Gen. Cooperation Newslett., 67: 75-76.

Carvolho CHS, Bohorava N, Bordallo PW, Abreu LL, Valicente FH, Bressan W, Can Paiva E (1997). Type II callus production and plant regeneration in tropical maize genotypes. Plant Cell Rep., 17: 73-76.

Duncan DR, Williams ME, Zehr BE, Widholm JM (1985). The production of callus capable for plant regeneration from immature embryos of numerous Zea mays genotypes. Planta, 165: 322-332.

Emons AMC, Kieft H (1991). Histological comparison of single somatic embryos of maize from maturation of embryogenic calli of *Zea mays* L. Can. J. Bot. 71: 1349-1356.

Fahye JW, Reed JN, Ready TL, Pace GM (1986). Somatic embryogenesis from three commercially important inbreds of *Zea mays* L. Plant Cell Rep., 5: 35-38.

Fransz PF, Schel JHN (1991a). Cyto-differentiation during the development of friable embryogenic callus of maize (*Zea mays* L.) Can. J. Bot. 69: 26-33.

Fransz PF, Schel JHN (1994). Ultrastructural studies on callus development and somtic embryogenesis in *Zea mays* L. In Bajaj YPS(ed) Biotechnology in agriculture and forestry. Maize. Springer, Berlin Heidelberg. New York. 25: 50-65.

Green CE, Philips RL, Kleese RA (1974). Tissue culture of maize (*Zea mays*). Initiation, maintenance and organic growth factors. Crop Sci. 14: 54 - 58.

Green CE, Philips RL (1975). Plant regeneration from tissue culture of maize. Crop Sci. 15: 417-421.

Lecouteux C, Tessereau H, Florin B, Courtois D, Petiard V (1994). Preservation of somatic embryos of carrot (*Daucus carota* L.) by dehydration after pretreatment with sucrose. Hort. Abstr. 64(2): 1192.

Machii H (1993). In vitro growth of encapsulated adventitious buds in mulberry, *Morus alba* L. Hort. Abstr., 63(7): 4918.

Pareddy PR, Petolino (1990). Somatic embryogenesis and plant regeneration from immature inflorescence of several elite inbred of maize. Plant Sci., 67: 211-219.

Redenbaugh K, Fujii J, Viss P, Slade D, Kossler M (1986). Scale-up artificial seeds. Abstr. VI international congrass of Plant tissue and Cell Culture. Minnesota U.S.A. Pp. 25-36.

Redenbaugh K (1991). Encapsulation of somatic embryos. Synseeds: Application of synthetic seeds to crop improvement. 203-215.

Takahata Y, Brown DCW, Keller WA, Kaizuma N (1993). Dry artificial seeds and desiccation tolerance induction in microspore – derived embryos of broccoli. Plant Cell, Tiss. Org. Cult., 35: 121-129.

Optimization and development of regeneration and transformation protocol in Indian mustard using lectin gene from chickpea [*Cicer arietinum* (L.)]

V. V. Singh[1], Vandana Verma[1]*, Aniruddh K. Pareek[1], Monika Mathur[1], Rajbir Yadav[1], Poonam Goyal[1], Ajay Kumar Thakur[1], Y. P. Singh[1], K. R. Koundal[2], K. C. Bansal[2], A. K. Mishra[1], Arvind Kumar[1] and Sandeep Kumar[1]

[1]Directorate of Rapeseed-Mustard Research, Bharatpur, Rajasthan 321 303, India.
[2]NRCPB, IARI, New Delhi, India.

A simple and easy protocol for regeneration and genetic transformation of *Brassica juncea* (L.) Czern and Coss cv. Pusa bold has been developed. The optimum regeneration was found on combination of BAP (3.0 mg/l) and IAA (0.2 mg/l). The genetic transformation protocol has been standardized using *Agrobacterium tumefaciens* (gv3101). Cotyledonary petioles used as explant were pre-cultured for 48 h and than co-cultivated with Agrobacterium suspension for 2 days. Putative transformants were selected from selection medium containing 20 mg/l Kanamycine and 250 mg/l cefotaxime. Complete plant formed after placing shoots on rooting medium (0.5 mg/l IBA). Confirmation of integration of transgene was done by PCR using *nptII* gene primer.

Key words: Indian mustard, *Agrobacterium tumefaciens*, chickpea lectin gene, cotyledonary petioles, transformation.

INTRODUCTION

Indian mustard [*Brassica juncea* (L)] is a major oilseed crop of India and is grown in around six million hectares, primarily in the north-western part of the country during the winter season. It is also a potential crop in western prairies of Canada, China and Russia. Brassica oilseeds are susceptible to many biotic and abiotic stresses which cause moderate to high level (10 - 90%) losses depending upon infestation and crop growth stage. Although the use of pesticides is quite effective, their excessive and injudicious use leads to imbalances in agro-ecosystem and health hazards (Singh, 2008). Transgenic plants expressing bacterial Bt genes and plant derived insecticidal lectin and protease inhibitor genes provide effective and ecologically sustainable option for insect control (Datta and Koundal, 2003). Plant lectins are a group of carbohydrate binding proteins present in high amount in legume seeds and the toxicity of plant lectins against a wide range of insects is well established (Peumans and Van Damme, 1995). Keeping these facts in view, an *in vitro* regeneration and transformation of *B. juncea var.* Pusa Bold was attempted using chickpea lectin gene construct.

MATERIAL AND METHODS

Plant materials and culture conditions

Seeds of Indian mustard var. Pusa bold were obtained from the Germplasm Unit of DRMR, Bharatpur, where cultivars are being maintained. Approximate one hundred Seeds were surface sterilized with 0.1% $HgCl_2$ for 5 - 6 min and rinsed 2 - 3 times (for 2 - 3 min each time) with distilled water. Sterilized planted seeds were then germinated in culture tubes on half MS medium (Murashige and Skoog, 1962) with 3% sucrose and solidified with 0.8% agar (Himedia). Among aseptically planted seeds, ninety were germinated at 25°C ± 2 in 16 h photoperiod and 1500 lux intensiy. The cotyledonary petioles were excised from 5 day old seedlings and planted on the regeneration medium with their abaxial surface in contact. While plating the explants, it was ensured that the cut end of the petiole remained either in contact or slightly immersed

---

*Corresponding author. E-mail: bandhanbiotech@gmail.com, vvs_71@yahoo.com

in the growth media.

Bacterial strain and binary plasmid vector

The *Agrobacterium* strain were grown overnight in minimal medium (Chilton et al., 1974) that is liquid YEM (Yeast Extract Manitol) at 28°C with continuous shaking under appropriate antibiotic selection condition (20 mg/l rifampicin and 50 mg/l kanamycin). The disarmed *A. tumifecience* strain gv 3101, carrying the binary plasmid vector pBinAR were used in preliminary transformation experiments. This vector mobilized into *A. tumifecience* strain gv3101.

Plant regeneration

To optimize various regeneration parameters for different explants (stems, hypocotyls, petioles, cotyledons and cotyledonary petioles), optimum concentration of hormones for shoot and root regeneration was used. BAP (0.0 - 5.0 mg/l) and IAA (0.0 - 0.5 mg/l) was used for shoot regeneration and IBA (0.0 - 1.0 mg/l) was used for root regeneration. All parameters were tested one after the other, in sequential order. The optimized conditions determined in earlier experiments were used in subsequent experiments. The above parameters (or treatments) were tested in the order as stated below.

For testing of first parameters different explants such as stems from regenerated shoots and hypocotyls, petioles, cotyledonary petioles, cotyledons were used from 5 day old germinated seedlings and transferred on MS media containing different combinations of shoot regenerating hormones. After 2 - 3 weeks, regenerated healthy, individual shoots from explants were excised and transferred on rooting media supplemented with different concentration of IBA (0.0 - 1.0 mg/l). Well developed rooted plants were kept for hardening and acclimatization. Nine plants were kept directly in soil rite: field soil mixture (1:0/1:1/3:1) and three plants were given pretreatment of liquid culture. Initially all plants were hardened in culture room, where they were covered with perforated poly bags. After 20 - 25 days when they had sufficient growth to sustain environmental conditions, they were transferred to the glass house for complete acclimatization.

Plant transformation

Agrobacterium strain gv3101 carrying the binary plasmid vector pBinAR was used for transformation. The vector contains *nptII* (neomycin phosphotransferase) for the selection of transformants. To carry out genetic transformation, cotyledonary petioles (with 2 mm petiole) were precultured on MS medium supplemented with 2, 4-D, BAP and IAA for a period of 48 hrs (Figure 1) (Chakrabarty et al., 2002). After 48 hrs, these petioles were co-cultivated with *Agrobacterium* culture, which was grown overnight in 50 ml liquid YEM containing antibiotics (rifampicin 20 mg/l and kanamycin 50 mg/l) at 28°C (Figure 2). The O.D of culture was maintained at 0.5 with an optimized incubation period (1 - 2 s to 30 min) of explant and *Agrobacterium* suspension. The cut ends of cotyledonary petioles were dipped in the *Agrobacterium* suspension from 1 - 2 s to 30 min and then transferred on co-cultivation medium (for 48 h) containing BAP and IAA. After co-cultivation explants were washed with double distilled water (2 - 3 times for 2 - 3 min) and treated with cefotaxime antibiotic solution (250 mg/l) for 5 - 6 min then blotted on sterile filter paper to remove any traces of bacterial contamination and then placed on selection medium with BAP (3.0 mg/l), IAA (0.2 mg/l) containing cefotaxime (250 mg/l) and kanamycin (20 mg/l). Shoot induction started after 15 - 20 days of co-cultivation. Regenerated shoots were excised and transferred to shoot elongation medium. Healthy shoots with an average height of

Figure 1. Pre-culturing of cotyledonary petioles of var. Pusa Bold.

Figure 2. Co-cultivation of pre cultured Cotyledonary petioles

1.5 cm were transferred to the rooting medium, after 3 - 4 weeks when roots grown sufficiently they were transferred to hardening medium containing soil rite. With this method primary transgenic were hardened to get T1 seeds.

Molecular analysis of putative transformants total genomic DNA was isolated from leaves of *in vitro* growing putative transformants and control plants using modified SDS DNA extraction method (Dellaporta et al., 1983). Gene specific primers *npt II* were used to amplify integrated transgene of 700bp. Each PCR reaction was carried out in 25 µl reaction mixture containing 1.5 mM $MgCl_2$, 200 µM dNTPs, (1.25 U) Taq DNA polymerase, 50 ng genomic DNA and oligonucleotide primers. The sequence of these oligonucleotide primers is as follows: Forward primer (F): 5'GAG GCT ATT CGG CTA TGA CTG 3'.Reverse primer (R): 5'ATC GGGAGC GGC GAT ACCGTA3'.

RESULTS AND DISCUSSION

Regeneration

Present study was conducted on *B. juncea* var. Pusa Bold with a view to optimize regeneration and transformation protocol using cotyledonary petioles. In this study, cotyledonary petioles, hypocotyls, petioles of 5 day old seedlings and stems from back cultured shoots were used and inoculated on MS medium with varying concentrations of BAP (0.0 - 5.0 mg/l) and IAA (0.0 - 0.5 mg/l). The optimum regeneration frequency of 58.65% was obtained on combination of BAP and IAA (Table 1). Although hypocotyls and stems showed 60 - 70% regeneration frequency and comparatively gave good response than cotyledonary petioles, but in further experiments of transformation, both stems and hypocotyls gave very poor transformation frequency (3 - 5%). Therefore cotyledonary petioles were used as an explant in transformation experiments. Our results are in conformity with earlier reports of regeneration obtained in *Brassica juncea* var. PJK and Rai-1 using cotyledonary petioles (Sharma et al., 1990; Babu et al., 2003). In contrast, genetic transformation in *B. juncea* was done by using hypocotyl as an explant source and successfully transformed RLM198, Pusa Bold, Skorospieha II, Pusa Jaikisan cultivars of Indian mustard (Babu et al., 2003). Presence of BAP was found to be critical in this study as reported earlier in Indian mustard. In absence of BAP, there was no shoot regeneration. At BAP concentration of 1.0 mg/l and 2.0 mg/l, regeneration frequency of 5.78 and 21.50 was observed. Maximum shoot regeneration was observed at a BAP concentration of 3.0 mg/l when supplemented with 0.2 mg/l IAA (Table 2). Beyond optimum shoot regeneration level there was decline in regeneration frequency. In addition, BAP at concentration of 3.0 mg/l gave maximum number of shoots (6-7 per explants). In contrast to this, lower range (2 - 4 shoots/explant) was obtained at other concentration. These results are also in accordance with earlier reports of Babu et al. (2003) and Sharma et al. (2004) in *B. juncea*. Since best results were achieved by using cotyledonary petioles as explants on MS medium containing 3.0 mg/l BAP and 0.2 mg/l IAA. This combination was used for transformation study. Regenerated shoots were kept on root inducing medium having different concentration of IBA (Table 3).

Plant transformation

In the present study, cotyledonary petioles were dipped

Table 1. Shoot regeneration frequency in *Brassica juncea* cv Pusa Bold at various combinations of BAP & IAA.

BAP/IAA	0.0	0.1	0.2	0.3	0.4	0.5
0.0	0.0	0.0	0.0	0.0	0.0	0.0
1.0	10.25	13.24	19.75	13.00	8.95	10.95
2.0	22.63	19.43	26.18	21.73	17.14	22.85
2.5	21.30	24.50	32.25	23.50	20.15	19.52
3.0	47.25	50.10	58.65	52.21	43.50	25.56
3.5	30.19	34.45	32.11	28.50	23.60	21.63
4.0	22.85	20.36	28.45	21.85	24.65	19.25
4.5	13.62	10.95	19.52	16.34	15.64	13.55
5.0	9.06	14.52	10.23	13.26	14.26	8.08

for 1 - 2 s in *Agrobacterium* suspension and co-cultivated for 48 h on co-cultivation media. This treatment which was given for 1 - 2 s showed higher transformation frequency of 15 - 20%. After a period of 12 - 15 days, cut tips of cotyledonary petioles initiated shoot regeneration and started differentiating into green or white color shoots. The transformation frequency was calculated on the basis of the number of green shoots obtained out of the total number of shoots formed on the selection media. Transformation frequency varying between 15 - 20% was obtained from 10 independent batches, each having approximately 95 explants. Green shoots thus obtained were separated and sub cultured on same media.

Kanamycin level (20 mg/l) was maintained in sub culturing and rooting to rule out any possibilities of escapes. Multiplied shoots of 3 - 4 cm length were excised and cultured on medium containing IBA (0.5 mg/l) for rooting. Rhizogenesis started after 2 - 3 weeks of inoculation and shoots having well developed roots was obtained in 3 - 4 weeks (Figure 3). The rooted plant-lets were then transferred to pots filled with hardening media that is soil rite (Figure 4). However, rooting in putative transformants was found to be difficult. This may be due to insertion of that lectin gene in root forming zone. Insertion resulting into hormonal imbalances might be responsible for preventing root induction. Half MS supplemented with IBA was showing high frequency of root induction than normal MS. While on basal medium it was showing less frequency of root induction and non-fibrous roots. Therefore, in all our experiments half MS medium containing IBA was used for root induction from regenerated shoots. During hardening of putative transformants, they were pretreated with liquid culture so that they able to acclimatize rapidly according to environmental conditions. Further during hardening process soil rite: field soil (3:1) ratio found to be most appropriate for plants. Transformants were hardened and acclimatized to get T1 seeds. Molecular analysis of putative transformations was performed to confirm the presence of chickpea lectin gene using *nptII* primers. Results were showing that some of transformants succeed (l-4 in Figure 5) to get that gene of 0.7 kb.

Table 2. Transformation table of *Brassica juncea* cv Pusa Bold.

Cultivars	No. of explants Co-cultivated	Total no of shoots developed on selection media	Total no. of green shoot	No. of explant on rooting	No. of explant for hardening
Pusa Bold	95	65	12	3	2
	80	40	7	2	3
	105	75	15	3	1
	95	75	14	1	1
	110	66	10	1	0
	70	56	11	0	3
	66	43	8	4	2
	101	65	2	2	0

Table 3. Effect of various concentrations of auxins (IBA) on formation of roots.

IBA (mg/l)	% of Root Induction	No. of roots/shoot
0	20	3
0.1	40	2
0.2	30	4
0.3	50	5
0.4	40	4
0.5	80	12
0.6	60	4

Figure 4. Hardening of putative transformants.

Figure 3. Rooting in putative transformants

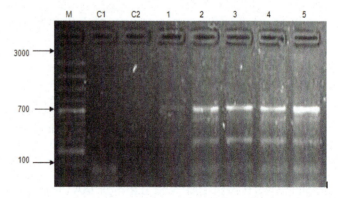

Figure 5. Molecular analysis of putative transformants through specific primers (M = Marker, C = Controls, 1 - 5 Putative Transformants).

Present study clearly demonstrates *in vitro* regeneration and *Agrobacterium* mediated transformation of *B. juncea* variety Pusa Bold. The present process of transformation is simple and easy in operation. However, selection of appropriate explant, choice of *Agrobacterium* strain with proper media and hormone combination are the major determinants for the success for any trans-formation and regeneration system. The development of aphid resistant transgenic *Brassicas* will definitely be a boom to mustard growing farmers of the country.

ACKNOWLEDGEMENT

Authors gratefully acknowledge the financial grant and gene construct received under Network project on Transgenics in Crops, ICAR, New Delhi. We are also thankful to Dr. K.C. Bansal, Coordinator, NPTC and Dr. Rekha Kansal, NRCPB, New Delhi for their all round help and support to carry out this experiment.

REFERENCES

Babu BS, Brijmohan SB, Sabharwal V, Chaudhary A, Lakshmikumaran M (2003). *In vitro* regeneration & genetic transformation of *B. juncea* via Agrobacterium using cotyledonary petiole explants. *Brassica*. 5(1-2): 16-23.

Chakrabarty R, Viswakarma N, Bhat SR, Kirti PB, Singh BD, Chopra VL (2002). Agrobacterium mediated transformation of cauliflower: Optimization of protocol and development of Bt-transgene cauliflower. Indian Acad. Sci. 27: 495-502.

Chilton MD, Currier T, Farrand S, Bendich A, Gordon MP, Nester EW (1974). *Agrobacterium tumifecience* DNA and PS8 bacteriophage DNA not detected in crown gall tumors. Proc .Natl. Acad. Sci. USA. 71: 3636-3672.

Datta S, Koundal KR (2003). Genetic transformation of *B. juncea* (l.) using Insecticidal Lectin gene from cowpea (*Vigna unguiculata* (L. Walp). *Brassica* 5 (1-2): 33-40.

Dellaporta SL, Wood J, Hicks JB (1983). Plant DNA minipreparation version II. Plant Mol.Biol.Rep. 1: 19-21.

Murashige T, Skoog F (1962). A revised medium for rapid growth and bioassays with tobacco tissue culture. Physiol. plant 15: 473-497.

Peumans WI, Van Damme ELM (1995) Lectins as plant defense proteins. Plant Physiol. 109: 347-352.

Sharma KK, Bhojwani SS (1990) Histological aspects of in vitro root and shoot differentiation from cotyledon explants of *Brassica juncea* (L.) Czern. Plant Sci. 69: 107-214.

Sharma M, Sahni R, Kansal R, Koundal KR (2004). Transformation of Oilseed mustard *B. juncea*. var. PJK with Snowdrop lectin gene. Ind .J. Biotech. 3: 97-102.

Singh D (2008). Breeding for Aphid Resistance in Rapeseed–Mustard. In Sustainable production of oil seeds. A.Kumar, J.S.Chauhan and C.Chattopadhyay (eds).Agrotech Academy Publishing Udaipur (India), pp: 142-154.

Identification of PCR-based DNA markers flanking three low phytic acid mutant loci in barley

R. E. Oliver[1]*, C. Yang[2], G. Hu[1], V. Raboy[1] and M. Zhang[2]

[1]USDA-ARS, Small Grains and Potato Germplasm Research Unit, Aberdeen, ID 83210, USA.
[2]Hebei Institute of Food and Oil Crops, Shijiazhuang, Hebei 050031, People's Republic of China.

Phytic acid (PA) is the most abundant form of phosphorus (P) in cereal grains. PA chelates mineral cations to form an indigestible salt and is thus regarded as an antinutritional agent and a contributor to water pollution. Grain with low phytic acid (lpa) genotypes could aid in mitigating this problem. In barley, more than 20 lpa mutant lines have been isolated, representing at least 6 different genetic loci. These mutants have significantly reduced levels of seed PA, which are largely replaced by inorganic P, a form readily digestable by animals. Use of lpa lines in breeding has proved a practical approach for improvement of phosphorus nutrition in barley. Efficient utilization of these loci in marker-assisted selection breeding programs requires identification of closely-linked, high-throughput molecular markers. Here we report development of flanking, PCR-based markers for 3 major lpa loci in barley: lpa1-1 (M422), lpa2-1 linked locus (M640), and a locus linked to the myo-inositol 1-phosphate synthase (MIPS) gene (M678). In addition, marker position accuracy in the MIPs region has been improved by detection and elimination of marker redundancy.

Key words: Barley breeding, SSR marker, low phytic acid (lpa), grain nutrition.

INTRODUCTION

In barley (*Hordeum vulgare* L.) and other small grains, approximately 70-80% of total seed phosphorus (P) is stored as phytic acid (PA; *myo*-inositol-1,2,3,4,5,6-hexakisphosphate) (Ockenden et al., 1997; Raboy, 1997). PA plays critical roles in seed development and germination, DNA repair and mRNA export (Hanakahi et al., 2000; York et al., 1999), cell signaling (Menniti et al., 1993; Sasakawa et al., 1995; Lemtiri-Chlieh et al., 2000) and anti-oxidation (Graf et al., 1987). PA is not desirable for grain nutrition since it chelates essential cationic minerals, including calcium, iron, and zinc, forming a mixed salt known as phytate and reducing availability of minerals to human and non-ruminant animals. PA is the primary storage form of P in cereal grains; however, phytate is indigestible to non-ruminant animals, resulting in dietary defi-

*Corresponding author. E-mail: gongshe.hu@ars.usda.gov.

Abbreviations: cM; Centi-Morgan, **PA;** phytic acid, **MIPs;** *myo*-inositol 1-phosphate synthase, **Pi;** Inorganic phosphate, **STS;** Sequence-tagged-site, **SSR;** Simple sequence repeat, **RFLP;** Restriction Length Polymorphic Fragment, **ISSR;** inter simple sequence repeats and **Ins;** Inositols.

in dietary deficiency of P and accelerated eutrophication of waterways (Erdman, 1981; Sharpley et al., 2001). Nutrient impacts of phytate have prompted development of low phytic acid (lpa) mutants in crops such as maize, barley, rice, and soybean (Raboy et al., 2000; Shi et al., 2005; Larson et al., 1998; Larson et al., 2000 and Wilcox et al., 2000). These loss-of-function mutations decrease synthesis of PA from P; consequently, reductions in PA are compensated for by near-equivalent increases in bioavailable forms such as inorganic P (Pi) (Raboy et al., 2000). Thus, detrimental effects of phytate are alleviated without affecting overall seed P levels. Availability of these mutations and the potential benefits of their deployment have made development of *lpa* cultivars a key objective in feed barley breeding programs, providing further impetus to explore the genetics of this trait.

Mutations causing *lpa* can occur at any of the several steps in a complex biosynthetic pathway. Synthesis of PA requires conversion of glucose-6-phosphate to inositol-3-phosphate (Ins (3) P_1), catalyzed by *myo*-inositol 1-phosphate synthase (MIPS). The Ins ring is subsequently phosphorylated to synthesize PA (Loewus, 1990). Since the MIPS reaction is the sole source of Ins, which is necessary for all downstream reactions, mutation and re-

Table 1. Primer pairs for PCR-based markers tested for polymorphism between barley cvs. Harrington, Steptoe and Morex Sequence information was obtained from the GrainGenes website (http://wheat.pw.usda.gov). Italicized marker names are wheat SSRs; all others are barley SSR and STS markers. Marker names in bold exhibited polymorphism.

Locus	Marker	Forward sequence	Reverse sequence
lpa1	MSU21	TGGTCTTTCATGTACCTACC	TGTGTCATCAAGCACAACCA
	ABC153	TTCATCATCATCGTCATCGTG	CCTCTGCCGCTGGAACTA
	ABC252	CACAAGGCTCAAAACATAAC	AAGCTCACCAAGTCCCAGTC
	ABC165	CAATGACTTCAAGGGGTCTG	TCCATACCATTCCCATCTAA
	EBmag793	ATATATCAGCTCGGTCTCTCA	AACATAGTAGAGGCGTAGGTG
	EBmac415	GAAACCCATCATAGCAGC	AAACAGCAGCAAGAGGAG
	HVCSG	CACTTGCCTACCTCGATATAGTTTGC	GTGGATTCCATGCATGCAATATGTGG
	Bmac0144b	TACGTGTACATACTCTACGATTTG	ACTTATTCTGCATCCTGGGT
	EBmatc0039	TAGTCTCTTCATTTATACCATCACC	CATGCTGATCCCCCTTCT
	Bmac0216	GTACTATTCTTTGCTTGGGC	ATACACATGTGCAAAACCATA
	ABG317	CATGATGGGTCAAGCTCTGT	AACTCTGGGTGGTTTGTGAA
	cMWG660	CTGAACCCACAAGAGCAGAA	CCCAGCCAAAGCTGGTTTTT
	WMC175	CTCAGTCAAACCGCTACTTCT	CACTACTCCAATCTATCGCCGT
	MWC243	CGTCATTTCCTCAAACACACCT	CCGGCAGATGTTGACAATAGT
	Barc59	GCGTTGGCTAATCATCGTTCCTTC	AGCACCCTACCCAGCGTCAGTCAAT
	Barc101	GCTCCTCTCACGATCACGCAAAG	GCGAGTCGATCACACTATGAGCCAATG
	Cfd73	GATAGATCAATGTGGGCCGT	AACTGTTCTGCCATCTGAGC
lpa2	Bmag0011	ACAAAAACACCGCAAAGAAGA	GCTAGTACCTAGATGACCCCC
	MWG2031.2	TGTGACCTGTCAGACTGTTCAAGTT	AGCCAAGCATATCCTTCACTGACTA
	MWG889	CCCTGAATTCACGCGTTATT	GAATGAGCAACTACCGCATA
	Bmag120	ATTTCATCCCAAAGGAGAC	GTCACATAGACAGTTGTCTTCC
	GBM1126	AGCAGATGATTCCCCAGATG	GCCCACGGTGTAGATGTCTT
	WMG2259	TGATGGGTTCGCAAAGACG	TCCCTCTATGACATGGGCG
	GBM1492	GGAAGGAGACGAACACCAAA	AGGAGATCGAGCACGTAGGA
	GBM1419	CGTCACGCCACTCACCTC	CTTGAAGTCGGAACCCATGT
	GBM1174	TCTGGAAGAGGAAGGTGAGC	TTCCTCTTACCGTTCTTCGC
	ABG320	GATCCAACAGCAAGGAAAGA	AGACGAGTGGACACATGATG
	Bmag4	GTTTCCCATGCGACGTTC	GATGACGATTGATAGGTGT
	Bmac135	ACGAAAGAGTTACAACGGATA	GTTTACCACAGATCTACAGGTG
	Bmac31	AGAGAAAGAGAAATGTCACCA	ATACATCCATGTGAGGGC
	EBmac565C	ATTTGAATGTCCAACAGAATC	AATTGATAAGTTACTGACACACG
	Bmac0035	TCTCATCATTTTTTTGGGTGG	TGTGACCAAATACAAGAGGCC

duced expression of MIPS causes reduced PA synthesis. The MIPS gene, which exists as a single copy in barley (Larson and Raboy, 1999), has been mapped to chromosome 4H and M678 is a mutant closely linked to this gene.

Besides MIPS, other lpa genotypes result from mutation of kinases in later steps of the PA pathway. More than 20 lpa mutants have been identified in barley, representing at least 6 different loci, more than any other species to date (Hu et al. manuscript submitted). Barley lpa1-1 (M422) and lpa2-1 (M1070) have been mapped to chromosomes 2H and 7H, respectively, and result in 50 and 70% reductions in PA (Larson et al., 1998). In lpa1-1, PA is replaced by a molar equivalent of Pi; in lpa2-1, PA is replaced by both Pi and inositol pentakisphosphate, an intermediate in PA biosynthesis. Barley lpa3-1 (M635) and M955 are non-allelic but linked on chromosome 1H, and exhibit 75 and 95% reductions in PA, both with a corresponding increase in Pi. Other mutants include M640 on chromosome 7H (non-allelic to lpa2-1), and M499 and M2080, both of which are unmapped and could be non-allelic to identified loci (Hu et al. manuscript submitted). Pyramiding of these genes could be useful in generating cultivars with minimal PA.

Since PA is inversely related to Pi, lpa mutants result in a High Inorganic P (HIP) phenotype, which can be identified in a laboratory assay. Since different lpa mutations may have different effects on agronomic traits, combinations of different mutant genes could potentially optimize agronomic benefits in plants. Efficient use of lpa genes in breeding and selection, however, requires a rapid and reliable method for differentiation of individual loci be-

Table 1. Contd.

MIPS	Bmag369	CACTAGGCACCAATGACTG	ATCGAAAATCTTAGCTTTGG
	EBmac757	GTGTCTTTTTCACTTCCTTTG	TCTTCACTGTTGAGATGATGA
	EBmac755	AGCCTTGTGTATCAGGACA	CTGCTGGTGTTCTCTAAAAGT
	Bmac0064	CTGCAGGTTTCAGGAAGG	AGATGCCCGCAAAGAGTT
	Bmac582	GCCACATATGCACCCTAGTG	CATGGGGTAGTTTGTGCCTT
	Bmag217	AATGCTCAAATATCTATCATGAA	GGGGCTGTCACAAGTATATAG
	MWG799	TGCAAACTTGATGGCAGGCC	TCGGCGGCCTTGAGGTTGC
	Bmag767	AACTTACCTTCATATGTTGTGG	GAACACTATGATTCCATACGTC
	AWBMS0022	CCACTTCAAAGGCTTCCACA	CCGGAGAGTTGCTAATCT
	Bmac375	CCCTAGCCTTCCTTGAAG	TTACTCAGCAATGGCACTAG
	ABG472	CCGCGTACGCGAATCTGAGT	GCCCAGCTAGGTCGACAATA
	EBmac701	ATGATGAGAACTCTTCACCC	TGGCACTAAAGCAAAAGAC
	EBmac635	TGCTGCGATGATGAGAACT	TAGGGTAGATCCGTCCCTATG
	BCD453b	AGATTTGTACAACTCAACGGATATCA	AGCTCAAGCCTATTAGGATTCGG
	EBmac679	ATTGGAGCGGATTAGGAT	CCCTATGTCATGTAGGAGATG
	EBMac788	TAACTTACTTTATATCCATGGCA	ATGATGAGAACTCTTCACCC
	Bmag138b	ACCAGGAGGAATGAGAGAG	AATAAACCTTGAGACGATGG
	Bmac577	TCATACAGAAGCCCACACAG	TGCATGTTCATTCTAGACAGG
	EBmag781	CTATTTTCTAATGCTTGGACC	TGTCTAGTTCATCATCATTGC
	EBmac658	GTATGCAAGTGTAGGTGTGTG	CATGGGTTTACCCACATAC
	GBM1509	CAATCGTTGTCCAGAACCCT	GGCCGACAAATATGCTTCAT
	Bmac84	CTTGTGCCCTTTGATGCAC	CATAACTTGAGGATGTGTGTGACA
	GBM1299	GATCCCCCTAAAAGCAGACC	CTGCCTAGTCCCTGCATCTC
	GBM1220	GCTACCAGAACCCAGGAACA	TGAGCAACCTGAAACTGTCG
	GBM1448	GTATGACACCCGATCCATCC	CAAAATTTGGGACCTGAGGA
	GBM1338	ACGCACAGATACGTACACGC	GCCCCTCCTAGAACACAACA
	Bmag0714B	ATTCCTTATAGAGACACACGC	TTCTCTCCAACAATAAGAAGC
	EBmac691	TTAACAGAGGGCATTGGT	TCCTTTTCTCCATTTGAGTT
	Bmac181	ATAGATCACCAAGTGAACCAC	GGTTATCACTGAGGCAAATAC
	HVM68	AGGACCGGATGTTCATAACG	CAAATCTTCCAGCGAGGCT
	WMS6	CGTATCACCTCCTAGCTAAACTAG	AGCCTTATCATGACCCTACCTT

cause phenotypic assays can not distinguish individual plants for their genotype compositions. Although several lpa-linked RFLP, SCAR and ISSR markers have been identified (Larson et al., 1998; Roslinsky et al., 2007), high-throughput performable and regular PCR-based flanking markers are necessary to reduce undetected recombination between the gene and marker and to facilitate marker-assisted selection. In this investigation, it was reported that identification of flanking, PCR-based markers for 3 major lpa loci: lpa1-1, M640 in the lpa2 region, and the M678 in the MIPs gene region. Those markers are particularly in seedling stage selections in breeding program.

MATERIALS AND METHODS

Plant material

Mutations were induced via sodium azide treatment of barley cv. Harrington using methods described elsewhere (Nilan et al., 1973). Mapping populations were developed by crossing mutant lines M 422 (lpa1-1), M640 (lpa2 region) and M678 (MIPs region) to 'Steptoe.' An additional populations of M640 x 'Morex' and M678 x 'Morex' were also developed and used to detect additional polymerphism at the lpa2-1 locus. Populations were advanced in field plots in Aberdeen, ID and seeds were selected from F3 families for homozygosity testing of phosphorus phenotype in 20 individual seeds of each family. Pooled tissue sample from 20 homozygous seedlings of each family was used to represent the corresponding F2 individuals in DNA extraction and in genetic analysis.

Genotypic analysis

Homozygosity of F3 families at each mutant locus was determined using the single-seed HIP assay, in which Pi is detected colorimetrically (Chen et al., 1956; Raboy et al., 2000). A deep blue color indicated a high concentration of Pi, characteristic of a homozygous mutant genotype. The lpa mutations are recessive (Larson et al., 1998); thus, a 1:2:1 phenotypic ratio is still expected in seed from heterozygous plants. F2 plants were harvested individually and the harvested seeds were F3 families. PA content was measured in 20 seeds per F3 family and families were considered homozygous if assay results were uniform.

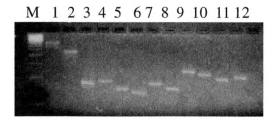

Figure 1. Polymorphisms between DNA samples of barley cvs. Harrington and Steptoe for 6 SSR markers flanking 3 lpa loci. M: 100 bp DNA ladder (Bio-Rad). Odd-numbered lanes were Harrington DNA; even-numbered lanes are Steptoe. PCR products from 2 DNA samples using the same primer pairs were loaded side by side. The order of DNA markers from left to right is MSU21, Bmag415, EBmac701, Bmag714B, Bmag120, and AWBMS0022.

Table 2. Mapping results of flanking DNA markers at 3 *lpa* mutant loci. In each locus, the mapping populations and number of homozygous families were specified. Proximal and distal markers and their genetic distances from each locus were reported.

Locus area	lpa1 (2H)	M640 (7H)	M678 (4H)
Mutant allele and mapping population used	M422 x Steptoe*	M640 x Steptoe* M640 x Morex*	M678 x Steptoe
No. of Homozygous family used	188	51	86 for EBmac701 41 for Bmag714B
Proximal marker	EBmac415	Bmag120	EBmac701
Recombinant	28	7	16
Genetic distance	28/188 = 14.9 cM	7/51 = 13.7 cM	16/86 = 18.6 cM
Distal marker	Msu21	AWBMS0022	Bmag714B
Recombinant	18	6	8
Genetic distance	18/188 = 9.69 cM	6/51 = 11.8 cM	7/41 = 17.0 cM

SSR and STS analysis

DNA was extracted using a cetyl trimethylammonium bromide (CTAB) protocol essentially as described elsewhere (Jackson et al., 2006). Molecular markers known to be near *lpa* loci were selected based on previous mapping information and the 2005 bar-ley consensus map (GrainGenes, http://wheat.Pw.usda.gov; Larson et al., 1998; Ramsay et al., 2000; Roslinsky et al., 2007; Varshney et al., 2007). Marker names and sequences screened in this study are summarized in Table 1. Primers of MSU21 amplifying non-specific bands were redesigned to generate simpler banding patterns using Primer3 program (http://frodo.wi.mit.edu) in these cases, the new primer sequences are shown under the same name in Table 1. All markers are either STS or SSR; several are homologous to RFLP or ISSR markers of the same name.

PCR reaction conditions were uniform for all markers, with each 25 μl reaction volume containing 50 ng template DNA, 1 μl each primer (10 μM), 2.5 μl 10X buffer containing the manufacture recommended Mg concentration, 1 μl dNTPs (2.5 mM for each nucleotide), and 1 U Taq polymerase (RedTaq, Sigma, St. Louis, MO). The thermal profile was 94°C for 3 min; 39 cycles of 94°C for 30 s, 52°C for 30 s, and 72°C for 1 min; and a 4°C hold. Reaction products were analyzed on 3% SFR high resolution agarose gels (Amresco, Solon, Ohio) stained with ethidium bromide and run in 1X TAE buffer. A 100 bp DNA ladder (Bio-Rad, Hercules, CA) was used as size marker in each gel. Polyacrylmide gel electrophoresis (PAGE) was conducted according to the protocol published previously (See et al., 2002).

Linkage analysis

Banding patterns were scored on each homozygous family and compared to the genotypes of both parents. Mutant or wild-type scores were also compared to phenotypic results from the HIP colorimetric assay for each individual family. Genetic distances between molecular markers and lpa loci were based on the percentage of recombinant phenotypes within each mapping population. Positions of flanking markers were determined based on the recombinant patterns detected in the individuals of mapping populations and compared to the published barley genetic maps.

RESULTS

Marker polymorphism

Although markers were selected based on known proximity to mutant loci, only a fraction of the primers evaluated were polymorphic between Harrington and Steptoe. For lpa1, 17 primer pairs were screened, including 12 barley markers and 5 wheat SSRs located in the same

Table 3. Comparison of primer sequences of EBmac701 and EBmac788. Bold letters indicate the same primer sequence in 2 markers. Italicized letters indicate the overlapping sequence in primers of the 2 markers. Information in this table was obtained from the GrainGenes database.

Primer name	Forward sequence	Reverse sequence	Product size	Repeat domain
EBmac701	**ATGATGAGAACTCTTCACCC**	*TGGCA*CTAAAGCAAAAGAC	149	$(AC)_{23}$
EBmac788	TAACTTACTTTATATCCA*TGGCA*	**ATGATGAGAACTCTTCACCC**	168	$(TG)_{23}$

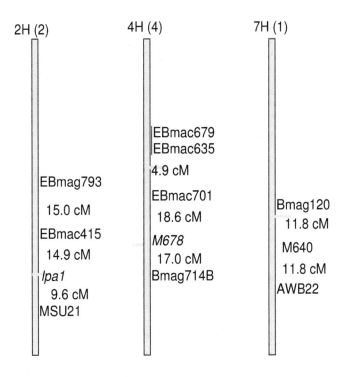

Figure 2. Genetic linkage maps of 3 barley lpa loci, showing flanking marker loci and genetic distance (cM) distances. Names of lpa loci are in bold. Markers and genes in the maps only illustrate the relative positions on chromosomes but not the actual locations.

in the same region of the corresponding wheat chromosome (Somers et al., 2004) (Table 1).

Three markers (MSU21, EBmag793 and EBmac415) were polymorphic (Figure 1). Preliminary test results using 44 individual DNA samples indicated that EBmag793 and EBmac415 are on the same side of the mutant locus but the former was further away (data not shown). MSU21 detected totally different recombinants compared to EBmac415, indicating that it was on another side of the gene. Therefore, only EBmac415 and MSU21 were used as flanking markers for genetic mapping in the selected populations.

Twenty-four primer pairs were evaluated for M640; four were polymorphic (Bmag0011, Bmag120, GBM1419, AWBMS0022). Mapping data indicated that Bmac0011 and Bmag120 were on the same side of the M640 gene but Bmac0011 was farther away. GBM1419 was mapped on the same location as Bmag120. Therefore, Bmag120 was selected as a flanking marker on one side of the gene. AWBMS0022 was the only marker that detected a different recombination pattern from the other 3 markers.

Bmag120 and AWBMS0022 were selected as flanking markers for the M640 gene. Polymorphism for AWBMS 0022 was a little difficult to detect using SFR agarose; therefore, this genotype was confirmed on a polyacrylamide gel. For the M678 in the MIPs region, 22 primers were tested, of which 4 were polymorphic (EBmac701, EBmac635, EBmac679, Bmag0714B). EBmac701, EB mac635, and EBmac679 were mapped to the same location and Bmag714B was mapped to the other side of the M678 allele or gene in the MIPs locus. In this study, it was selected EBmac701 and Bmag714B as flanking markers for this locus.

Mapping of flanking markers

Flanking markers selected for each locus were genetically mapped in corresponding mapping populations (Table 2 and Figure 2). Using 188 homozygous families from an M422 x Steptoe population, the closest proximal marker for lpa1 was EBmac415, linked at 14.9 cM, on the distal side; Msu21 was the closest, at a distance of 9.6 cM. Orders of proximal and distal markers were decided based on the published barley 2006 consensus map in lpa1 locus (Marcel et al., 2007) at lpa1. M640 was previously mapped to the lpa2 area but its phenotype was obviously different from lpa2, indicating that it may be a different gene (Hu et al. manuscript submitted).

The mutant name of M640 was used to refer to the locus. Data from 2 populations were combined to select flanking markers for lpa2: polymorphism was identified between M640 and Steptoe as well as between M640 and Morex. The proximal marker, Bmag120, was linked at 13.7 cM; the distal marker, AWBMS0022, was linked at 11.8 cM. The orders of proximal and distal markers in the lpa2 area were decided based on the published map of barley Barque73 x CPI (GrainGenes, http://wheat.pw.us da.gov; Hearnden et al., 2007).

Mapping of the MIPS linked locus utilized a population of 41 families for the marker Bmag714B from M678 x Steptoe and 86 families for EBmac701 from both M678 x Steptoe and M678 x Morex populations. At 18.6 cM, EB mac701 was the closest proximal marker; Bmag714B, at 17.0 cM, was the closest distal marker. The proximal and distal orders of EBmac701 and Bmag714B were decided based on the same Barque73 x CPI map (GrainGenes, http://wheat.pw.usda.gov; Hearnden et al., 2007).

SSR marker redundancy on chromosome 4H

Barley consensus maps from 2003 and 2005 differ in pri-

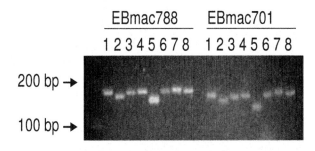

Figure 3. Comparison of amplification profiles of EBmac788 and EBmac701 using 8 DNA samples. Lane 1-8 represents the DNA of barley cvs. Harrington, Steptoe, Morex, CDC Alamo, Waxbar, Baronesse, barley germplasm Oregon Wolfe Barley Dominant, and barley cv. Azhul. PCR products were separated on 3% SFR agarose

mer content and marker position; however, both maps include EBmac701 and EBmac788 as discrete markers on chromosome 4H (GrainGenes, http://wheat.pw.usda.gov). Marker screening of the M678 in the MIPS region in this project initially included both markers. Results, however, indicate that EBmac788 is not a discrete marker but is a duplicate of EBmac701 in the opposite orientation. Comparison of primer sequences revealed that EBmac701F is identical to EBmac788R (Table 3). Flanking primers in each pair shared a five-base sequence on one end, with unique sequences on opposite ends. Thus, the product size of these markers differs by 19 bp, although resultant genotypes are identical. To confirm the equivalence of these markers, amplification profiles were compared using DNA from barley cultivars Harrington, Steptoe, Morex, CDC Alamo, Waxbar, Baronesse, Azhul, and barley germplasm Oregon Wolfe Barley Dominant (Figure 3). Profiles were identical for the 2 markers.

To further clarify the marker order in this area, EBmac 635 and EBmac679 were mapped in the same mapping population of M678 x Steptoe. Results indicated that these 2 markers are proximal to EBmac701 at about 4.9 cM distance (Figure 2). Since these 2 markers were mapped between EBmac701 and EBmac788 (Ramsay et al., 2000), the correct mapping position for EBmac701/EBmac788 should correspond to EBmac788 in the published map.

DISCUSSION

Considerable interest has centered on development of lpa grain as a solution to dietary micronutrient deficiency and environmental pollution. Feed barley with lpa could provide a nutritious and cost-effective alternative to current feedstuffs (Li et al., 2001; Sugiura et al., 1998; Veum et al., 2002). Hence, lpa has become a significant criterion in development of new feed barley cultivars. Recently, barley lines with lpa genes have been released (Bregitzer et al., 2008) and there is strong evidence for upcoming popular utilization of the lpa mutant genes or alleles in feed barley development. Lower phytate and higher inorganic phosphorus content in lpa mutations not only provides more available P for non-ruminant animals but also reduces phytate contamination in ground water. The contribution of lpa lines to decreased water contamination may be more significant because a clean environment will extensively affect people's lives in a positive way. With better knowledge of the beneficial impacts of lpa genes or alleles, incorporation of those genes in cultivar development will be enhanced.

Efficient selection of this trait in barley breeding programs could be facilitated by flanking DNA markers. Marker assisted selection has been proved a rapid and effective approach to enhance the selections when genotypes of seedlings are required. Using markers is critical where more than one gene need to be pyramided because phenotype assessments for each individual will be difficult. While the data presented here coincided with mapping results in published literature (Larson et al., 1998; Roslinsky et al., 2007), development of flanking markers at 3 major lpa loci may enable more rapid progress in practical use of these mutations. Flanking DNA markers will assure the presence of a specific gene or allele in the individuals because false positive will be very rare.

The genetic explanation is very simple: double recombination between 2 flanking DNA markers occurs at a much lower frequency compared to recombination between one and a single marker. The markers identified in this study are PCR-based, and thus amenable to high-through put genotyping. For the majority of markers, polymorphism was unambiguously visualized in agarose gels, a further advantage for rapid laboratory analysis. Markers with difficult-to-detect polymorphism in the agarose gel, such as AWBMS0022, can be validated by sequencing based genotyping facility or use of polyacrylamide gels. Development of flanking DNA markers for Lpa loci in this report is only the start. Results provided a base for further fine-tuning the markers for tighter linkage and easier detection. Even though markers were developed in limited barley cultivars, testing of markers near the ones identified in this study in different barley lines should enable identification of useful ones.

In addition to lpa marker development, this study has improved the marker accuracy in the MIPS region of chromosome 4H. Global mapping efforts may generate multiple instances of marker redundancy, such as that detected for EBmac701 and EBmac788. The marker redundancy will result in misinterpretation of genetic maps because the redundant marker will show different mapping positions due to the different mapping populations used. The consensus map has to incorporate different maps because of the difficulty of re-testing all the markers in the same population. Inaccurate mapping positions or redundancy may be more problematic in gene fine mapping or map-based gene cloning. Therefore, it may be a good idea for the marker end-users to verify published marker positions and distances in the population of interest before using them. Publishing the confirmed data will benefit the barley marker community.

ACKNOWLEDGEMENTS

We thank Dr. Jianli Chen for the critical review of the manuscript. This research was funded by USDA CRIS project number 5366-21310-003-00D.

REFERENCES

Bregitzer PP, Raboy V, Obert DE, Windes J, Whitmore J (2008). Registration of 'Clearwater' low-phytate hulless spring barley. J. Plant Regist. 2: 1-4.

Chen PS, Toribara TY, Warner H (1956). Microdetermination of P. Anal. Chem. 28: 1756-1758.

See D, Kanazin V, Kephart K, Blake T (2002). Mapping Genes Controlling Variation in Barley Grain Protein Concentration. Crop Sci. 42: 680-685.

Erdman J (1981). Bioavailability of trace minerals from cereals and legumes. Cereal Chem. 58:21-26.

Graf E, Epson KL, Eaton JW (1987). Phytic acid: A natural antioxidant. J. Biol. Chem. 262: 11647-11650.

Hanakahi LA, Bartlet-Jones M, Chappell C, Pappin D, West SC (2000). Binding of inositol phosphate to DNA-PK and stimulation of double-strand break repair. Cell 102: 721-729.

Hearnden P, Eckermann P, McMichael G, Hayden M, Eglinton J, Chalmers K (2007). A genetic map of 1000 SSR and DArT markers in a wide barley cross. Theor. Appl. Genet. 115: 383-391.

Jackson EW, Avant JB, Overturf KE, Bonman JM (2006). A quantitative assay of *Puccinia coronata* f. sp. *Avena* DNA in *Avena sativa*. Plant Dis. 90:629-636.

Larson SR, Raboy V (1999). Linkage mapping of maize and barley *myo*-inositol 1-phosphate synthase DNA sequences: Correspondence with a low phytic acid mutation. Theor. Appl. Genet. 99:.27-36.

Larson SR, Rutger JN, Young KA, Raboy V (2000). Isolation and genetic mapping of a non-lethal rice (Oryza sativa L.) low phytic acid 1 mutation. Crop Sci. 40: 1397-1405.

Larson SR, Young KA, Cook A, Blake TK, Raboy V (1998). Linkage mapping of two mutations that reduce phytic acid content of barley grain. Theor. Appl. Genet. 97: 141-146.

Lemtiri-Chlieh F, MacRobbie EAC, Brearley CA (2000). Inositol hexakisphosphate is a physiological signal regulating the K^+-inward rectifying conductance in guard cells. Proc Nat Acad. Sci. USA. 97:8687-8692.

Li YC, Ledoux DR, Veum TL, Raboy V, Zyla K (2001). Low phytic acid barley improves performance, bone mineralization, and phosphorus retention in turkey poults. J. Appl. Poult. Res. 10: 178-185.

Loewus FA (1990). Inositol biosynthesis. In: Morfe DJ, Boss WF, Loewus FA (eds) Inositol Metabolism in Plants. Wiley-Liss, New York pp.13-19.

Marcel TC, Varshney RK, Barbieri M, Jafary H , de Kock MJD, Graner A, Niks RE , de Kock MJ (2007). A high-density consensus map of barley to compare the distribution of QTLs for partial resistance to *Puccinia hordei* and of defence gene homologues. Theor. Appl. Genet. 114: 487-500.

Menniti FS, Oliver KG, Putney JW, Shears SB (1993). Inositol phosphates and cell signaling: New views of $InsP_5$ and $InsP_6$. Trends Biochem. Sci. 18: 53-56.

Nilan RA, Sideris EG, Sander C, Konzak CF (1973). Azide – A potential mutagen. Mutation Res. 17:142-144.

Ockenden I, Falk DE, Lott JNA (1997). Stability of phytate in barley and beans during storage. J. Agric. Food Chem. 45: 1673-1677.

Raboy V (1997). Accumulation and storage of phosphate minerals. In: Larkins BA, Vasil IK (eds) Cellular and Molecular Biology of Plant Seed Development. Kluwer Academic Publishers, Netherlands pp. 441-477.

Raboy V, Gerbasi PF, Young KA, Stoneberg SD, Pickett SG, Bauman AT, Murthy PP, Sheridan WF, Ertl DS (2000). Origin and seed phenotype of maize low phytic acid 1-1 and low phytic acid 2-1. Plant Physiol. 124: 355-368.

Ramsay L, Macaulay M, degli Ivanissevich S, MacLean K, Cardle L, Fuller J, Edwards KJ (2000).A simple sequence repeat-based linkage map of barley. Genetics 156:1997-2005.

Roslinsky V, Eckstein PE, Raboy V, Rossnagel BG, Scoles GJ (2007). Molecular marker development and linkage analysis in three low phytic acid barley (*Hordeum vulgare*) mutant lines. Mol. Breed 20: 323-330.

Sasakawa N, Sharif M, Hanley MR (1995). Metabolism and biological activities of inositol pentakisphsphate and inositol hexakisphosphate. Biochem. Pharmacol. 50: 137-146.

Sharpley AN, McDowell RW, Kleinman PJA (2001). Phosphorus loss from land to water: Integrating agricultural and environmental management. Plant and Soil 237: 287-307.

Shi J, Wang H, Hazebroek J, Ertl DS, Harp T (2005). The maize low-phytic acid 3 encodes a myo-inositol kinase that plays a role in phytic acid biosynthesis in developing seeds. Plant J. 42: 708-719.

Somers DJ, Isaac P, Edwards K (2004). A high-density microsatellite consensus map for bread wheat (*Triticum aestivum* L.). Theor. Appl. Genet. 109: 1105-1114.

Sugiura SH, Raboy V, Young KA, Dong FM, Hardy RW (1998). Availability of phosphorus and trace elements in low-phytate varieties of barley and corn for rainbow trout (*Oncorhynchus mykiss*). Aquaculture 170: 285-296.

Varshney RK, Marcel TC, Ramsay L, Russell J, Röder MS, Stein N, Waugh R, Langridge P, Niks RE, Graner A (2007). A high density barley microsatellite consensus map with 775 loci. Theor. Appl. Genet. 114: 1091-1103.

Veum TL, Ledoux DR, Bollinger DW, Raboy V, Cook A (2002). Low-phytic acid barley improves calcium and phosphorus utilization and growth performance in growing pigs. J. Anim. Sci. 80: 2663-2670.

Wilcox JR, Premachandra GS, Young KA, Raboy V (2000). Isolation of high seed inorganic P, low-phytate soybean mutants. Crop Sci. 40: 1601-1605.

York JD, Odom AR, Murphy R, Ives EB, Wente SR (1999). A phosphorlipase C-dependent inositol polyphosphate kinase pathway required for efficient messenger RNA export. Science 285: 96-100.

Intercropping and its application to banana production in East Africa: A review

George Ouma

Department of Botany and Horticulture, Maseno University, P. O. BOX 333, Maseno, Kenya.
E-mail: goumaoindo@yahoo.com or gouma@maseno.ac.ke.

Bananas are very important in Kenya for domestic consumption and export. They are extensively grown where they are mainly intercropped with short term crops. There has been an increase in the grower interest in using intercropping, growing two or more crops simultaneously on the same land in the development of new cropping systems for their land. Intercropping could reduce management inputs and result in sustainable systems that more effectively use and even potentially replenish natural resources used during crop production for long term management of farmland. While intercropping has been practiced more widely in the developing countries of Central America, Asia and Africa, developed countries have not adopted it well. Some benefits of intercropping to the grower are risk minimization, effective use of available resources, efficient use of labour, increased production per unit area of land, erosion control and food security. This paper discusses the effects of intercropping on pest and disease control, physiology of the crops grown, cultural practices such as date of planting, spacing and plant density, soil fertility and time of planting among other effects and lastly banana production in East Africa in relation to intercropping and declining soil fertility in banana-based cropping systems.

Key words: Sustainable, cultural, food security, crops, efficient, cropping system, management.

INTRODUCTION

Banana Production in East Africa

Bananas (Musa spp. AA) and plantains (Musa spp. AAB) are of major importance as staple food crops in much of Sub-Sahara Africa. The region as a whole produces nearly 30 million tonnes of the crop annually (Gold et al., 1999). Production has increased in recent years but this has been due to increased in area planted rather than increase in productivity. Between 1970 and 2001, the area under banana and plantain in Africa increased from 3.2 to 4.7 million hectares and in the same period yields marginally decreased from 6.16 tonnes per ha 5.99 tonnes per ha.

The largest producer and consumer in East Africa is Uganda which produced 9.5 million tonnes in the year 2001 (Reddy et al., 1992). The crop is produced almost exclusively by small scale farmers and is used almost entirely for home consumption. The type of banana grown is the cooking type (East African highlands banana). It is the main type grown in the East African Highlands. The constraints to banana production are pests, diseases and its perishability. These are the major constraints to banana production. Black sigatoka leaf spot disease (*Mycosphaerella fijiensis*) has spread throughout the region and all the banana cultivars are susceptible (Reddy et al., 1992). Another disease is *Fusarium oxysporum* (Cubense).

This is a soil borne fungus that affects banana production. The fungus is persistent in the soil for many years after infection rendering chemical control impossible.

Nematodes are also a problem in bananas. Nematodes (*Rodopholus similis, Pratylenchus* spp., *Melodygyne* spp., *Helicolylenchus, M*ultunctus and weevils (*Cosmopolites sordidus* are common and cause yield losses due to reduced nutrient up take and toppling of plants viruses such as banana streak virus are also constraints to banana production (Reddy et al.,1992).

Bananas are also very perishable causing losses between farm gate and market. There is need to improve post-harvest handling and storage techniques during periods of over production (Reddy et al., 1992).

Banana intercropping in East Africa

Intercropping is a very common cropping system in East Africa and it is practiced by majority of the farmers mainly due to declining land sizes and food security needs. There has been in grower interest using intercropping pos-sibly because it could reduce management inputs that result in sustainable systems that more efficiently use and even potentially replenish natural resources used during crop production for long term management of farm land. It has been used in the developing countries of Central America, Asia and Afica (Altier and Liebman, 1994) and its advantages are: risk minimization, effective use of available resources, efficient use of labour, increased crop productivity, erosion control, food security (Andrew and Kassam, 1976) and pest control (Wein and Smithson, 1979).

Bananas have perennial characteristics and may be grown on the same piece of land for up to 50 years. Cultivation is through clonal propagation. The usual spacing is 3 by 3 m. When being established crops like beans, coffee, maize and sweet potatoes are intercropped with the young banana plants. When the land was still plentiful, the intercrop would be phased out after a year (two cropping seasons) and farmers would start mulching the bananas. Only a few tree crops such as ficus (*Ficus nataliensis*), Jack fruit (*Artocarpus heterophillus*), and pawpaw (*Carica papaya*) remained in the plantation mainly to serve as wind breaks. In Tanzania coffee-banana cropping system is the most widespread farming practice characteristic of the Kilimanjaro region, Mbeya, Kagera and Arusha areas (Anon, 2008). The benefit to adopting this system versus a pure coffee system is that it offers higher returns to the small holder (Anon, 2008). Bananas can also be inter planted with coconut of any age between 8-25 years but palms of more than 25 years are more suitable for intercropping since the light transmitted increases with increase in age and it is ideal for raising perennial crops (Jodha, 1979; Grossman and Quales, 1993). This happens at the Kenyan coast. Intercropping banana with annual crops can be remunerative. Farmers with limited resources have traditionally multicropped their lands to minimize risks associated with growing a single crop and to ensure more stable subsistence in terms of food nutrition and possible income. Intercrops which can easily be raised in banana plantation at early stages of growth are radish, cauliflower, cabbage, spinach, chili, brinjal, yam and cucurbitaceae crops are grown as intercrops. (Cassava/banana combination is one of the most efficient cropping systems (Bekunda and Woomer, 1996, 1999).

In Uganda bananas are intercropped with pineapple as a source of food and income for the family while pineapples are exported (Grossman and Quales, 1993; Jodha, 1979). Due to the increase in land pressure recently in East Africa intercropping of bananas is now carried out in old plantations and some other crops inter planted between banana plants are vanilla, solanacea crops, fruit trees and sugarcane (Bekunda and Woomer, 1996).

Effects of banana intercropping

Several studies have been conducted to investigate the effect of various intercrops on the performance of bananas with respect to yield, growth and pest incidence. For example there are attempts to investigate bananas in Agro forestry systems. This has been undertaken by small holder farmers with small pieces of land to meet their wood, fodder and food needs (Akeampong, 1995 and 1999). In a study conducted in Burundi to study the effects of nine trees species planted at 4 × 8 m on the yield of bananas planted at 4 × 4 m and beans planted at a density of 1000 plants per hectare., the results showed that none of the trees affected the yield of bananas and beans at their early growth stage. The legume intercrops had no effect on banana yield.

Therefore, land use efficiency of small holder farms in East Africa can be increased by incorporating food and or fodder legumes into banana cropping systems (Ddungu, 1987) Akeampong et al., 1999). In the same study when banana was intercropped with three densities of *Grevilla robusta* of 208, 313 and 0.25 trees per hectare, it was reported that after three and a half years wood volume *of G. robusta* was highest, but banana and bean yields were unaffected (Ddungu,1987). *Cedrela serrata* was found to be the best tree to be intercropped in a banana/bean intercrop (Akeampong et al., 1995; Akeampong et al., 1999). As far as pest management in banana intercropping is concerned it has been reported that the number of *C. sordidus*, banana weevil was lowest in banana intercropped with millet. Yields losses were as high in the intercropped banana and mulched ones (Raman, 2006).

Similar findings were also reported by other workers (Raman et al., 2006). Leguminous crops in the *genera Canavalia muzinna* and *Tephrosia vogelli* have been reported to be having repellent or insecticidal properties against the banana weevils, *C. sordidus* and nematodes (*R. similis*) when intercropped with bananas (McIntyre et al., 1981). In one study it was tested if weevil and nematode populations can be affected under banana intercropped with the legumes such as *Canavolia ensiforms*, *Mucuna pruviens* and *T. vogelli*. It was reported that they caused an attack by banana root neorosis whose incidence was highest in *T. vogelli* intercrop than the banana mono culture. Banana yields were unaffected by the legumes showing that they can be intercropped with bananas (Mcintyre et al., 1981). Similarly banana intercopping with sweet potatoes has reduced the incidence of the root lesion nematode (*Pratylenctius goodey*) in heavily infected banana plants around Lake Victoria Basin (Bekunda and Woomer, 1996). Banana is mostly grown by small scale farmers on an average land size less than two hectares. Nutrient depletion on small holder farms has been cited as the main reasons for the decline in banana yields (De and Singh, 1979). Soil fertility in East Africa is mainly replenished by the use of organic crop residues (Delvaux, 1996) but many banana residues are not returned to the soil because there have been many alternative uses (Rubaihayo, 1991; Rogers and Dennis,1993).

Inorganic fertilizers could be used to supplement the organic inputs. Unfortunately these have a supply constraint that hindering their uses (Rubaihayo, 1991). According to (Rubaihayo 1991)) soil fertility depletion in small scale farms is largely a consequence of socio economic and policy constraints and distortion. Farmers in East Africa use the following organic resources to replenish soil fertility in banana based cropping systems namely banana stalks and leaves at pruning upon harvest, a practice that may contribute to banana weevil, stem nematode and sigatoka fungal attacks. Banana stalks are also applied as mulches (Bekunda and Woomer, 1996). Farmers apply a wide range of additional resources to banana based cropping systems including field crop residues and burned residues and on farm manures, intercropping with legumes may also be as strategy to offset fertility depletion (Bekunda and Woomer, 1996). With respect to light particularly where Banana is interplanted with big agroforestry trees, it has become adapted to such shaded conditions and its yields are not affected (Senevirathing et al., 2008). Further, some nutrients are absorbed better under intercropping systems (Kurien et al., 2006).

Conclusion

Intercropping is a very beneficial cropping system in East Africa because of its advantages of increasing food security and reducing poverty and increasing soil fertility. It is increasingly becoming important in East Africa where land scarcity is increasing.

Banana is a very important crop in East Africa but due to the fact that small scale farmers also require food security, it should be intercropped with many of the annual crops being grown by farmers to achieve this.

REFERENCE

Akyeampong EL Hitimana S, Franzel Munyemana PC (1995). The agronomic and economic performance of banana, bean and tree intercropping in highlands of Burundi: an interm assessment. Agrofor. Sys. 31(3): 199 -200.

Akyempong EL, Hitimana E, Munyemana TPC (1999). Multistrata Agroforestry with beans, bananas and Grevillea robusta in the highlands of Burundi. Exp. Agric. 35: 357-369.

Altier MA, Leibman M (1994). Insect, weed and plant disease management in multiple cropping systems In Francis CA (ed) Multiple cropping systems, Macmillan Company, New York p. 383.

Andrew DJ, Kassam AH (1976). The importance of multiple cropping systems in increasing world food supplies: In RI Papendick PA, Sanchez GB Triplett (Eds). Multiple cropping Am. Soc. Agron. Spec. Publ. 27. pp. 171-200

Anonymous (2008). A review of Coffee-Banana Based cropping Systems in Tanzania Economics.

Bekunda M, Woomer PL (1996) Organic resources management in banana-based cropping systems of the Lake Victoria Basin, Uganda. Agric. Ecosys. Environ 59(3): 171-180.

Bekunda M (1999). Farmers responses to soil fertility decline in banana-based cropping systems in Uganda. Managing Africa's soils. No. 4 Russell Publishers, Nottingham 2

De R, Singh SP (1979). Management practices for intercropping systems. Proc. Intl. Workshop on intercropping. 10-13 Jan. 1979, Hyderabad, India. Intl. Crops Res. Inst. For the semi-arid tropics Pantancheru, India. pp. 17-21.

Delvaux B (1996) Soils in bananas and plantains. Gouven, S. (Ed) London, U. K. Chapmans Publishers. pp. 230-253.

Ddungu JCM (1987). Regional needs for banana and plantain. Improvement in East Africa. In: Banana and plantain strategies Parsley GJ, De Longhe, East Africa (Eds) pp. 36-38.

Gold CS, Karamura EB, Kiggunda A, Baganda F, Abera AMK (1999). Geographic shifts in highland banana (Musa group AAA-E. A.) Production in Uganda. Afr. Crop Sci. J. 7: 223-298.

Grossman JW, Quarles (1993). Strip intercropping for Biological control. IPM Practitioner April. pp. 1-11.

Jodha NS, (1979) Intercropping in traditional farming systems. Proc. Intl. Workshop on intercropping 10-13 Jan. 1979, Hyderabad, India Intl. Crops.

Kurien S, Kumar S, Nemkavil V (2006). Relative efficiency of 32p uptake in a banana-based Intercroppings system. Fruits 6(1): 353-366.

Lima AF, Lopez LH (1979) Plant popu-lation and spatial arrangement study on intercropping maize and beans P. vulgaris in North East Brazil. Proc. Int. Workshop. On intercropping, 10-13 Jan. 1979, Hyderabad, India, Intl. Crops Res. Inst. For the Semi-Arid. Tropics. Pantancheru, India, pp. 41- 45.

McIntyre BC, Gold I, Kashaija H, Ssali Night E, Bwamiki D (2001) Effects of legume Intercrops on soil borne pests, biomass, nutrients and soil water in banana. Biol. Fertil. Soils 34(5): 342-348.

Rahman M, ZRahman MH, Haqu ME, eKalar MH, Naber SL (2006). Banana-based intercropping system in Northern part of Bangladesh. J. Agron. 5(2): 228-231.

Reddy KC, Visser P, Buckner P (1992). Pearl millet and cowpea yields in sole and intercrop systems and their after effects and soil and crop productivity. Field Crops Res. 28: 315-326.-19.

Roger F, Dennis RD, (1993). Developing an effective Southern pea and sweet corn intercrop system. Hort. Technol. 3 (2): 178-183.

Rubaihayo PR (1991) Banana-based cropping systems research. A report on Rapid Rural Appraisal sun production, Research Bulletin No. 2 Makerere University.

Senevirathing AMK, Stirling CM, Rogrigo VH (2008). Acclimation of photosynthesis and growth of banana (Musa spp.) to natural shade in the humid tropics. Exp. Agric. 44: 301-312.

Wein HC, Smithson JB, (1979) The evaluation of genotypes for inter cropping. Proc. Intl. Workshop on intercropping 10-13 Jan. 1979. Hyderabad, India. Intl. Crop Res. Inst. For the Semi-Arid Tropics, Patancheru, India, pp. 105-110.

A study on stigma receptivity of cytoplasmic-nuclear male-sterile lines of pigeonpea, *Cajanus cajan* (L.) Millsp.

R. H. Luo[1], V. A. Dalvi[1], Y. R. Li[1] and K. B. Saxena[2]

[1]Crop Genetic Improvement and Biotechnology Laboratory, Guangxi Academy of Agricultural Sciences, Nanning, P.R. China 530007.
[2]International Crops Research Institute for the Semi-Arid Tropics (ICRISAT), Patancheru, A.P. India.

Stigma receptivity in pigeonpea [*Cajanus cajan* (L.) Millsp.] was studied using 2 male-sterile lines (ICPA 2039 and ICPA 2043) under field conditions. An experiment was conducted to observe the stigma receptive period at Nanning (22°N 108°E) in Southern China. The study revealed that the stigma was receptive 48 h before flower opening and continued up 4 days after flower opening. The peak stigma receptivity was on the day of flower opening with 84 and 86% pod set after hand pollination in the male-sterile lines ICPA 2039 and ICPA 2043, respectively. This floral stage is characterized by the corolla to calyx length ratio of 2.75 ± 0.075 (ICPA 2039) and 2.60 ± 0.283 (ICPA 2043). The long time span of stigma receptivity in pigeonpea encourages insect-aided natural out-crossing. This information will help breeders to carry out hybridization activity with high success rates. Also the long receptivity may facilitate more seed yield in isolated seed production blocks.

Key words: CMS lines, pigeonpea, stigma receptivity.

INTRODUCTION

Pigeonpea (*Cajanus cajan* (L.) Millsp.) is a commercially important legume crop. In recent past, hybrid breeding is considered as an effective tool in this crop (Stakstad, 2007). In crop plants like pigeonpea where male-sterility system is recently developed, it is necessary to study the environmental effect on stigma receptivity duration. Stigma receptivity is an important factor to have higher success rate of natural out-crossing. The literature reports some studies on this aspect but very few with male-sterile lines. With the development of the cytoplasmic-nuclear male-sterile (CMS) lines, it is necessary to study this aspect in detail to enhance the utilization of CMS technology.

For successful commercialization of any hybrid, easy seed production method is a pre-requisite, which is dependant on insect behavior in the particular location, stability of male-sterile line, and duration of stigma receptivity (Saxena et al., 2006). This paper emphasizes the research need of stigma receptivity in China. Already some research has been done in India on pigeonpea (Prasad et al., 1977) and other crops such as silk oat (Kalingnire et al., 2000) and bullelgrass (Shafer et al., 2000). This research paper confirms the previous findings of stigma receptivity in pigeonpea but at different environmental conditions. This will help in standardization of the stigma receptivity time at different locations.

MATERIALS AND METHODS

For the present study the seeds of cytoplasmic-nuclear male-sterile (CMS) lines ICPA 2039 and ICPA 2043 (along with its maintainers) were sown in isolation during 2008 rainy season (Crop duration from June - January) in field at Guangxi Academy of Agricultural Sciences, Nanning (22°49 N), China. To study the stigma receptivity hand pollinations were carried out on male-sterile plants using pollen of the maintainer line at different stages of flower buds. As the female parent is male-sterile, there was no need of emasculation for controlled pollinations and this avoided chances of accidental self- or cross-pollination due to isolation.

To identify appropriate bud size for maximizing pod set after hand

*Corresponding author: E-mail: vijay_dalvi79@rediff.com

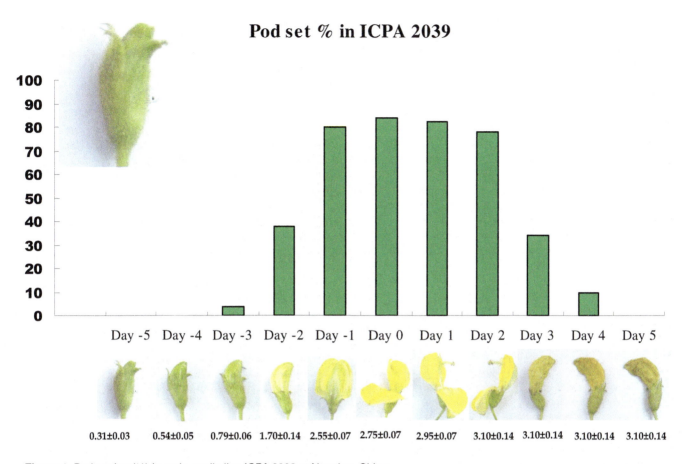

Figure 1. Pod setting (%) in male-sterile line ICPA 2039 at Nanning, China.

pollinations, 5 young buds with petals just emerging (Figure 1, inset) were selected randomly on 5 male-sterile plants and photographs were taken at 24 h intervals starting 0930 h. The procedure used by Dalvi and Saxena (unpublished) was used for this study. The day when the flowers opened was designated as 'Day 0' and it took 5 days for the selected young buds to open and another 5 days to drop from the pedicel base. The initially selected bud stage was designated as 'Day 5' and the subsequent stages after each 24 h were designated as 'Day -4', 'Day -3', 'Day -2', and 'Day -1'. Similarly, the stages after flower opening were designated as 'Day +1', 'Day +2', 'Day +3', 'Day+4' and 'Day +5' (Figure 1). To develop a visual bud selection index for hybridization, the lengths of corolla and calyx were measured on each day and their ratios were estimated. In case of ICPA 2039 these indices ranged from 0.31 ± 0.034 (Day -5) to 3.10 ± 0.141 (Day +5) and at 'Day 0' stage this ratio was 2.75 ± 0.075. Similar indices were recorded in case of ICPA 2043 male-sterile line. The Day-5, Day 0 and Day+5 indices were 0.27 ± 0.025, 2.60 ± 0.283 and 2.90 ± 0.141, respectively.

11 male-sterile plants each of ICPA 2039 and ICPA 2043 were selected randomly and one plant was assigned for pollinating one stage of bud. In male-fertile plants of the maintainer line the pollen dehiscence started a day before flower opening that is, at Day -1 stage and the dehisced pollen grain remained intact on the anther lobes up to Day +1. During this period the pollen grains exhibited >95% viability when examined under microscope using 2% aceto-carmine solution. Therefore for pollinations fully developed but unopened flower buds were harvested from the respective maintainer lines and 50 pollinations were done at each stage (from Day -5 to Day +5). To minimize the possible effect of micro-environment on fertilization only 10 pollinations were done on each bud stage every day. The targeted pollinations were done during 21st - 25th December 2008. Each pollinated bud was tagged with a thread for identification and pod set was recorded 3 weeks after completing the pollinations. The pod set after hand pollinations was considered as indicator of stigma receptivity.

RESULTS

The present experiment revealed that it needs on average of about 11 days for a tiny bud to complete its life as flower. In this study we have examined 2 cytoplasmic-nuclear male-sterile lines viz. ICPA 2039 and ICPA 2043. Only 4% pod set was recorded in ICPA 2039 when the pollinations were made 72 h before anthesis and prior to this no pod set was observed indicating stigma was not receptive at that time. On the subsequent day, the pod set improved rapidly and it was highest (84%) when pollinations were made on the day of flower opening (Day 0) and it remained in the high regime for another 2 days with 82 and 78% pod set (Figure 1). Subsequently, the pod set declined with time and there was no pod set 96 h after flower opening (Day 0). In case of ICPA 2043 the pod set was considerable (14%) 72 h before flower opening and the maximum pod set was 86% on the day of flower opening (Figure 2). The pod set declined further

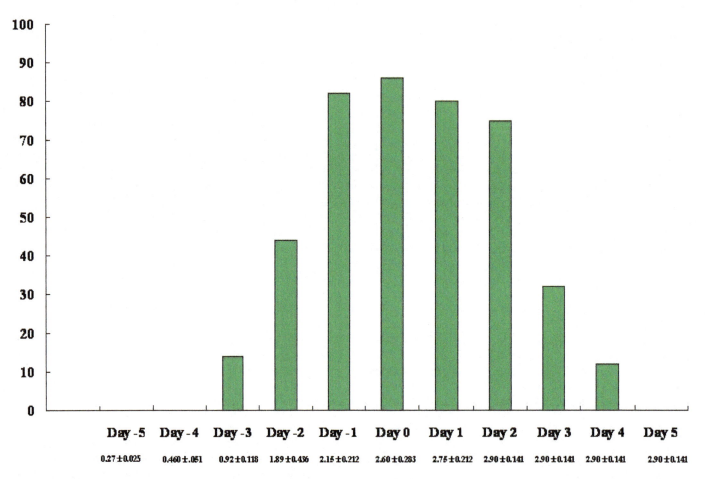

Figure 2. Pod setting (%) in male-sterile line ICPA 2043 at Nanning, China.

and no pod set was observed on Day +5. This experiment showed that in ICPA 2039 the stigma was receptive for 6 days. Such longer stigma receptive period in male-sterile lines will help for better pod set by hand pollination. Similarly, in ICPA 2043 the stigma was receptive for 6 days. There is need to study the effect of environmental factors such as temperature and humidity, whether these parameters have any effect on the duration of stigma receptivity. The mean temperature during the study period was 13°C with a mean relative humidity of 72%.

DISCUSSION

The previous studies predicted that the variation observed in pod setting at different stages could be attributed to the inherent developmental changes in stigma and embryo sac of the female flowers. The large variation observed in grain setting in silky oat (Kalinganire et al., 2000) was attributed to the changes in stigma and embryo sac structures while in buffelgrass (Shafer et al., 2000) it was due to the protogynous nature of the flower. In pigeonpea such studies are needed to understand the role of developmental changes in stigma and embryo sac in pod set. The decline in the pod set rate is attributable to the age of the flower.

The results of the present experiment showed that in pigeonpea the receptivity of stigma started 48 - 72 h before flower opening and continued to be receptive 96 h there after, but within this period a considerable variation for pod set was observed on different days. Dalvi and Saxena (unpublished) also observed a similar trend for stigma receptivity duration at Patancheru, India. The previous studies reported 68 h stigma receptivity before flower opening and 20 h after flower opening at Bihar, India by Prasad et al. (1977). The differences observed in various studies could be attributed to the differences in the methodology and/or the genotypes used in the studies and environmental conditions. From the study conducted at Patancheru, India it is concluded that for maximizing pod set the pollinations should be initiated a day before flower opening and be continued for three days. To select the appropriate floral buds for pollination the calyx to corolla index should be between 1.8 ± 0.80 to 3.0 ± 1.34. Our results are in accordance with this study.

The present study showed that at Nanning the stigma of ICPA 2039 remained receptive for a total of about 192 h (including the short period at Day -3). In case of ICPA 2043 the stigma receptivity duration was the same. Since honey bees (*Apis spp.*) visit pigeonpea flowers after they open and from this time the stigma remains receptive for 120 h and this period coincides with high activity of pollinating insects, which are responsible for cross pollination in this crop. The high yields recorded in the large-scale hybrid (male-sterile × male-fertile line) pigeonpea seed production studies under natural conditions (Saxena, 2006; IIPR, 2007) confirm this hypothesis. The information generated from this study can also be used to maximize the pod set when crosses are made between 2 male-fertile lines where emasculation of female flower is essential. Since pollen dehiscence starts a day before flower opening and maximum pod set is observed on the day of flower opening it may be recommended that for maximizing the pod set emasculations be done at Day -2 stage and pollinations could be made either on Day -1 or Day 0 stages, provided humidity is not a constraint.

REFERENCES

IIPR (2007). Enhancing yield and stability of pigeonpea through heterosis breeding. Progress Report 2006 – 2007. Indian Institute of Pulses Research, Kanpur.

Kalinganire A, Hardwood CE, Slee MU, Simons AJ (2000). Floral structure, stigma receptivity, and pollen viability in relation to protandry and self-incompatibility in silky oat (*Grevillea robusta*). Ann. Bot. 86: 133-148.

Prasad S, Prakash R, Haque Md F (1977). Floral biology of pigeonpea. Tropical Grain Legume. 7: 12.

Saxena KB (2006). Seed Production systems in pigeonpea. Patancheru 502 324 Andhra Pradesh, India: International Crops Research Institute for the Semi-Arid Tropics. 76: 92-9066-490-8.

Saxena KB, Kumar RV, Latha KM, Dalvi VA (2006). Commercial pigeonpea hybrids are just a few steps away. Ind. J. Pulses Res. 19(1): 7-16.

Shafer GS, Burson BL, Hussey MA (2000). Stigma receptivity and seed set in Protogynous bullelgrass. Crop Sci. 40: 391-397.

Stakstad E (2007). The plant breeder and the pea. The Science 196-197.

Grain filling rate is limited by insufficient sugar supply in the large-grain wheat cultivar

Guohua Mi*, Fanjun Chen and Fusuo Zhang

The Key Lab of Plant Nutrition, MOA, College of Resources and Environmental Sciences. China Agricultural University, Beijing, 100193, China.

Wheat grain yield can be limited by source, sink or by both. Inconsistency of the previous results reflects the interactions between genotypes and environments. In north China where the hot, dry wind was frequent during grain filling, new winter wheat cultivars with large grains suffer from loss of grain weight quite often because the grain filling rate (GFR) is low. In the present study, the carbohydrate assimilation and utilization was investigated to possible role of carbon assimilation and utilization in limiting grain filling rate by comparing two winter wheat cultivars, a large-grain cultivar Jing9428 (slow GFR), and a small-grain cultivar CAU3291 (fast GFR). It was shown that there was no significant difference in net photosynthetic rate and leaf area index during grain filling between the two cultivars. However, soluble sugar concentration in stems of Jing9428 decreased much more sharply than CAU3291 when the linear grain growth phase began suggesting an insufficient supply of photosynthates for grain filling. In grains, the ratio of starch to sucrose, glucose and fructose content of Jing9428 was significantly lower than CAU3291. It was suggested that insufficient photosynthate supply rather than weak sink strength was the main reason limiting grain filling rate of the large-grain type wheat cultivars in north China.

Key words: Winter wheat, grain filling, source-sink relationship, carbohydrate.

INTRODUCTION

Wheat grain yield can be limited by source (the supply of assimilates) or sink (the capacity of the grains to accumulate assimilates), or both (Evans et al., 1975). The results of different studies by manipulating source-sink relationships during grain filling through removing leaves or grains, and shading have shown inconsistent conclusions (Winzeler et al., 1989; Jedel and Hunt, 1990; Savin and Slafer, 1991; Slafer and Savin, 1994; Cruz-Aguado et al., 1999). For example, Cruz-Aguado et al. (1999) reported that wheat growth in the tropics is more source limited than in temperate areas. While in Australia, Slafer and Savin (1994) suggested that during post-anthesis period, grain yield of wheat is either sink-limited or co-limited by both source and sink but never source-limited. This inconsistency reflects the genotype and environment interactions in the availability of assimilate for grain growth (Ma et al., 1990).

In winter wheat growing region of North China, two environmental factors limit wheat yield. The short spring season limits grain number per spike. The frequent hot dry wind during grain filling stage limits grain weight. Therefore, traditional wheat cultivars usually have many spike number, but few grains per spike and small grains. Comparing with cultivars bred in 1970's, new cultivars bred in 1990's have the same spike number per hectare, a little more grains per spike, but significant larger grain weight (about 36 mg/grain versus 45 mg/grain). Grain weight has been a most important factor in increasing yield potential (Liu and Meng, 1994). To increase wheat yield further (more than 6 ton/ha), some cultivars with even larger grains (more than 48 mg/grain) has been developed in late 1990's. However, grain weight potential of such cultivars was not always realized due to their low grain fill rate (GFR). Grain filling is quite often interrupted under unfavorable weather conditions during grain filling stage. According to source/sink theory, grain filling rate can be limited by source or sink, including sink size (grain number per hectare) and sink strength (example, enzym activity, phloem unloading, hormone control etc.) (Farrar, 1993). By using two wheat cultivars with different grain size and grain filling rate, this study aims to elucidate the possible role of photosynthate supply and utilization in

---

*Corresponding author. E-mail: miguohua@cau.edu.cn.

Table 1. Grain yield and yield components of two wheat cultivars.

Cultivars	Grain yield (g/m^2)	Spike number per m^2	Kernel number per spike	Grain weight (mg/grain)
CAU3291	516	527	29.1	35.6
Jing9428	467	443	28.8	43.1
T-Test	NS	NS	NS	*

*: significant at 0.05 levels.
NS: not significant.

limiting grain filling rate in winter wheat in northern China.

MATERIALS AND METHODS

Field site and experimental set-up

The study was conducted during the 2002 and 2003 growing seasons at the experimental station of the China agricultural University, Beijing. The soil was characterized by high fertility with organic matter of 2.32%, Nitrate-N of 10.8 mg/kg, Olsen-P of 18.6 mg/kg, Ammonium acetate-K (NH$_4$OAC-K) of 149 mg/kg and pH of 7.6. With the intensive management, 112.5 kg N/ha was applied as granular urea at seeding and 112.5 kg N/ha at stem jointing stage. Phosphorus was applied as diammonium phosphate (DAP) at the rate of 180 kg/ha (P$_2$O$_5$) at seeding. Irrigation was conducted when needed.

Two cultivars, CAU3291 and Jing9428 were used in the study. CAU3291 is a small-grain genotype with fast grain filling rate (GFR). Jing9428 is a big-grain genotype with slow GFR. Both cultivars were bred in the same ecological area around Beijing, and released in late 1990's. Seed was sown at a rate of 225 kg/ha in rows 20 cm apart. The experimental design was a randomized complete block with 4 replications and plot size of 21.6 m^2.

Sampling

On the 3rd, 9th, 15th, 19th and 24th day after anthesis, 20 randomly chosen plants were sampled from each plot. Grains, stems (including leaf sheaths) and leaf blades of the main stems were separated, dried, weighed and ground. Grains and stems were used for the determination of carbohydrates. In the last sample, the total plant (main stem plus tillers) was used to determine the average grain number per spike.

At maturity (26th day after anthesis when 50% of the spikes become yellow), the spike number per m^2 was determined by counting the total spike number in an area of 0.2 m^2 (two 1 m-long rows) in each plot. Grain yield and the average grain weight were recorded in plants cut at ground level for a 3 m^2 area from each plot.

Determination of sucrose, glucose, fructose and soluble sugar and starch

Samples of grain and stem were extracted with 80% ethanol (50□). Sucrose, glucose and fructose were analyzed by HPLC methods (Shimada LC-10A). The HPLC conditions were as follows: Waters 600 Pump, Waters 600 controller, Waters 474 Scanning Fluorescence Detector, Aminex ®HPX-87C Column. Soluble sugar was assayed by an anthrone method (Zhang, 1992). Starch in the residue was determined after extraction with perchloric acid (Bai and Tang, 1993).

Leaf area index (LAI) and net photosynthetic rate (Pn)

On the 0th, 14th, and 21st day after anthesis, an area of 400 cm^2 was sampled from each plot to determine the leaf area index. The leaf area was measured with a ruler (length x maximum width x 0.78) (Quarrie and Jones, 1977). On the same day, net photosynthetic rate of the flag leaves was measured using a portable photosynthesis system (BAU-1, a joint product of Beijing Analytical Instrument Factory and the Department of Agronomy, China Agricultural University). This is a closed system that measures changes in CO$_2$ concentration over time. Measurements were made between 9:30 to 12:00 a.m. when the photosynthetically active radiation (PAR) was 1021 ± 56 µmol m^{-2} s^{-1} and the leaf chamber temperature was 27 ± 2°C. Three replicate plants were measured in each plot. Measurements were made at ambient CO$_2$ concentration.

Statistics

Data were analyzed by using the SAS statistics program (V6.03) (SAS Institute Inc., 1985-1987).

RESULTS

Yield and yield components

Wheat yield is determined by spike number, grain number and grain weight. CAU3291 got higher yield which was greatly contributed by its relatively more spikes per hectare, and more grains per spike (Table 1). However, its grain weight is significantly lower than that of Jing9428. The result was typical for these two cultivars when there was hot, dry weather during grain filling stage. Though the grain weight of Jing9428 was much higher than that of CAU3291, it is lower than the potential value which was reported as about 48 mg/grain in favorable weathers. Therefore the grain fill process was analyzed to clarify the reason.

Grain fill

Grain growth of wheat can be divided into three phases, (1) the initial lag phase; (2) the apparent linear phase; and (3) the maturation phase (Herzog, 1986). In the present experiment, no significant difference was found between CAU3291 and Jing9428 during the initial lag phase (Figure 1). Then, grain filling rate CAU3291 increased much quickly than that of Jing9428 in the linear phase (Figure 1). On the 19th day after anthesis, the grain of CAU3291 reached the maturation phase, while that of Jing9428 kept linearly growth. It seemed that the grain of Jing9428 stopped growth suddenly on day 24 after

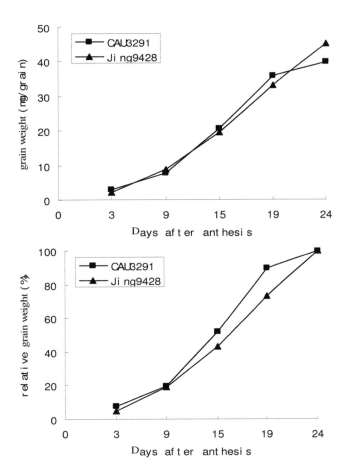

Figure 1. Grain growth process of two wheat cultivars Jing9428 and CAU3291. Upper, absolute grain weight (l.s.d. (0.05) = 1.32); lower, relative grain weight (l.s.d. (0.05) = 3.5%).

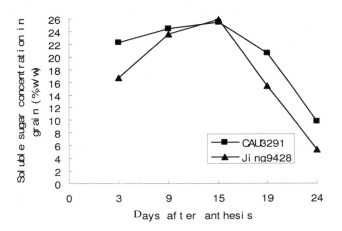

Figure 2. Soluble sugar concentration in stems and leaf sheathes of two wheat cultivars Jing9428 and CAU3291 during grain filling stage (l.s.d. (0.05) = 2.04).

Jing9428 kept linearly growth. It seemed that the grain of Jing9428 stopped growth suddenly on day 24 after anthesis without reaching the maturation phase, possibly because of the hot weather. So it was necessary to analyze the physiological limitation accounting for the grain growth pattern of Jing9428.

Carbohydrate supply and utilization

In general, factors controlling grain growth may include two aspects, one is carbohydrate supply from the current photosynthate or sugar stored in vegetative organs (mainly stems and leaf sheaths) pre- or/and post-anthesis (source), the other is carbohydrate utilization in grain (sink). Leaf area index (LAI) can be a measurement of source size. In this experiment, no difference was found in LAI of these two cultivars during grain filling stage (Table 2). The soluble sugar storage in stems and leaf sheaths of both cultivars increased to the maximum on day 15 post-anthesis then decreased with the linear growth of grains (Figure 2). However, sugar concentration in stems and leaf sheaths of Jing9428 decreased more sharply after day 15, suggesting an insufficient supply of current photosynthate for grain filling.

In general, the concentration of sugar and reduced sugar (glucose, fructose) in grains of Jing9428 was lower than that of CAU3291, though not significantly (Figure 3). Also, starch synthesis in grains of Jing9428 was slower than that of CAU3291 (Figure 4). The ratios of starch to sugar, glucose, and fructose content in grains indicate conversion rate of sugar to starch (Table 3). It was found that the ratios of starch to sugar, glucose, and fructose content in grains of CAU3291 were significantly lower (from the 9th day after anthesis and onward) than in Jing9428. These data suggest the insufficient supply of carbohydrates to grains of Jing9428 in comparison to CAU3291.

DISCUSSION

Slafer et al. (1994) reported that breeders have increased wheat yield potential mainly through increasing the number of grains per m^2 (sink size) rather than through increasing individual grain mass. While in east China where the climate is much favorable for wheat growth, wheat cultivars with big spike (numerous, larger grains per spike) has been successful in getting yield as high as 9 ton/ha (Fu and Li, 1998). In North China, however, the grain growth of such cultivars may be limited by the short grain filling duration due to the hot, dry weather. As found in the present study, grains of Jing9428 did not reach their potential weigh because the grain filling rate was slow so that the grain fill process failed to finish in the given short maturation period (Figure 1). Thorne (1985) suggested that factors within the grains, rather than out of them, may limit grain filling. If the sink strength of Jing9428 was weak, it might have exerted a negative feedback on carbon partitioning to the grains. As a result, soluble sugar concentration in stems should have kept a high level during the rapid grain growth. We did find such

Table 2. Leaf area index (LAI) and net photosynthesis rate (Pn, mg CO_2 dm^{-2} h^{-1}) during grain filling of two wheat cultivars Pn was measured under atmosphere CO_2 concentration from 09:30 to 12:00 when the PAR was 1021 ± 56 μmol m^{-2} s^{-1}. Data are the average of three replicates with standard deviation.

Cultivars	At anthesis		14 days after anthesis		21 days after anthesis	
	LAI	Pn	LAI	Pn	LAI	Pn
CAU3291	4.4 ± 0.7	24.5 ± 3.1	4.6 ± 0.5	27.2 ± 2.7	2.5 ± 0.5	10.2 ± 2.1
Jing9428	4.1 ± 1.1	26.3 ± 2.4	4.1 ± 0.6	28.1 ± 3.1	2.8 ± 0.8	15.9 ± 3.4

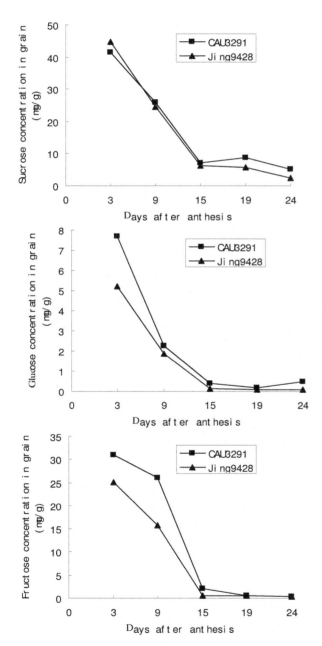

Figure 3. Sucrose (upper), Glucose (middle) and Fructose (lower) concentrations in grains of two wheat cultivars during grain filling stage (l.s.d.(0.05) is 2.07, 1.63, and 2.83 for sucrose, glucose, and Fructose, respectively).

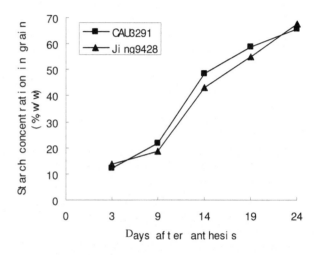

Figure 4. Starch accumulation in grains of two wheat cultivars (l.s.d.(0.05) = 2.94%).

phenomenon in the previous study in which the big-spike, stay-green cultivar LZ953 accumulated large amount of soluble sugar in stems (Mi et al., 2002). However, it seemed not the same case as in Jing9428 in which soluble sugar concentration decreased sharply during the linear grain growth phase (Figure 2). Bonnett and Incoll (1993) demonstrated in barley that there was an earlier loss of carbohydrate from the stem of plants when the incident radiation was halved and the photosynthetic input considerably reduced. Therefore, the slow grain filling rate in Jing9428 may be contributed to insufficient photosynthate supply rather than sink strength. Indeed, the LAI and Pn of leaves in Jing9428 were almost the same as in CAU3291 (Table 2), though the grain weight potential of the former was much more than that in the later (Figure 1, Table 1).

Consequently, the soluble sugar supply at the late grain filling stage was insufficient in Jing9428 in comparison to CAU3291 (Tab. 3). Blade and Baker (1991) also reported that large-seeded cultivars are more sensitive to assimilated supply.

In general, there are two ways to increase carbohydrate supply during grain filling, that is, longer leaf area duration (LAD) or higher net photosynthesis rate (Pn). Stay-green has been proven an effective trait in high-

Table 3. Ratio of starch to sucrose, glucose, and Fructose in grains of two wheat cultivars.

		Days after anthesis				
		3	9	14	19	24
Starch/sucrose	CAU3291	3	8	68	67	124
	Jing9428	3	8	69	95	266
	T-Test	NS	NS	NS	**	**
Starch/glucose	CAU3291	16	96	1274	3271	1339
	Jing9428	27	99	3923	5505	6765
	T-Test	NS	NS	**	**	**
Starch/Fructose	CAU3291	4	8	231	920	1526
	Jing9428	6	12	771	1059	2255
	T-Test	NS	NS	**	**	**

**: significant at 0.05 level, NS: not significant.

yielding corn breeding (Lee and Tollenaar, 2007; Mi et al., 2007). In wheat, high-yielding big-spike cultivars which had a longer LAD were also developed in East China where the weather is suitable for a long grain filling period (about 40 days) (Jiang and Li, 1993). In North China, however, the hot dry wind during late grain filling stage may kill the plants before the grains reach normal maturity. So, increasing Pn should be a promising way to improve such big-grain cultivars as Jing9428.

ACKNOWLEDGEMENTS

The study is supported by the NKBRSF Project of China (G2009CB118607) and the Natural Science Foundation of China (No. 39600088). F.Z. is supported by the Program for Changjiang Scholars and Innovative Research Team in Universities (IRT0511).

REFERENCES

Bai B, Tang X (1993). Assay techniques in Plant Physiology. China Science and Technology Publisher. Beijing.

Blade SF, Baker RJ (1991). Kernel weight response to source-sink changes in spring wheat. Crop Sci. 31: 1117-1120.

Bonnett GD, Incoll LD (1993). Effects on the stem of winter barley of manipulation the source and sink during grain-filling. II. Changes in the composition of water-soluble carbohydrates of internodes. J. Exp. Bot. 44: 83-91.

Cruz-Aguado JA, Reyes F, Rodes R, Perez IP, Dorado M (1999). Effect of source to-sink ratio on partitioning of dry matter and ^{14}C-photoassimilates in wheat during grain filling. Ann. Bot. 83: 655-665.

Evans LT, Wardlaw FT, Fischer RA (1975). Wheat In: Evans L.T. (ed.) Crop Physiology: Some Case Histories. Cambrideg University Press. Cambridge.

Farrar JF (1993). Sink strength: what is it and how do we measure it? Plant, Cell Environ. 16: 1013-1046.

Fu ZL, Li HQ (1998). Investigation on reasons for super yield of winter wheat in the Yellow River area of China. Tritical Crops. 18: 48-51.

Herzog H (1986). Source and sink during the reproductive period of wheat: Development and its regulation with special reference to cytokinins. Paul Parey Scientific Publishers. Berlin and Hamburg.

Jedel PE, Hunt LA (1990). Shading and thinning effects on multi-and standard-floret winter wheat. Crop Sci. 30: 128-133.

Jiang H, Li J (1993). Comparison of maturing phases of high yield wheat varieties. J. Shandong. Agric. Univ. 24: 437-445.

Lee EA, Tollenaar M (2007). Physiological Basis of Successful Breeding Strategies for Maize Grain Yield. Crop Sci. 47 (S3): S202–S215.

Liu JX, Meng FH (1994). Change of yield component traits of winter wheat cultivars and the breeding target for the future in Beijing area. Beijing Agric. Sci. 12: 11-13.

Ma YZ, MacKown CT, Van Sanford DA (1990). Sink manipulation in wheat: Compensatory changes in kernel size. Crop Sci. 30: 1099-1105.

Mi G, Tang L, Zhang F, Zhang J (2002). Carbohydrate storage and utilization during grain filling as regulated by nitrogen application in two what cultivars. J. Plant Nutri. 25: 213-229.

Mi GH, Chen FJ, Zhang FS (2007). Physiological and genetic mechanisms for nitrogen-use efficiency in maize. J. Crop Sci. Biotech. 10: 57-63.

Quarrie SA, Jones HG (1977). Effect of abscisic acid and water stress on development and morphology of wheat. J. Exp. Bot. 28: 192-203.

SAS (1987). SAS/STAT Guide for Personal Computers, 6th editon. Cary, NC, SAS Institute.

Savin R, Slafer GA (1991). Shading effects on the yield of an Argentinian wheat cultivar. J. .Agric. Sci. 116 : 1- 7.

Slafer GA, Satorre EH, Andrade FH (1994). Increase in grain yield in bread wheat from breeding and associated physiological changes. In: Slafer G.A. (ed.) Genetic Improvement of Field Crops. Marcel Dekker, New York.

Slafer G.A, Savin R (1994). Source-sink relationships and grain mass at different positions within the spike in wheat. Field Crops Res. 37: 39-49.

Winzeler M, Monteil PH, Nosberger J (1989). Grain growth in tall and short spring wheat genotypes at different assimilate supplies. Crop Sci. 29: 1487-1491.

Zhang X (1992). Determination of soluble sugar. In Zhang, X. Eds., Study Methods for Plant Physiology (in Chinese). Liao Ning Science and Tecnhnology Publisher, Shenyang, China.

Physicochemical and functional characteristics of cassava starch in Ugandan varieties and their progenies

Ephraim Nuwamanya[1], Yona Baguma[2]*, Naushad Emmambux[3], John Taylor[3] and Rubaihayo Patrick[1]

[1]Department of Crop Science, Makerere University Kampala, P. O. Box 7062, Kampala, Uganda.
[2]National Crops Resources Research Institute (NaCRRI), P. O. Box 7084, Kampala, Uganda.
[3]Department of Food Science, Faculty of Agriculture, University of Pretoria, South Africa.

Due to poor root quality traits in improved, disease resistant cassava (*Manihot esculenta* Crantz.) varieties and hence low acceptability among farmers, a study was undertaken to improve these varieties by crossing them with disease susceptible, farmer preferred local landraces. Five improved varieties and four local landraces were used and hybridisations among them were made in a poly-cross nursery block. Starch isolated from the nine cassava varieties and their F_1 progenies was analysed for physicochemical and functional properties. Significant differences were observed between varieties, progenies and within the F_1 progenies. The amylose content ranging between 19.0 - 25.0% was negatively correlated to swelling power and solubility but positively correlated to starch content. Average starch granule sizes ranged between 7.0 - 12.0 μm, though smaller granules ranged between 2 - 6.9 μm and large granules between 13 – 20 μm. Granules were mainly truncated in shape and similar across varieties and their progenies. Individual parents had peak viscosity, set back viscosity and viscosity at breakdown higher than the progenies suggesting inherent genetic and biochemical differences among parents used in the study. Variations were also observed in the parents and progenies for starch swelling power, solubility and starch content on dry basis. Starch associated molecules such as proteins and lipids did not vary significantly but dietary fibre significantly ($P< 0.05$) varied both in parents and F_1 families. Significant correlations ($r > 0.45$) were observed among starch properties including swelling power and breakdown viscosity. Based on these results, selections for lines with different starch quality and quantity properties can be made among the F1 families for future dietary and industrial uses.

Key words: Progenies, starch granules, viscosity, physicochemical properties.

INTRODUCTION

In recent years, cassava has received more attention as a root crop not only for its resistance to abiotic stresses (Chavez et al., 2005; Baguma, 2004) but also its high productivity with considerable starch yield (up to 30% of the fresh root or 80% of root dry matter) and purity (Ceballos et al., 2006; Benesi, 2005). Cassava has been grown in Uganda since the 1860s with the recurrent introduction of varieties to mitigate possible losses of germplasm due to viral disease outbreaks that have always threatened the crop (NARO, 2005). Such introductions have not always suited the preferences of local consumers despite their resistance to diseases and/ or pests hence the continued need to provide cultivars which are resistant to disease and/or up to the taste of the local consumers (Baguma, 2004). With the added importance of cassava starch in industry, these cultivars are required to be able to produce starch suitable for

---

*Corresponding author. Email: ybaguma@naro-ug.org, bgmyn@yahoo.co.uk.

suitable for dietary and industrial applications. Thus different breeding programmes aimed at producing cultivars with high quality starch producing cassava roots have been initiated in Uganda (NARO, 2005). Starch constitutes the main component of the cassava root (Ceballos et al., 2006) and thus plays an important role in the use of cassava as a food and industrial crop. Cassava starch has been studied and characterised for its different properties such as granule structure, pasting properties and functional properties such as swelling power and solubility (Zaidul et al., 2007; Gomes et al., 2005; Charles et al., 2004). Studies have shown that cassava starch granules are truncated with various shapes and sizes ranging from 2 - 40 µm (Moorthy, 2002) and displays an A-type x-ray pattern (Tukomane et al., 2007). The starch has pasting properties typical of other tuber and root starches with low amounts of proteins, lipids and fibre (Charles et al., 2004). In particular, this study was aimed at detailing the different starch attributes of cassava that can be exploited for a number of emerging starch industries. The production of bio-ethanol requires crops with appreciable amounts of starch and/or sugar which can be easily turned into ethanol by fermentation. Production of cultivars with high sugar concentrations in addition to starch would thus reduce the cost of production of bio-ethanol (Ceballos et al., 2008). Other industries that require cassava with novel starches include the production of glucose syrups and in the emerging construction and mining industry where cassava starch is required because of its good viscosity properties. In addition, the food and dietetics industry require improved starches ideal for delivering health benefits to people since starch forms a major part of the nutritional components among most communities (Morell and Mathews, 2005). A good example for this cause is the production of resistant starch which passes through small intestines into the large bowel where it is fermented to produce a range of products such as short chain fatty acids that are important in the prevention of cancer (Topping and Cliffton, 2001). Starches with a low glycemic index are also important in health (Baguma, 2004). These starches release glucose at a slow rate hence important in helping patients with type II diabetes (Morell and Myers, 2005). For increased diversity in terms of progeny combinations produced and seed number, a poly-cross nursery block was used. The poly-cross approach has been used in a number of crops such as sugar cane, maize, (KoLliker, 2005), sweet potato (Ernest et al., 1993) and cassava to increase on the diversity, yield and maximize the number of progenies that could be presented among each of the progeny parents. This necessitated its use in this study where screening of starch properties from four land races and five improved cultivars including two Nigerian landraces and three IITA lines and their progenies was undertaken. The physico-chemical and functional properties of starch from progeny lines in each of the female parents used in the poly cross were accessed and the relationships among these properties determined to see how they affect each other.

MATERIALS AND METHODS

Field experiments were set up at the National Crops Resources Research Institute (NaCRRI), Namulonge, in Central Uganda. Namulonge has served as the screening site for many cassava properties, justifying its suitability for this experiment (Otim-Nape et al., 1998). Parental lines included locally grown varieties namely; Bao, Bamunanika, Kakwale and Nyaraboke preferred by farmers due to their good root qualities but are yet, low yielding. These were recipients for gene conditioning high yield and other superior root qualities. Donor parents included introductions from IITA (Nigeria) namely NASE 10, NASE 12 and 95/SE-00036 characterized by a high yield potential (20-35 kg/ha), and Nigerian landraces TME 5 and TME14 that combine high dry matter content, high yielding tuber potential and resistance to the dreaded cassava mosaic disease.

Due to insufficient knowledge on the specific combining abilities and flowering habits of the selected parents, a poly-cross mating design was used to maximize hybridization and seed production. In doing so, every individual had an equal chance of mating with each other in the population (du Ploy, 1985). To prevent pollen from undesired sources, the crossing block containing 100 plants per parent was isolated from the nearest cassava field by a distance of 100 m. At harvest, seeds in each family were collected and then bulked to form half-sib families. In the following year, seeds were pre-germinated in a high humidity chamber at a 30 °C. Pre-sprouting was necessary because seeds often have a dormancy period of a few months after maturity and require relatively high temperatures (30 - 35 °C) for optimum germination (Ceballos et al., 2004). Adequate soil moisture and freedom from weeds was maintained to ensure high and uniform germination of the seeds. The seedlings were maintained in the nursery for four weeks. Clonal evaluation consisted of nine half-sib families consisting of 1,077 progenies that were planted using family replication procedure (Jaramillo et al., 2005).

Harvesting and collection of root samples

1077 progenies constituting 10 progeny families and their respective parents were harvested at 12 months after planting; two roots were randomly collected per progeny and prepared for starch extraction by peeling and washing with distilled water.

Starch extraction

Native cassava starch extraction was carried out using a method described by Benesi (2005). A hundred grams of the fresh tuberous cassava roots were washed, peeled, and homogenized with 100 ml of 1 M NaCl to aid the release of starch from the solution using a Waring blender.

The mixture was stirred with a stirring rod for 2 min and filtered using a triple cheese (muslin) cloth. The filtrate was allowed to stand for 1 h to facilitate starch sedimentation and the top liquid was decanted and discarded. 200 ml of distilled water was added followed by centrifugation at 3,000 g for 10 min. The starch was air-dried on aluminium pans at room temperature for 24 - 36 h and stored in plastic air tight containers at room temperature. The extracted starch from each of the progeny families for a particular parent was bulked before analyses.

Determination of moisture content

Moisture content was determined according to (Benesi, 2005) with modifications. 1.50 g of starch was dried in a forced air oven for 3 h at 105°C. The samples were transferred to a dessicator and allowed to cool to room temperature and the difference in the weight of starch was used to calculate the apparent moisture content.

Determination of protein and nitrogen content

The percentages of total nitrogen and protein were determined by Dumas combustion method in a Nitrogen/Protein analyser (Leco Model FP-528). 0.15g of the cassava starch sample was used and the total protein content was calculated using a protein conversion factor of 6.25.

Determination of starch content

The starch content was determined using a Megazyme total starch assay kit based on the AOAC method 996.11 by enzymatic hydrolysis of starch (0.1 g) using amylase/amyloglucosidases and quantification of glucose using glucose oxidase/peroxidase reagent.

Determination of amylose content

Amylose content was determined from 0.5 g starch (dry weight basis) using the Megazyme amylose/amylopectin assay kit by selective quantitative precipitation of amylopectin with concanavalin A (Con A), quantitative estimation of amylose on hydrolysis using amylase/amyloglucosidases and estimation of glucose by glucose oxidase/peroxidase assay.

Determination of total dietary fibre in extracted starch

Total dietary fibre was determined using the Megazyme available carbohydrate and dietary fibre assay kit based on the AOAC method 985.29 by deffating, drying, hydrolysis and deproteinisation of the 1.0 g of starch sample. The resultant solution was treated with ethanol (four times the volume of the sample) to precipitate the fibre and remove depolymerised proteins and glucose. The residue was filtered and washed with 78% ethanol followed by 95% ethanol and 70% acetone to remove any organic compounds left in the mixture. It was then dried in an air forced oven and weighed.

Determination of granule structure and size of cassava starch

Starch samples (starch powder) were mounted on SEM stubs with adhesive tape and coated with gold. Scanning electron micrographs were taken by a JOEL JSM-840 microscope (JOEL, Tokyo, Japan). The accelerating voltage was 5 KV and the magnification used was ×4, 000 and ×5, 000. The granule size (diameter) was obtained using the Image Tool software Version 3.0 for windows (UTHSCSA, 2002).

Pasting properties of starch

Starch pasting properties were evaluated using a Rapid Viscosity Analyser (RVA model 3D, New Port Scientific, Sydney, Australia). Starch (2.5 g, dry basis) was suspended in distilled water and the total weight adjusted to 28 g. The sample was equilibrated at 50°C for 1 min, heated 92°C in 7.5 min at a rate of 5.7°C per min, held at 92°C for 5 min, cooled to 50°C in 7.5 min at a rate of 5.7°C/min and held again at 50°C for 1 min. From the resulting pasting curve, temperature at initial viscosity increased (pasting temperature), peak viscosity (PV), time to peak viscosity (P_t), hot paste viscosity (HPV), breakdown viscosity (BV), final viscosity (FV) and setback viscosity (SV) were recorded by the rapid viscosity analyser.

Swelling powers of the starch molecules

The swelling power of the starch granules was determined according to Charles et al. (2004) at different temperatures ranging from 30 - 80°C. Swelling power was recorded as the ratio in weight of the wet sediment to the initial weight of dry starch.

Statistical data analysis

Quantitative data analysis was carried out using GENSTAT discovery Edition 3 (VSN International). Means were calculated for each of the progeny families and the analysis of variance (ANOVA) was used to test for the difference within and among the clones and parental lines at 5% level of significance (p = 0.05). Relationships among different starch characteristics were analyzed using correlation coefficient and regression analysis.

RESULTS AND DISCUSSION

Proximate analysis of cassava starch

Results for proximate analysis of cassava starch from the parents and their progenies are presented in Table 1. The average starch moisture content among the parental lines ranged from 14.04 to 16.66% as compared to the progenies in which it ranged from 13.78 to 15.37%. The extracted starch had low levels of protein on dry basis with averages ranging from 0.28 - 0.52% among the parents and from 0.28 - 0.35% among the progenies. Protein content values obtained were lower than the values reported by Rodríguez-Sandoval et al. (2008) and Ceballos et al. (2006). Protein effects on starch properties depend on its content with high protein negatively affecting the pasting properties (Moorthy, 2002). The dietary fibre ranged from 0.02 - 0.56% among the parental lines and 0.28 - 0.93% among the progenies with significant (P < 0.05) effects on a number of pasting properties such as breakdown viscosity, peak viscosity, hot paste viscosity, pasting temperature and peak time. Increase in dietary fibre results in reduction in peak viscosity and increase in the pasting temperature. On the other hand, the lipid content averaged between 0.12 - 0.38% among the parents and 0.19 - 0.38% among the progenies lying in the range suggested by Moorthy (2002).

Increased lipid content improves starch textural properties and leads to viscosity stability hence improving the quality properties of starch (Moorthy, 1985). The low levels of starch associated compounds indicate the easiness with which cassava starch can be extracted from the roots.

Starch content

The results for starch and amylose content are presented

Table 1. Proximate analyses of cassava starch (dry basis) from the parents and their progenies.

Parent cultivar	MC^1 (%)	MC^2 (%)	$Protein^1$ (%)	$Protein^2$ (%)	$Fibre^1$ (%)	$Fibre^2$ (%)	$Lipid^1$ (%)	$Lipid^2$ (%)
Bamunanika	$16.49^a \pm 0.03$	$14.75^a \pm 0.04$	$0.35^a \pm 0.020$	$0.35^a \pm 0.004$	$0.37^a \pm 0.021$	$0.75^a \pm 0.017$	$0.38^a \pm 0.08$	$0.26^a \pm 0.08$
Bao	$16.47^a \pm 0.11$	$15.37^a \pm 0.19$	$0.52^b \pm 0.015$	$0.30^b \pm 0.012$	$0.37^a \pm 0.021$	$0.86^a \pm 0.017$	$0.22^a \pm 0.06$	$0.19^a \pm 0.08$
Kakwale	$14.62^b \pm 0.06$	$14.81^b \pm 0.38$	$0.29^a \pm .0003$	$0.29^a \pm 0.019$	$0.35^a \pm 0.014$	$0.28^b \pm 0.042$	$0.12^b \pm 0.11$	$0.38^a \pm 0.07$
Nyaraboke	$16.66^a \pm 2.00$	$14.91^a \pm 1.04$	$0.32^a \pm 0.007$	$0.32^b \pm 0.006$	$0.06^b \pm 0.014$	$0.69^a \pm 0.099$	$0.38^a \pm 0.06$	$0.28^a \pm 0.05$
95/SE/00036	$16.34^a \pm 0.24$	$13.78^a \pm 0.26$	$0.28^c \pm 0.009$	$0.31^b \pm 0.009$	$0.02^b \pm 0.007$	$0.59^a \pm 0.114$	$0.17^a \pm 0.08$	$0.28^a \pm 0.07$
NASE 10	$14.77^b \pm 1.44$	$15.31^b \pm 0.09$	$0.32^a \pm 0.007$	$0.31^b \pm 0.050$	$0.17^c \pm 0.000$	$0.33^b \pm 0.118$	$0.14^b \pm 0.09$	$0.34^a \pm 0.06$
NASE 12	$14.79^b \pm 0.76$	$14.08^a \pm 0.30$	$0.31^a \pm 0.012$	$0.28^a \pm 0.009$	$0.13^c \pm 0.021$	$0.58^a \pm 0.182$	$0.26^a \pm 0.11$	$0.35^a \pm 0.02$
TME 5	$14.04^c \pm 0.04$	$13.82^a \pm 0.25$	$0.31^a \pm 0.011$	$0.30^a \pm 0.004$	$0.56^d \pm 0.021$	$0.93^c \pm 0.171$	$0.37^a \pm 0.09$	$0.31^a \pm 0.03$
TME 14	$14.97^c \pm 0.12$	$14.97^a \pm 0.50$	$0.35^d \pm 0.011$	$0.33^a \pm 0.029$	$0.54^d \pm 0.028$	$0.49^d \pm 0.154$	$0.17^a \pm 0.08$	$0.22^a \pm 0.08$

[1] Results for the parental lines; [2] Results for the progenies; [a] Mean values of triplicate analyses in a column with the same superscript are not significantly different at 5%. MC^1 = moisture content of the parents; MC^2 = moisture content of the progenies

in Table 2. Among the parental lines, the starch content averaged from 70.36 - 89.90% while it was between 73.48 - 93.85% among the progenies. Significant differences were observed among the parents and the progenies in terms of starch content with the TME and 95/SE/00036 showing significantly ($P < 0.05$) lower starch content amongst the parents and showing significantly ($P < 0.05$) higher starch contents in case of the progeny families.

Amylose content of cassava starch

Results for the amylose content in cassava starch are presented in Table 2. The amylose content among the parental lines ranged from 23.01 - 26.98% while among the progenies, it ranged from 19.69 - 26.63%. There were no significant ($P > 0.05$) differences observed in amylose contents of different parents. A similar observation was reported by Moorthy (2002). The amylose content did not have significant ($P > 0.05$) effects on starch pasting properties although it was positively correlated to the pasting temperature and the peak time. The progenies however, showed significant ($P < 0.05$) differences with progenies of Nyaraboke showing significantly ($P < 0.05$) lower amylose content than the other families. Amylose content is important in almost all starch properties with low amylose contents leading to increased relative crystallinity of starch due to the reduced amorphous regions within the starch granule (Tukomane et al., 2007). Amylose content also affects the retrogradation properties of starch where high amylose starches have increased retrogradation tendencies caused by the aggregation of amylose which acts as nuclei during the process amylopectin retrogradation (Rodriguez-Sandoval et al., 2008). The influence of amylose on the pasting properties depends on its leaching out of the amylopectin network during heating into the solution affecting the starch's viscoelastic properties (Charles et al., 2004). Increase in amylose content leads to increase in the pasting temperature (Novel-Cen and Betancur-Ancona, 2005) due to the prolonged escape of amylose out of the amylopectin network during the gelatinisation of starch leading to prolonged swelling of starch granules (Moorthy, 2002) hence increasing the temperature required to form a starch paste.

Starch granule structure and morphology

Results showing scanning electron micrographs for starch granule shapes are presented in Figure 1. The micrographs showed granules with varying shapes in both the parents and progenies with the granules being characteristically kettle-drum shaped/truncated and some shapes ranging from oval to polygonal and round. Different surface morphologies were observed in the different granules and these ranged from few rough surfaced to dominantly smooth surfaced granules. Some granules had surface pores or fissures. These are important in the hydrolysis of starch as they aid the release of amylose hence important in starch solution properties (Tukomane et al., 2007). The top surface of truncated granules was either convex, biconvex and in some cases with various convex pits presented. The granule size in the parental lines ranged from 8.14 - 10.77 µm and between 8.03 - 9.36 µm in the progenies (Table 3). Granule size had a trimodal range with small

Table 2. Amylose content and starch content of the parents and the F_1 progenies clones.

Parental cultivars	Starch content % (db)		Amylose %	
	Parent	Progenies	Parent	Progenies
Bamunanika	74.84[a]±2.54	75.12[a]±1.62	25.90[a]±0.25	24.51[a]±1.51
Bao	78.49[a]±1.80	75.09[a]±1.65	25.16[a]±0.59	22.77[b]±0.06
Kakwale	81.79[b]±0.13	80.94[b]±1.08	25.72[a]±0.08	24.04[a]±0.18
Nyaraboke	81.76[b]±0.96	88.56[c]±0.09	25.28[a]±1.98	19.69[c]±1.96
95/SE/00036	89.90[c]±1.03	93.85[d]±1.93	23.01[a]±1.05	26.63[d]±1.23
NASE 10	85.38[d]±1.11	73.92[a]±2.34	23.64[a]±0.03	20.49[c]±0.89
NASE 12	76.12[a]±3.43	76.43[a]±4.47	26.98[a]±0.68	23.21[a]±0.98
TME 5	77.27[a]±0.42	74.29[a]±1.39	24.60[a]±3.19	24.09[a]±1.58
TME 14	70.36[e]±0.40	73.48[a]±0.67	23.44[a]±2.63	26.04[d]±1.12

[a]Mean values of duplicate analyses in a column with the same superscript are not significantly different at 5%.

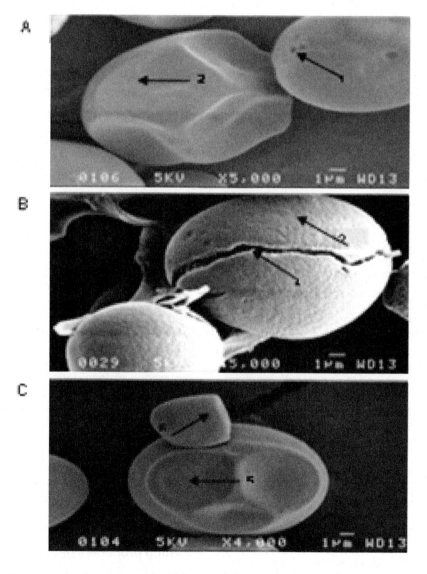

Figure 1. Structure of cassava starch granules showing the granule shapes and different surface morphologies A= TME 5 parent, B= Bao parent, C = Bamunanika parent. 1 = Surface pores on a smooth granule. 2 = Smooth surfaced granules, 3 = Rough surfaced granules 4 = Fissure in the granule 5 = Convex pits on the top surface of the granule, 6 = small granule on top of large granule.

Table 3. Granule size analysis across different parents and their progenies, size ranges and their trimodal distribution.

Variety/Family	Size[1]	Range[1]	Granule shape[1]	N	A%	B%	C%
Bamunanika	10.72[a]±3.12	1.0-19.5	truncated, polygonal, oval	108	11.11	65.28	23.61
Progenies	9.06[a]±3.09	2.0-16.0	truncated, polygonal	131	20.61	67.18	23.61
Bao	10.77[a]±3.19	5.0-18.5	truncated, polygonal	166	8.38	68.86	22.75
Progenies	8.06[a] ±2.86	1.5-14.5	truncated, polygonal	114	32.75	64.65	2.58
Kakwale	8.14[a] ±2.86	2.0-15.2	truncated, polygonal, oval	118	26.95	70.21	2.83
Progenies	8.97[a] ±3.07	2.9-15.4	truncated, polygonal oval	140	26.56	65.63	7.81
Nyaraboke	9.92[a] ±2.66	5.0-17.0	truncated, polygonal round	105	12.28	76.32	11.40
Progenies	8.03[a] ±2.90	1.8-14.8	truncated, polygonal	114	34.29	60.00	5.71
95/SE/00036	10.29[a] ±3.13	3.1-19.9	truncated, polygonal, oval	185	15.91	69.89	14.21
Progenies	9.36[a] ±3.30	2.5-18.9	truncated, polygonal, oval	107	24.29	65.42	10.28
NASE 10	8.33[a] ±3.26	1.4-18.8	truncated, polygonal	127	30.21	61.87	7.91
Progenies	9.35[a] ±3.21	1.4-16.7	truncated, polygonal	109	29.21	59.55	11.23
NASE 12	10.03[a] ±2.68	3.0-18.6	truncated, oval	98	7.84	82.35	9.80
Progenies	9.11[a]±2.49	2.4-14.8	truncated, oval	102	18.81	76.23	4.95
TME 5	8.51[a] ±3.08	1.7-15.2	truncated, oval rounded	108	32.28	59.84	7.84
Progenies	8.73[a] ±2.93	2.6-15.1	truncated, oval, rounded	107	29.59	61.23	9.18
TME 14	9.61[a] ±2.22	3.5-15.5	truncated, polygonal, oval	103	7.92	84.16	7.92
Progenies	8.97[a] ±2.72	2.1-16.2	truncated, polygonal, oval	122	21.95	69.11	8.94

[a] Mean values of n analyses in a column with the same superscript are not significantly different at 5%; N: number of granules observed/measured in a particular size range and shape; Trimodal distribution shown as the percentage where A= small sized granules from 1-6.9 µm, B= medium sized granules of 7-12.9 µm, C= large granules of 13-20 µm.

with small granules ranging from 2-6 µm and composing of 7.8-34.3% among the parents and the progenies, middle sized granules ranging from 7-12 µm (59.6-84.2%) and large granules ranging from 13-20 µm (2.6-23.6%). The trimodal range observed may be attributed to the harvest time and the growing conditions for cassava (Sriroth et al., 1999) where granules of starch obtained from plants grown in the dry season has predominantly smaller granules. Due to the observed differences in size distributions among the progenies and their parents, variations were observed in their starch pasting proper-ties and swelling power. Most of the parents had a low percentage of small granules compared to the progenies. However, the percentage of middle sized granules was higher in most of the parents compared to their respective progenies. This could have resulted into the differences observed in the pasting characteristics of the progenies compared to the parents. Based on the granule morphology and size variations, cassava starch can be produced to suit various uses such as in textile applications. In par- in the progenies it ranged between 170.79-226.54 RVU. The high average peak viscosity observed among cassava starches analysed reflect the low amylose content in cassava compared to wheat and maize starches (Zaidul et al., 2007). Starches with low amylose content gelatinize easily with consequent leaching out of amylose and rapid increases in viscosity. This is especially common in starches from potato and other tuber and root crops, which display high viscosities with large differences between the peak and final viscosities (Noda et al., 2006). In the parental lines, the average peak viscosity was low in local landraces compared to the introduced varieties with the exception of NASE 12 (Table 4). Low peak viscosity translates into good cooking properties (Moorthy, 2002) hence the farmers preference of local landraces especially Kakwale in this case. The hot paste viscosity (HPV) was generally lower applications. In particular the increased percentage of small granules among different progenies increases the potential for the use of this starch in the bio-ethanol industry (Ceballos et al., 2008). Starch granules are also important in the characterisation of different botanical sources of starch (Moorthy, 2002) and are a major focus in the modification of starch used in laundry (Varavinit et al., 2007).

Pasting properties of cassava starch

Results for cassava starch viscosity and pasting properties are shown in Table 4. The peak viscosity of the parental lines ranged from 253.01 - 344.96 RVU while in the parental lines (66.33 - 124.30 RVU) than in the proge-

Table 4. Rheological and pasting properties of cassava starch from nine parental varieties and their progenies.

Variety/Family	Peak (RVU)	HPV (RVU)	Break (RVU)	Final (RVU)	Setback (RVU)	PTi	Peak time (Min)
Bamunanika	290.21^a±7.01	73.33^a±5.66	216.88^a±1.35	156.57^a±1.84	73.67^a±6.01	66.93^a±0.74	5.44^a±0.05
Progenies	204.17^b±0.12	124.75^b±7.31	79.42^b±7.19	182.92^b±5.89	57.99^b±1.17	68.65^b±0.35	7.77^b±0.14
Bao	292.89^a±4.75	66.33^a±7.07	222.29^a±3.71	145.96^c±4.54	70.08^a±1.90	66.78^a±0.11	5.14^c±0.05
Progenies	170.79^c±0.4	81.75^{ca}±4.36	76.26^b±7.68	130.07^d±3.28	57.25^b±1.64	68.81^b±0.29	7.51^b±0.03
Kakwale	253.01^d±3.43	79.96^c±0.06c	168.56^c±2.98	193.36^e±0.06	63.71^c±1.58	70.4^b±1.27	5.97^d±0.05
Progenies	222.05^e±0.5	106.71^d±2.53	115.34^d±2.00	159.71^a±6.42	86.67^d±2.59	66.73^a±0.04	6.57^e±0.05
Nyaraboke	298.13^a±3.25	73.88^a±2.06	223.25^a±2.59	154.71^a±6.42	85.84^d±2.59	69.98^b±0.04	5.97^d±0.05
Progenies	226.54^e±0.41	94.84^d±0.12	129.7^e±3.37	172.79^b±6.19	73.46^a±3.13	66.73^a±0.04	7.50^b±0.14
95/SE/00036	336.55^f±0.18	80.34^c±0.23	256.21^f±0.41	162.75^a±1.41	82.42^d±1.65	63.28^c±0.04	4.24^f±0.04
Progenies	221.59^e±4.48	79.54^c±1.12	142.04^g±0.41	141.92^c±1.77	62.38^c±0.64	64.90^d±0.00	4.67^g±0.00
NASE 10	303.75^a±1.88	80.46^c±5.01	223.29^a±3.13	152.34^a±5.18	71.88^a±0.18	68.45^b±0.07	5.53^h±0.00
Progenies	221.46^e±0.05	97.67^e±0.19	120.34^d±4.37	159.63^a±2.03	67.28$_e$±0.51	65.80^a±0.35	5.13^c±0.00
TME 5	282.6^a±5.05	124.30^b±2.65	216.04^a±1.36	227.17^f±2.59	89.17^d±0.00	66.23^a±0.00	5.75^i±0.03
Progenies	195.55^b±1.95	127.34^b±7.66	93.34^i±2.35	178.67^b±2.60	76.81^a±2.66	68.05^b±0.00	6.90^j±0.24
TME 14	344.96^h±2.77	80.54^c±3.24	264.42^j±0.47	167.54^a±2.41	87.00^d±2.41	64.75^d±0.00	5.35^k±0.05
Progenies	203.17^b±5.66	99.25^e±3.29	103.92^k±2.36	165.42^a±6.72	66.08^e±3.54	68.45^b±0.00	6.70^j±0.04

^{a}Values with the same superscript in the column not significantly different at 5%. HPV= hot paste viscosity; PTi= pasting temperature; RVU= rapid viscosity units.

progenies (79.54 - 127.34 RVU) with significant (P< 0.05) differences observed among the parents and their progenies. Similar results were observed for the break down viscosity in the parental lines (168.56 - 264.42 RVU) and the progenies (56.13 - 142.04 RVU) while the final viscosity ranged from 145.96 - 227.17 RVU in the parental lines and 130.07 - 182.92 RVU in the progenies. The low final viscosities observed compared to the peak viscosities indicate the low tendency of cassava starch to retrograde (Moorthy, 2002). The set back viscosity was generally higher in the parents (63.71 - 89.17 RVU) than in the progenies (56.04 - 86.67 RVU). Starch viscosity is important in the characterisation of starch and the differences observed provides an opportunity for selection of cultivars from the F_1 progenies for industrial and food uses. The pasting temperature of starch from the different parental lines ranged from 63 - 69°C and was similar to that of the progenies where it ranged from 64.90 to 68.81. Cassava starch has low pasting temperature (average 68°C) hence, it forms pastes much easier compared to starches with high pasting temperatures such as potato (average 72°C) (Moorthy, 2002) and rice (average 69.5°C) (Cameron et al., 2007).This is due to the low stability of cassava starch granules on heating which makes them loose their molecular structure easily (Novelo-Cen and Betancur-Ancona, 2005). The peak time was low among the parental lines (4.24 - 5.97 min) compared to the progenies (4.67 - 7.77 min) hence their ability to form pastes much easier than the progenies. Significant (P < 0.05) differences were observed in the different pasting properties of the progenies with wider variations observed among them. Such differences can be attributed to the differences in the size distributions of starch granules (Table 3) and give more opportunity for selection of starch with different uses compared to the parents. The starch pasting curves for the different varieties and progenies are presented in Figures 2a and b. The results showed an outstanding feature for starch obtained from the progenies where the pasting curve had no clear peak with a 'shoulder' at attainment of peak viscosity suggesting differences in the crystalline nature of starch and the general structure of amylopectin component of starch in the different progenies compared to the different parents.

Swelling power of cassava starch

The results of starch swelling power at different temperatures are presented in Table 5. An average of 2 fold increase was observed with a 10°C change in temperature. At higher temperatures (>70°C), a sudden increase in swelling power was observed. This may be attributed to the disruption of starch granules at higher temperatures and consequent release of all the amylose from the amylopectin network (Charles et al., 2007).

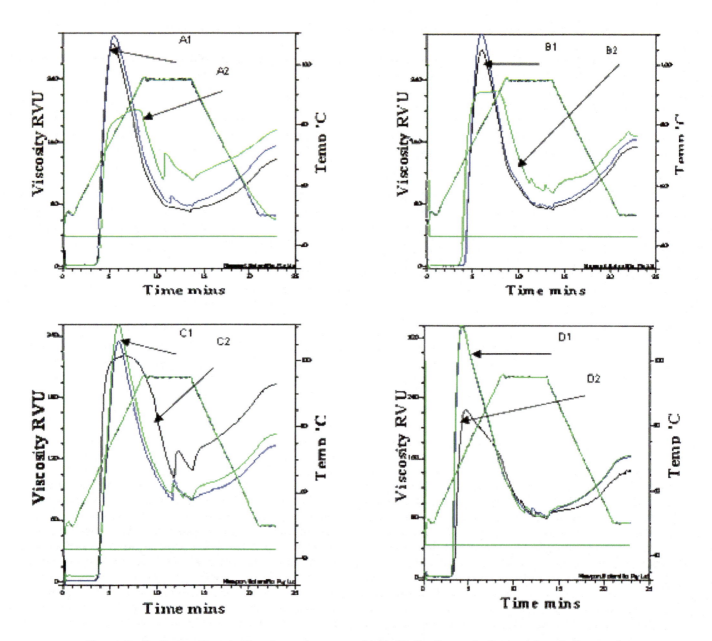

Figure 2a: Pasting profiles of different parents compared with F1 Families and other varieties. A=Bamunanika; B=Nyaraboke; C=Kakwale; D= 95/SE/00036. X1= Parent, X2= Progeny.

Significant (P< 0.05) differences were observed in swelling power at 30°C between the parents and progenies with the progenies showing an average high swelling power. At higher temperatures the differences generally disappeared. Uptake of water by starch granules results into progressive swelling as temperature increases (Charles et al., 2007). Swelling power is an important parameter especially in characterisation of starches from different botanical origins which display different swelling powers at a given temperature (Moorthy, 2002; Charles et al., 2007). It also affects both the eating quality of cassava roots and the use of starch in a number of industrial applications (Moorthy, 2002). High swelling power results into high digestibility and ability to use starch in solution suggesting improved dietary properties and the use of starch in a range of dietary applications.

Correlations among the physicochemical and functional properties of cassava starch

The results for the interrelationships between various starch properties are presented in Table 6. The amylose content was positively correlated to the starch content (r

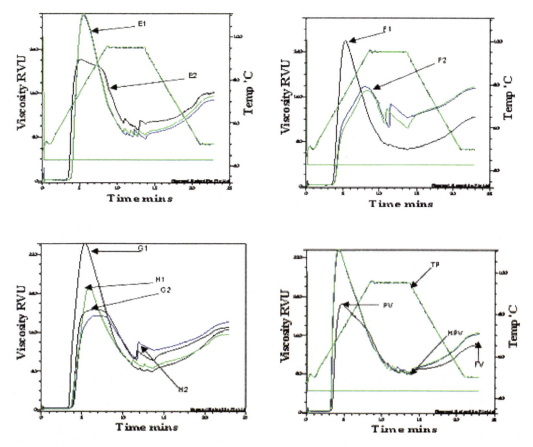

Figure 2b: Pasting profiles of different parents compared with F1 Families and other varieties.
E= NASE 12; F=NASE 10; G= TME 14; H= TME 5. X1= Parent, X2= Progeny. PV= Peak Viscosity, HPV = Hot Paste Viscosity or Trough Viscosity, FV= Final Viscosity, FV-HPV= Set back viscosity, PV-HPV = Break down viscosity TP = Temperature profile.

Table 5. Cassava starch swelling power at different temperatures.

Variety/Progenies	30 °C	40 °C	50 °C	60 °C	70 °C	80 °C
Bamunanika	$1.52^a \pm 0.22$	$2.63^a \pm 0.71$	$5.95^a \pm 0.28$	$7.87^a \pm 0.36$	$10.21^a \pm 0.57$	$16.16^a \pm 0.75$
Progenies	$2.12^b \pm 0.11$	$3.64^a \pm 0.22$	$7.21^b \pm 0.89$	$9.91^b \pm 0.09$	$11.18^a \pm 0.73$	$16.89^a \pm 1.61$
Bao	$1.54^a \pm 0.29$	$3.33^a \pm 0.61$	$6.39^a \pm 1.39$	$7.85^a \pm 0.36$	$10.79^a \pm 0.61$	$15.76^a \pm 0.96$
Progenies	$2.73^b \pm 0.51$	$3.64^a \pm 0.52$	$6.79^a \pm 0.42$	$9.09^b \pm 0.14$	$12.24^a \pm 1.43$	$15.05^a \pm 0.09$
Kakwale	$1.32^a \pm 0.19$	$3.74^a \pm 0.37$	$6.39^a \pm 0.31$	$8.72^a \pm 0.34$	$11.61^a \pm 0.10$	$15.66^a \pm 0.55$
Progenies	$2.04^b \pm 0.39$	$6.01^b \pm 1.25$	$7.04^a \pm 0.26$	$9.98^b \pm 0.27$	$14.59^b \pm 1.44$	$18.59^a \pm 2.48$
Nyaraboke	$1.68^a \pm 0.45$	$3.59^a \pm 0.65$	$6.21^a \pm 0.68$	$8.73^a \pm 0.34$	$12.11^a \pm 0.25$	$16.32^a \pm 0.74$
Progenies	$2.05^b \pm 0.03$	$4.24^b \pm 0.88$	$7.11^a \pm 0.42$	$9.39^b \pm 0.25$	$10.51^a \pm 1.52$	$16.05^a \pm 1.06$
95/SE/00036	$1.95^a \pm 0.09$	$3.85^a \pm 0.07$	$6.54^a \pm 0.02$	$8.31^a \pm 1.16$	$11.32^a \pm 0.85$	$15.67^a \pm 0.74$
Progenies	$1.94^b \pm 0.11$	$4.54^b \pm 0.96$	$5.62^a \pm 1.47$	$9.75^b \pm 0.62$	$12.78^a \pm 1.88$	$15.15^a \pm 1.37$
NASE 10	$1.57^a \pm 0.18$	$3.10^a \pm 0.06$	$5.97^a \pm 0.23$	$9.11^b \pm 0.25$	$12.62^a \pm 0.98$	$17.14^a \pm 1.37$
Progenies	$2.00^b \pm 0.29$	$5.40^b \pm 0.54$	$7.97^a \pm 0.12$	$10.77^b \pm 0.01$	$12.54^a \pm 1.01$	$20.79^b \pm 0.91$
NASE 12	$1.62^a \pm 0.25$	$3.73^a \pm 0.07$	$5.95^a \pm 0.32$	$8.16^a \pm 0.32$	$12.99^a \pm 1.83$	$17.08^a \pm 0.76$
Progenies	$2.36^b \pm 0.01$	$4.00^b \pm 0.52$	$6.74^a \pm 0.46$	$9.92^b \pm 0.31$	$13.29^a \pm 1.61$	$18.01^a \pm 2.99$
TME 5	$1.26^a \pm 0.10$	$3.15^a \pm 0.14$	$6.17^a \pm 0.63$	$7.53^a \pm 0.27$	$11.27^a \pm 0.93$	$16.82^a \pm 1.14$
Progenies	$1.92^b \pm 0.57$	$6.31^b \pm 1.16$	$7.46^a \pm 0.26$	$9.11^{ab} \pm 0.79$	$12.83^a \pm 2.09$	$16.29^a \pm 2.14$
TME 14	$1.27^a \pm 0.18$	$3.18^a \pm 0.08$	$5.95^a \pm 0.31$	$8.59^a \pm 0.37$	$10.38^a \pm 1.85$	$17.33^a \pm 0.93$
Progenies	$2.38^b \pm 0.01$	$5.57^b \pm 1.13$	$7.60^a \pm 0.33$	$8.89^a \pm 0.03$	$13.61^a \pm 1.17$	$15.74^a \pm 1.77$

[a]Mean values with the same superscript are not significantly different at 5%.

Table 6. Correlation matrix between the different starch parameters.

	Amy	Ash	BV	CF	FV	GS	Lpd	MC	PV	PT	Pt	Ptn	SP	SV	Sol	SC	HPV	WSM
Amy	1.00																	
Ash	0.06	1.00																
BV	-0.27*	0.13	1.00															
CF	0.32	0.25*	-0.38*	1.00														
FV	0.18	0.61*	-0.29*	0.08	1.00													
GS	-0.30*	0.46*	0.10	-0.49*	0.16	1.00												
Lpd	0.18	0.73*	0.09	-0.45*	0.45*	0.35*	1.00											
MC	-0.54*	-0.64*	-0.01	-0.28*	-0.29*	-0.35*	-0.37*	1.00										
PV	-0.20	0.32*	0.88*	-0.54*	0.10	0.24	0.35*	0.01	1.00									
PT	0.38*	0.53*	-0.62*	0.42*	0.51*	0.03	0.40*	-0.60*	-0.56*	1.00								
Pt	0.45*	-0.22	-0.73*	0.49*	0.44*	-0.60*	-0.21	0.17	-0.57*	0.38*	1.00							
Ptn	-0.28*	-0.09	0.27	-0.17	0.21	0.24	-0.38*	0.18	0.35*	-0.43*	-0.09	-0.09						
SP	-0.73*	-0.01	0.57*	-0.81*	-0.30*	0.36*	0.20	0.54*	0.59*	-0.63*	-0.66*	-0.66*	1.00					
SV	0.23	0.23	0.49*	-0.32*	0.14	-0.23	0.54*	-0.03	0.57*	-0.07	-0.16	-0.16	0.20	1.00				
Sol	-0.74*	0.18	0.24	0.08	0.26*	0.32*	-0.24	0.10	0.23	-0.11	-0.26*	-0.26*	0.58*	0.25	1.00			
SC	0.52*	0.31*	0.17	-0.05	-0.28*	0.17	0.23	-0.64*	0.04	0.14	-0.27*	-0.27*	-0.48*	0.28*	-0.23	1.00		
HPV	0.35*	0.45*	-0.58*	0.18	0.87*	0.17	0.50*	-0.35*	-0.22	0.71*	0.49*	-0.01	-0.44*	-0.26*	-0.08	-0.55*	1.00	
WSM	0.10	0.25*	-0.47	0.32*	0.43*	-0.48*	0.33*	0.04	-0.39*	0.68*	0.55*	-0.44*	-0.29	0.10	-0.02	-0.24	0.71*	1.00
														0.24	-0.06	0.32*	0.45*	0.51*

*n= Significant correlations at 5%. BV= breakdown viscosity, CF= crude fibre, FV= final viscosity, GS= granule size, MC= moisture content, PV= peak viscosity, PT= peak temperature, Pt= peak time, SP= swelling power, SV= set back viscosity, Sol= Solubility, SC= starch content, HPV= hot paste viscosity, WSM= cold water soluble materials, Lpd = lipid content, Ptn = protein content.

= 0.522) and the peak time (r = 0.450) while it was negatively correlated to swelling power (r = -0.733) and moisture content (r = -0.536) suggesting its importance in starch solution and pasting properties. However, no significant relationships were observed between amylose content and other pasting parameters including peak and final viscosity. There were positive correlations between the ash content and peak time (r = 0.533), hot paste (r =0.451) and final viscosity (r = 0.607). This could be attributed to the presence of phosphorus and minerals such as calcium, which impart high viscosity to starch and increases its gel strength (Moorthy, 2002). The negative correlation between dietary fiber and peak viscosity (r = -0.544) explains the reductive effect of fiber on the viscosity of starch. Dietary fibre was also positively correlated to peak time (r = 0.486) and pasting temperature (r = 0.422) as expected since fibre acts as a barrier to the free swelling of starch granules hence increasing the pasting temperature and peak time during pasting (Moorthy, 2002). Thus the high swelling power observed in cassava starch is as a result of low fibre contents associated with it. Swelling power was negatively correlated to peak time (r = -0.659), pasting temperature (r = -0.629) and crude fibre (r = -0.805) suggesting the low peak time and low crude fibre associated with cassava starch which has a high swelling power (Charles et al., 2007), and positively correlated to peak viscosity (r = 0.588) explaining the increase in the swelling power with increase in the pasting properties of starch (Rickard et al., 1991). On the other hand, cold water solubility was positively correlated to peak time (r = 0.549) and the pasting temperature (r = 0.684).

ACKNOWLEDGEMENTS

This work was supported under the BIO-EARN program, with a grant from the Swedish International Cooperation Agency/Department for Research Cooperation (Sida/SAREC).

REFERENCES

Baguma Y (2004). Regulation of starch synthesis in cassava. Doctoral Thesis, Swedish university of agricultural sciences, Uppsala, Sweden.

Benesi IR (2005). Characterisation of Malawian cassava germplasm for diversity, starch extraction and its native and modified properties. PhD Thesis, Department of Plant Sciences, University of the Free State, South Africa pp. 74-123.

Ceballos H, Sanchez T, Morante N, Fregene M, Dufour D, Smith A, Denyer K, Perez J, Calle F, Mestres C (2006). Discovery of an Amylose-free Starch Mutant in Cassava. (*Manihot esculenta* Crantz J. Agric. Food Chem. 55(18): 7469-7476.

Charles A, Huang T, Lai P, Chen C, Lee P, Chang Y (2007). Study of wheat flour–cassava starch composite mix and the function of cassava mucilage in Chinese noodles. Food Hydrocolloids 21: 368-378.

Cameron K, Wang Y, Moldenhauer A (2007). Comparison of Starch Physicochemical Properties from Medium-Grain Rice Cultivars Grown in California and Arkansas Starch/Stärke 59: 600-608.

Charles A, Chang Y, Ko W, Sriroth K, Huang T (2004). Some Physical and Chemical Properties of Starch Isolates of Cassava Genotypes. Starch/Stärke 56: 413-418.

Ch´avez A, S´anchez T, Jaramillo G, Bedoya J, Echeverry J, Bola¯nos E, Ceballos H, and Iglesias C (2005). Variation of quality traits in cassava roots evaluated in landraces and improved clones, Euphytica 143: 125-133.

Gomes A, Mendes da Silva C, Ricardo N (2005). Effects of annealing on the physicochemical properties of fermented cassava starch (polvilho azedo). Carbohydrate Polymers 60: 1-6.

Jaramillo G, Morante N, Perez J, Calle F, Ceballos H, Arias B, Belloti A (2005). Diallele analysis in cassava adapted for the mild altitude valleys environment. Crop sci; 45, 1058-1063

Moorthy S (2002).Tuber crop starches. Tech Bulletin No. 18 CTCRI, Trivandrum.

Moorthy S, Mathew G, Padmaja G, (1993). Functional properties of the starchy flour extracted from cassava on fermentation with a mixed culture inoculum. J. Sci. Food Agric., 61: 443-447.

Moorthy S (1985). Effect of different types of surfactants on cassava starch properties. J. Agric. Food Chem. 35: 1227-1232.

National Agricultural Research Organisation (NARO) (2005). Cassava research program www.naro.go.ug.

Noda T, Fujikami S, Mura H, Fukushima M, Takigawa S, Endo M (2006). Effect of potato starch characteristics on the textural properties of Korean-style cold noodles made from wheat flour and potato starch blends. Food Sci. Technol. Res. 12: 278-283.

Novelo-Cen L, Betancur-Ancona D (2005). Chemical and Functional Properties of *Phaseolus lunatus* and *Manihot esculenta* Starch Blends, Starch/Stärke 57: 431-441.

Rickard J, Asaoka M, Blanshard J (1991). The physicochemical properties of cassava starch. Trop. Sci. 31: 189-207.

Rodríguez-Sandoval E, Fernández-Quintero A, Cuvelier G, Relkin P, Bello-Pérez L, (2008). Starch Retrogradation in Cassava Flour from Cooked Parenchyma. Starch/Stärke 60: 174-180.

Sriroth K, Santisopasri V, Petchalanuwat C, Kurotjanawong K, Piyachomkwani K, Oates C (1999). Cassava starch granule structure-function properties; Influence of time and conditions of harvest on four cultivars of cassava starch. Carbohydr. Polym. 38: 161-170.

Tukomane T, Leerapongnun P, Shobsngob S, Varavinit S (2007). Preparation and Characterization of Annealed- Enzymatically Hydrolyzed Tapioca Starch and the Utilization in Tableting. Starch/Stärke 59: 33-45.

Varavinit S, Paisanjit W, Tukomane T, Pukkahuta C (2007). Effects of Osmotic Pressure on the Crosslinking Reaction of Tapioca Starch. Starch/Stärke 59: 290–296.

Zaidul I, Norulaini N, Omar A, Yamauchi H, Noda T (2007). RVA analysis of mixtures of wheat flour and potato, sweet potato, yam, and cassava starches. Carbohydrate Polymers 69: 784-791.

du Ploy C (1985). Progress and limitations in breeding of the sweet potato (*Ipomoea batatas*) in South Africa ISHS Acta Hortic. 194: 77-82.

Ceballos H, Fahrney K, Howeler R, Ospina B (2008). Cassava Potential for Bioethanol. Symposium --Biofuels in Developing Countries: Opportunities and Risks 625-628.

Combining ability for maize grain yield in *striga* endemic and non-endemic environments of the southern guinea savanna of Nigeria

G. Olaoye and O. B. Bello*

Department of Agronomy, University of Ilorin, Ilorin, Nigeria.

Combining ability studies for maize grain yield and other agronomic characters were carried out using ten open-pollinated maize varieties and their 45 F_1 hybrids in a *Striga hermonthica* (Del.) Benth endemic zone (Shonga) and non-endemic zone (Ilorin) in Kwara State, Nigeria, during the 2005 cropping season. Both general combining ability (GCA) and specific combining ability (SCA) effects for *Striga* related characters such as *Striga* shoot counts, syndrome ratings, flowering *Striga* shoots and barren maize plants were generally low, suggesting the role of additive and dominant gene action in tolerance to *S. hermonthica* (Del.) Benth. Parents Acr 94 Tze Comp5 and Tze Comp3 C2 had significant ($p < 0.05$) positive GCA effects for grain yield and other agronomic characters in both *Striga* endemic and non-endemic environments respectively. Crosses Tze Comp3 C2 x Hei 97 Tze Comp3 C4, Tze Comp3 C2 x Acr 94 Tze Comp5 and Ak 95 Dmr - Esrw x Acr 94 Tze Comp5 had significant ($p < 0.05$) positive SCA effects for grain yield only in *Striga* endemic environment. These parents and hybrids appeared to have gene pools for *S. hermonthica* tolerance that can be manipulated and used to develop promising hybrids for early maturity and high grain yield across the Southern Guinea Savanna ecology.

Key words: *Striga hermonthica*, tolerance, combining ability, grain yield, Nigeria.

INTRODUCTION

Striga infestation is one of the most serious constraints to cereals production by smallholder farmers in sub-Saharan Africa. Infestation usually results in substantial yield losses, averaging more than 70% of *Striga* free environment (Kim, 1991). Much of the damage occurs before *Striga* emerges from the ground and the degree of damage depends on susceptibility of the cultivar, *Striga* species, level of infestation, and any additional stress in the host's environment (Shinde and Kulkarni, 1982; Vasudeva Rao et al., 1982; Basinki, 1995). Of the five *Striga* species, *S. hermonthica* (Del.) Benth and *S. asiatica* (L.) Kuntze are the most noxious weeds threatening 44 million hectares of agricultural land in Africa (Sauerborn, 1991). *S. hermonthica* (Del.) Benth infestation in particular constitutes a serious threat to maize production in the savanna ecologies of Nigeria. Breeding for tolerance/resistance to *S. hermonthica* (Del.).

Benth offers a viable option for the management of this weed and is economically compatible with the low-cost input requirement of the subsistence farmers in controlling *Striga* (Ramaiah, 1986; Kim et al., 2002). Available data suggests that *Striga* resistance is controlled by a relatively few genes with additive effects (Shinde and Kulkani, 1982; Vasudeva Rao et al., 1982). Kim (1994b) observed a considerable variability in the resistance of maize varieties to *S. hermonthica*. Findings from two independent studies, (Mumera and Below, 1996; Gurney et al., 2002) revealed that identification of *Striga* resistance maize genotypes should focus on the ability of ear sink to successfully compete with *Striga* for assimilates. However, maize breeding programmes designed for the development of commercial maize hybrids and improved maize genotype tolerant to *Striga* parasites usually requires a good knowledge of combining ability of the genetic materials to be used. Kim (1994a) used combining ability approach to study the genetics to maize tolerance of *S. hermonthica* in 10 inbred parents under *S. hermonthica* infestation in Mokwa, Nigeria. The results showed that such study was highly suitable for the development of *Striga* tolerant maize genotypes. Mumera and Below

---

*Corresponding author. E-mail: obbello2002@yahoo.com.

(1996) reported that counts of Striga emerged plants differed by more than three folds between the most and least susceptible genotypes with early maturing types generally being the most resistant. Kim et al. (2002) reported that tolerant open pollinated varieties (OPVs) produced 2.0 - 2.5 times the yield of susceptible varieties, especially under high Striga infestation. Also, Kim and Adetimirin (1997) studied responses of tolerant and susceptible maize varieties to timing and rate of nitrogen fertilizer under S. hermonthica infestation in southern guinea savanna ecology of Nigeria. Their results revealed that among all the tolerance factors studied, the most important component for Striga management was genetic tolerance. Hence, development of S. hermonthica tolerant genotypes appears feasible and promising in these agro-ecological zones.

Striga infestation has reached an endemic status not only in the northern guinea savanna, but also in the Southern Guinea Savanna (SGS) of Nigeria. It constitutes a serious threat to maize production and farmers are being compelled to abandon their farmlands to Striga, or change to production of less susceptible crop. The objectives of this study therefore were (1) to assess both general and specific combining abilities of ten open pollinated maize varieties for maize grain yield and other agronomic characters in Striga endemic and Striga free environments and, (2) to identify open pollinated varieties and hybrids that combined tolerance to S. hermonthica with grain yield and suitable agronomic traits for use in commercial hybrid maize production in Striga endemic zones of the Southern Guinea Savanna of Nigeria.

MATERIALS AND METHODS

The genetic materials used for this study comprised of 10 open pollinated maize varieties, which have been developed for yield and adaptation to biotic and abiotic stress factors. They are also early to medium maturing white cultivars with maturity period of 90 to 100 days. The varieties were obtained from the International Institute of Tropical Agriculture (IITA), Ibadan, Nigeria. Partial diallel crosses were made between the 10 open pollinated maize varieties during 2004 cropping season at the Teaching and Research farm, University of Ilorin, Nigeria. The resultant 45 F_1 hybrids were harvested, processed and stored in the cold room prior to field evaluation. The field study was carried out in Striga endemic (Shonga) and Striga free (Ilorin) environments both in the Southern Guinea Savannah (SGS) of Kwara State of Nigeria during the raining season of 2005. The trial was laid out in a Randomized Complete Block Design with four replicates. Entries which included the hybrids and parents were made in two-row plots of 5 x 1.5 m each and planted at inter-row spacing of 75 cm and within row spacing of 50 cm to enhance a plant population of about 53,555 stands per hectare. Three seeds were planted on a hill but were later thinned to two at three weeks after planting (WAP). NPK 20-10-10 fertilizer was applied at the rate of 80 kgNha^{-1} in split doses immediately after thinning and at 6 WAP. In the Striga endemic zone, Striga related parameters such as Striga shoot counts at 10 WAP and Striga syndrome ratings using a Scale 1 - 9 as described by Kim (1994a) were collected. Others included: number of flowering Striga plants as well as barren maize plants. At Striga endemic and non-endemic fields, agronomic parameters such as maize establishment plant count, days to anthesis and silking and plant height were also measured. Plant height was measured from soil level to the base of the tassel. Days to 50% silking (number of days from planting to when 50% of the population have silked) as well as days to 50% pollen shed (number of days from planting till the time 50% have shed pollen) were recorded. Anthesis-silking interval was estimated as the difference between days to pollen shed and silking. Maize grain yield (t/ha) was also measured after adjusting for moisture at harvest. Data collected were subjected to separate diallel analyses using Griffing (1956) Method II (parents and crosses together), Model I (fixed effects). General and specific combining abilities (GCA and SCA) were computed using SAS (1999) for the 10 parents open pollinated varieties (OPVs) and their 45 F_1 hybrids with respect to Striga related and maize agronomic characters.

RESULTS

General and specific combining ability effects for Striga parameters

ANOVA for GCA effects of parents for Striga related traits in the Striga endemic environment are presented in Table 1. GCA effect for Striga shoot count was very low probably due to high tolerance of the parents to S. hermonthica emergence. The highest GCA effect for this trait was 95.50 from Ak 95 Dmr-Esrw as against the least value of 0.22 in parent Tze Comp4-Dmr Srbc2. GCA effect was also low among the parents for number of flowering shoots at 12 weeks after maize was planted, with Hei 97 Tze Comp3 C4 and Acr 97 Tze Comp3 C4 having the least effects. However, Ak 95 Dmr-Esrw was significant for this trait. GCA effect for number of barren maize plants was very low in most of the parents with parents Tze Comp3 C2 and Hei 97 Tze Comp3 C4 having the least GCA effects, while Acr 90 Pool 16-Dt had significant GCA effect for this character. GCA effect for Striga syndrome rating was also generally low with parents Tze Comp3 C2 and Acr 97 Tze Comp3 C4 recording the least effects. However, significant GCA effect was observed for Striga syndrome rating in Acr 90 Pool 16-Dt.

Specific combining ability (SCA) effects for Striga related parameters are presented in Table 2. SCA effects for both Striga shoot count (that is, emergence and infestation) and number of flowering Striga plants were generally low. However, highly significant SCA effect was observed in crosses Tze Comp4 C2 x Tze Comp3 C2, Acr 97 Tze Comp3 C4 x Tze Comp3 C2, Hei 97 Tze Comp3 C4 x Acr 94 Tze Comp5, Acr 94 Tze Comp5 x Ak 95 Dmr-Esrw and Tze Comp3 C2 x Ak 95 Dmr-Esrw for Striga shoot count at 8 WAP. Similarly, SCA effect for number of flowering Striga plants at 12 weeks after Maize was planted was significant for hybrids Acr 90 Pool 16-Dt x Acr 94 Tze Comp5, Acr 90 Pool 16-Dt x Tze Comp3 C2, Tze Comp4 C2 x Tze Comp3 C2, Acr 97 Tze Comp3 C4 x Tze Comp3 C2, Hei 97 Tze Comp3 C4 x Ak 95 Dmr-Esrw and Acr 94 Tze Comp 5 x Ak 95 Dmr-Esrw. Values recorded in respect of the remaining crosses were very low and non-significant.

Non-significant SCA effects were observed in the hybrids for number of barren maize plants (Table 3).

Table 1. Estimate of GCA effects for *Striga* related parameters in *Striga* endemic environment (Shonga) of southern guinea savanna of Nigeria.

Parents	*Striga* shoot count	Number of flowering *Striga*	Number of barren maize plants	*Striga* syndrome rating
Acr 90 Pool 16-Dt	35.36	13.94	137.48*	1120.08**
Tze Comp 4-Dmr Srbc2	0.22	1.50	40.71	6.51
Tze Comp4 C2	15.95	10.69	13.97	141.37
Acr 97 Tze Comp3 C4	11.27	0.01	122.89	1.44
Hei 97 Tze Comp3 C4	0.23	1.05	0.81	7.45
Acr 94 Tze Comp5	35.27	14.13	39.98	49.65
Tze Comp3 Dt	11.07	7.56	277.18	76.31
Tze Comp3 C2	34.78	13.87	0.06	0.35
Ak 95 Dmr-Esrw	95.50	70.04**	17.96	11.22
Tze Msr-W	1.02	2.05	1.13	8.21

*, ** Significant at < 0.05 and < 0.01 levels of probability respectively.

However, SCA effects were significant for *Striga* syndrome ratings in crosses Tze Comp4-Dmr Srbc2 x Acr 97 Tze Comp3 C4, Tze Comp4-Dmr Srbc2 x Tze Comp3 Dt, Tze Comp4-Dmr Srbc2 x Tze Comp3 C2, Tze Comp4 C2 x Acr 97 Tze Comp3 C4, Hei 97 Tze Comp3 C4 x Tze Comp4 C2 and Tze Comp3 Dt x Ak 95 Dmr-Esrw.

General and specific combining ability effects for maize grain yield and related traits

Estimates of GCA effects for grain yield and agronomic traits in *Striga* endemic and *Striga* free environments are presented in Table 4. GCA effects for maize agronomic characters in *Striga* endemic and *Striga* free environments differed significantly in the parents. Parent Acr 90 Pool 16-Dt recorded significant GCA effects only for maize establishment count and maize grain yield in both *Striga* endemic and *Striga* free environments. Parent Acr 94 Tze Comp5 exhibited significant GCA effects for days to pollen shed and grain yield in *Striga* free environment, and also had significant GCA effects for both days to silking and grain yield in *Striga* endemic environment. Tze Comp3 C2 only showed positive and significantly high GCA effects for both anthesis-silking interval and grain yield in both *Striga* endemic and *Striga* free environments. Acr 94 Tze Comp5 only had significant GCA effect for plant height also showed significant effect for grain yield in *Striga* endemic environment.

GCA effects for maize grain yield in *Striga* endemic environment were generally low in many of the parents. However, Acr 94 Tze Comp5 and Tze Comp3 C2 exhibited high GCA effects for maize grain yield and some of the agronomic characters in both *Striga* endemic and *Striga* free environments. Acr 90 Pool 16-Dt and Tze Comp4 Dmrsrbc2 recorded high GCA effects for grain yield in *Striga* free environment.

SCA effects for maize establishment count (Table 5) in *Striga* endemic and non-endemic environments were highly significant in hybrid Tze Comp4 C2 x Acr 97 Tze Comp3 C4. Acr 90 Pool 16-Dt x Hei 97 Tze Comp3 C4 had significant SCA effects for days to 50% pollen shed and grain yield in *Striga* free environment, while SCA effects for days to 50% silking in *Striga* endemic and non-endemic environments were highly significant in cross Acr 90 Pool 16-Dt x Hei 97 Tze Comp3 C4. Conversely, non-significant SCA effects were recorded for anthesis-silking interval and plant height. Yield assessment in *Striga* endemic environment showed significant effects in crosses Tze Comp3 C2 x Hei 97 Tze Comp3 C4, Tze Comp3 C2 x Acr 94 Tze Comp5 and Ak 95 Dmr-Esrw x Acr 94 Tze Comp5. Hybrids Tze Comp4 C2 x Acr 97 Tze Comp3 C4, Acr 90 Pool 16-Dt x Hei 97 Tze Comp3 C4 had significant SCA effects for maize grain yield and also for maize establishment count and flowering traits respectively in *Striga* free environment.

Table 2. Estimates of SCA effects for *Striga* shoot count (upper diagonal) and number of flowering *Striga* (lower diagonal) in F_1 hybrids 8 WAP in *Striga* endemic environment (Shonga, Nigeria).

Parent	Acr 90 Pool 16-Dt	Tze Comp4-Dmr Srbc2	Tze Comp4 C2	Acr 97 Tze Comp3 C4	Hei 97 Tze Comp3 C4	Acr 94 Tze Comp5	Tze Comp3 Dt	Tze Comp3 C2	Ak 95 Dmr-Esrw	Tze Msr-W
Acr 90 Pool 16-Dt	–	0.20	0.01	5.33	24.12	11.47	9.39	60.00	12.87	2.67
Tze Comp4-Dmr Srbc2	1.89	–	6.94	4.10	4.71	36.28	0.23	0.52	28.90	4.67
Tze Comp4 C2	0.28	2.58	–	1.94	1.18	41.84	9.30	110.12**	21.42	3.78
Acr 97 Tze Comp3 C4	1.05	3.64	0.25	–	2.14	54.71	32.47	105.57**	7.72	8.34
Hei 97 Tze Comp3 C4	8.60	0.01	3.99	0.01	–	105.15**	9.34	18.09	36.96	5.76
Acr 94 Tze Comp5	48.23*	29.27	23.58	34.00	64.39	–	1.30	56.89	77.67*	2.45
Tze Comp3 Dt	3.20	4.55	1.92	8.34	0.89	4.16	–	3.62	106.38**	6.46
Tze Comp3 C2	43.03*	0.31	62.90*	64.27**	7.11	3.28	1.31	–	3.63	11.62
Ak 95 Dmr-Esrw	10.04	20.84	15.01	5.32	39.72**	73.39**	30.56	32.31	–	7.53
Tze Msr-W	1.78	3.41	5.87	3.45	4.86	7.32	8.56	2.93	4.34	–

*, ** Significant at < 0.05 and 0.01 levels of probability respectively.

Table 3. Estimates of SCA effects for number of barren plants (upper diagonal) and *Striga* syndrome rating (lower diagonal) in F_1 hybrids 12 WAP in *Striga* endemic environment (Shonga, Nigeria).

Parent	Acr 90 Pool 16-Dt	Tze Comp4-Dmr Srbc2	Tze Comp4 C2	Acr 97 Tze Comp3 C4	Hei 97 Tze Comp3 C4	Acr 94 Tze Comp5	Tze Comp3 Dt	Tze Comp3 C2	Ak 95 Dmr-Esrw	Tze Msr-W
Acr 90 Pool 16-Dt	–	25.94	46.71	139.02	118.74	8.94	37.59	39.24	23.48	0.41
Tze Comp4-Dmr Srbc2	6.75	–	120.14	228.14	2.15	5.54	145.67	124.74	94.49	0.21
Tze Comp4 C2	101.23	48.43	–	323.35	177.16	1.24	4.46	2.51	316.07	0.42
Acr 97 Tze Comp3 C4	49.91	251.5*	268.57*	–	17.96	31.56	161.84	4.28	15.77	0.23
Hei 97 Tze Comp3 C4	3.27	37.51	251.11*	12.39	–	265.2	10.43	15.83	12.32	0.31
Acr 94 Tze Comp5	0.37	24.48	19.67	18.69	195.77	–	108.09	26.65	138.49	0.32
Tze Comp3 Dt	111.07	389.5*	157.31	99.72	6.32	5.53	–	0.82	393.32	0.17
Tze Comp3 C2	141.71	254.9*	12.31	21.67	3.26	21.07	11.51	–	290.64	0.43
Ak 95 Dmr-Esrw	171.09	7.59	6.39	2.26	23.45	62.07	397.5*	164.51	–	0.02
Tze Msr-W	7.59	8.58	4.52	32.58	25.84	21.23	3.45	251.00	2.02	–

*, ** Significant at < 0.05 and 0.01 levels of probability respectively.

Table 4. Estimate of GCA effects for maize grain yield and agronomic traits under Striga endemic (Shonga) and non-endemic environments (Ilorin) of southern guinea savanna of Nigeria.

Parent	Maize establishment plant count		Days to 50% pollen shed		Days to 50% silking		Anthesis-silking interval		Plant height		Grain yield	
	Striga endemic	Striga free	Striga endemic	Striga free	Striga endemic	Striga free	Striga endemic	Striga free	Striga endemic	Striga free	Striga endemic	Striga free
Acr 90 Pool 16-Dt	711.02*	384.05*	0.01	9.10	9.10	0.60	0.01	0.67	9.10	390.72	0.20	1.04*
Tze Comp 4-Dmr Srbc2	28.80	485.53*	16.16	88.41	88.41	24.50*	0.72	0.04	88.41	361.25	1.98	1.02*
Tze Comp4 C2	54.32	16.93	2.80	9.96	9.96	0.02	1.87	11.54	9.96	14.89	1.86	0.73
Acr 97 Tze Comp3 C4	16.08	37.36	0.55	153.56*	153.22*	47.03*	0.03	9.47	153.82*	34.44	5.75	0.53
Hei 97 Tze Comp3 C4	146.77	835.45*	0.49	0.95	0.95	0.24	0.06	13.04*	0.95	92.65	0.25	0.36
Acr 94 Tze Comp5	173.07	102.97	0.24	163.11*	163.43*	2.39	0.70	1.63	163.35*	167.33	66.11*	1.45*
Tze Comp3 Dt	327.83*	106.22	1.20	118.32*	118.31*	2.96	4.99	5.43	118.45*	45.00	4.04	0.06
Tze Comp3 C2	37.16	29.72	3.31	55.23	55.23	0.23	12.13*	20.08*	55.23	287.90	55.14*	1.07*
Ak 95 Dmr-Esrw	119.43	218.34*	0.14	9.29	9.29	0.06	1.83	0.05	9.29	25.85	7.88	0.19
Tze Msr-W	4.67	3.67	7.89	4.57	4.57	11.14	0.61	7.80	4.57	452.54	0.41	0.07

*, ** Significant at < 0.05 and < 0.01 levels of probability respectively.

DISCUSSION

General and specific combining ability effects for Striga parameters

In breeding for Striga tolerance, the lower the value obtained for Striga related parameters, the better the genotypes with respect to these traits. Significant GCA and SCA effects recorded in Striga endemic environment for Striga related parameters such as Striga shoot count, Striga syndrome ratings, flowering Striga shoots and barren maize plants, suggest differential reaction of the genotypes to Striga infestation. These results showed that both additive and non-additive gene effects played major roles in the inheritance of tolerance to the parasite both in the OPVs and hybrids respectively. Low GCA effects recorded for Striga shoot count and number of flowering Striga plants in the parents is also indicative of high tolerance to S. hermonthica emergence and survival, consequently, a reduction in the rate of Striga seed multiplication in the soil. The low GCA effects for Striga syndrome rating and number of barren maize plants similarly suggest their tolerance to S. hermonthica infestation. Parents Tze Comp4 Dmr Srbc2 and Tze Msr-W with very high GCA effects for Striga shoot count could be regarded as susceptible while Tze Msr-W and Acr 97 Tze Comp3 C4 with low GCA effects have good tolerance to S. hermonthica. Kim (1994a) had earlier reported low GCA effects for S. hermonthica emergence and host-plant response for most tolerant maize inbreds and high GCA effects for the susceptible. In the present study, additive gene action played a greater role in inheritance of tolerance to S. hermonthica (Del.).

Benth as previously observed by Adetimirin et al. (2001). Generally, the result obtained from our study showed that some parents were tolerant to S. hermonthica, but level of tolerance varied probably due to differences in genetic background among the parental populations used. These results also support the findings of Ransom et al. (1990) who observed that maize genotypes differed significantly in their tolerance to Striga asiatica infestation. This would suggest that a significant portion of Striga tolerance is derived from gene complexes (Kim et al., 1998), which may be best exploited in hybrid combinations where disruption through segregation would be minimized.

Significant SCA effects recorded for Striga related characters indicated differential response of the crosses to these Striga parameters. In other words, non-additive gene action played significant role in the inheritance of Striga tolerance in most of the crosses. The most resistant crosses are those involving Acr 94 Tze Comp5 and Tze Comp3 C2 which are both resistant Striga. In the earlier study, Kim (1991) reported that the highest level of tolerance to S. hermonthica was obtained from crosses involving two resistant parents while most of the susceptible hybrids were from crosses involving of susceptible × susceptible parents as

Table 5. Estimate of SCA effects of selected crosses for maize grain yield and agronomic traits in Striga endemic (Shonga) and non-endemic (Ilorin) environments of the southern guinea savanna of Nigeria.

Hybrid	Maize establishment plant count		Days to 50% pollen shed		Days to 50% silking		Anthesis-silking interval		Plant height		Grain yield	
	Striga endemic	Striga free	Striga endemic	Striga free	Striga endemic	Striga free	Striga endemic	Striga free	Striga endemic	Striga free	Striga endemic	Striga free
Acr 90 Pool 16-Dt x Hei 97 Tze Comp3 C4	3.00	82.94	5.14	9.25*	17.69*	27.05	0.17	0.76	24.98	8.10	1.07	2.03*
Acr 90 Pool 16-Dt x Tze Comp3 Dt	3.53	86.64	0.14	0.47	0.01	*	2.28	2.18	45.68	148.86	0.76	2.35*
Acr 90 Pool 16-Dt x Ak 95 Dmr-Esrw	1.40	0.15	1.05	1.98	5.83	0.01	2.05	0.27	2.45	11.16	0.49	1.87*
Tze Comp4 C2 x Acr 97 Tze Comp3 C4	234.44*	240.61*	0.30	0.24	0.41	9.94	3.48	0.05	152.99	263.99	0.36	1.07*
Tze Comp4 C2 x Hei 97 Tze Comp3 C4	60.84	58.90	0.27	0.96	0.06	2.09	2.67	1.62	47.37	163.05	0.14	1.19*
Acr 97 Tze Comp3 C4 x Hei 97 Tze Comp3 C4	37.25	43.90	0.44	0.01	0.24	4.82	8.18*	1.26	8.87	257.51	0.29	1.15*
Hei 97 Tze Comp3 C4 x Acr 94 Tze Comp5	62.56	4.30	0.02	2.06	1.07	0.50	0.01	1.27	24.00	329.55	0.09	1.72*
Hei 97 Tze Comp3 C4 x Tze Comp3 Dt	5.00	0.36	0.23	3.52	0.45	2.88	5.57	8.75	38.55	11.14	0.86	2.31*
Tze Comp3 C2 x Hei 97 Tze Comp3 C4	5.76	183.50	0.41	0.46	0.63	0.03	1.17	2.47	82.26	28.17	1.83*	0.01
Tze Comp3 C2 x Acr 94 Tze Comp5	8.70	2.12	4.14	5.33	0.36	0.06	7.90	3.37	42.23	143.65	3.67*	0.71
Ak 95 Dmr-Esrw x Acr 94 Tze Comp5	0.66	1.34	0.04	3.22	2.11	0.36	0.02	3.70	70.90	1.54	1.60*	0.46

*, ** Significant at < 0.05 and < 0.01 levels of probability respectively.

as observed with Tze Comp3 Dt in this study. The result suggests that genes for tolerance may be recessive since S. hermonthica tolerance appears more common in tolerance x tolerance crosses compared with tolerance x susceptible crosses.

General and specific combining ability effects for maize grain yield and agronomic traits

There were differential responses among the parent OPVs for maize agronomic characters in both S. hermonthica endemic and non-endemic environments. Low GCA effects recorded for maize grain yield in Striga endemic environment in many of the parents indicate poor general combination in terms of grain yield under heavy Striga infestation and lack of heterotic response for grain yield in many of the parents used. However, two parents (Tze Comp3 C2 and Acr 94 Tze Comp5-W), which exhibited high GCA effects for maize grain yield, will be suitable as parents for yield improvement in Striga endemic environ-ment. Badu-Apraku and Lum (2007) reported that varieties differed significantly in grain yield under both Striga endemic and Striga free conditions. The authors also identified Acr 94 Tze Comp 5-W as the most promising genotypes in terms of grain yield, reduced Striga damage and low Striga emer-gence. In an earlier study conducted in Abuja and Mokwa (Nigeria), Menkir et al. (2001); Badu-Apraku et al. (2008) independently reported low grain yield for most of parents used in the present study under Striga infestation. Both studies also identified OPVs Acr 94 Tze Comp5 and Tze Comp3 C2 as being superior for grain yield under Striga infestation which further confirmed their suitability as cultivar *per se* in *S. hermonthica* endemic environment as well as sources of genes for *S. hermonthica* tolerance and higher maize grain yield across the SGS ecology. Therefore, apart from their suitability as cultivar in *S. hermonthica* endemic environment, these two parents could be hybridized with other proven cultivars to increase grain yield in *Striga* endemic environment of Nigeria savannas.

The significant GCA effects for maize establishment count exhibited by many of the parents indicated that the present gene pools can be manipulated for better germination and survival especially since the environment is also drought-prone. Highly significant GCA effects recorded among some parents for flowering traits (days to 50% tasselling and silking) indicated late maturity of the parents, while those with low effects indicate earliness in maturity. The significant GCA effect recorded for Tze Comp3 C2 in both environments for anthesis-silking interval shows differential response of the parent to differences in environmental factors and also could be crossed with other promising genotypes to generate populations with early maturity and high yielding. Shanghi et al. (1983); Revilla et al. (1999) in independent studies also reported the importance of GCA effects for days to tasselling, silking and maturity in open pollinated varieties of maize.

Plant height is also an important trait to be considered in maize breeding especially since maize plant with high plant height could lodge easily. Significant GCA effects for plant height in parents Acr 94 Tze Comp5, Acr 97 Tze Comp3 C4 and Tze Comp3 Dt under *Striga* infestation, showed variability in plant height among these genotypes. Thus, Acr 94 Tze Comp5 and Tze Comp3 C2 which combined high maize grain yield with reduced *Striga* damage, plant height and anthesis-silking interval could be ideal cultivar *per se* or utilized in hybrid combinations for further testing in *Striga* endemic area to ascertain consistency in performance.

SCA effects for maize grain yield and other related characters in *Striga* endemic environment were generally low in many of the hybrids indicating poor specific combination for these traits under severe *Striga* infestation. However, crosses Tze Comp3 C2 x Hei 97 Tze Comp3 C4, Tze Comp3 C2 x Acr 94 Tze Comp5 and Ak 95 Dmr-Esrw x Acr 94 Tze Comp5 with significant SCA effects for maize grain yield, appeared to be ideal specific combiners for grain yield in *Striga* endemic environment. This suggests that non-additive gene effects played a major role in the expression of grain yield among crosses under *Striga hermonthica* infestation which is also similar to earlier report (Olakojo and Olaoye, 2005) of importance of non-additive gene action in the inheritance of tolerance to *S. lutea* infestation in the southwestern ecology of Nigeria. Therefore, these three hybrids could be utilized as sources of inbred line extraction for the development of high yielding varieties for cultivation in *Striga* endemic ecology of the Nigeria Guinea Savanna. Hybrid, Tze Comp4 C2 x Acr 97 Tze Comp3 C4 with significant SCA effects for maize grain yield and maize establishment count, could be crossed with other promising genotypes to generate populations with better germination, survival and high grain yield in the drought-prone ecology of the SGS. Acr 90 Pool 16-Dt x Hei 97 Tze Comp3 C4 on the other hand could be ideal for early season cultivation in the SGS, having exhibited significant SCA effects for earliness and maize grain yield.

The parents used in this study as well as the crosses generated exhibited different levels of significant GCA and SCA effects for *Striga* tolerant traits, maize agronomic traits and grain yield in *S. hermonthica* endemic and non-endemic environments. Several studies (Kim, 1994ab; Berner et al., 1995; Abreu, 1997; Akanvou et al., 1997; Lane et al., 1997) have also shown that both additive and non-additive gene effects are important in the inheritance of different *Striga* parameters and grain yield. However, the results from present study which corroborates earlier findings of Kim (1994a); Berner et al., (1995); Akanvou et al., (1997) which were conducted in West and Central Africa on the relative importance of GCA to SCA effects for the different *Striga* parameters, differed from those of Lane et al. (1997). For example, reports of earlier authors noted that GCA effects played important role in the inheritance of plant host damage while SCA effect was more important for *S. hermonthica* emergence. Lane et al. (1997) in their own study reported that both additive and non-additive gene effects played equal and important roles in the inheritance of *Striga* parameters. In other words, relative importance of GCA and SCA effects for *Striga* parameters may vary depending on population sampled or environment where the study was conducted.

Menkir et al. (2001) suggested the establishment of parallel breeding programme targeted for yield improvement in *Striga* endemic and *Striga* free environment. Therefore OPV parents Acr 94 Tze Comp5 and Tze Comp3 C2, besides being ideal as cultivar *per se*, represent new sources of *Striga* tolerance genes for future breeding of high yielding *Striga* tolerant maize varieties for *S. hermonthica* endemic ecologies of Nigeria's savannas. Acr 94 Tze Comp5 and three other parents which exhibited high GCA effect for grain yield in *Striga* free environment could form a parallel gene pool for development of future varieties for high grain yield and general adaptation to the Nigeria savannas.

REFERENCES

Abreu AFB (1997). Predicao do potencial jenetico de popeulacoes segregantes do feijoeiro utilizando genitores interraciais. Doctoral theis, universidade Federal de Lavras, Lavras, MG.

Adetimirin VO, Aken'Ova ME and Kim SK (2001). Detection of epistasis for horizontal resistance to *Striga hermonthica* in maize. Maydica 46: 27–34.

Akanvou L, Doku EV, Kling J (1997). Estimates of genetic variances and interrelationships of traits associated with *Striga* resistance in maize. Africa Crop Sci. J. 5: 1–8.

Badu-Apraku B, Fontem Lum A (2007). Agronomic Performance of *Striga* Resistant Early-Maturing Maize Varieties and Inbred Lines in the Savannas of West and central Africa. Crop Sci. 47: 737-748.

Badu-Apraku B, Fontem Lum A, Fakorede MAB, Menkir A, Chabi Y, Thed C, Abdulai V, Jacob S, Agbaje S (2008). Performance of Early Maize Cultivars Derived from Recurrent Selection for Grain Yield and *Striga* Resistance. Crop Sci. 48: 99-112.

Basinki JJ (1995). Witch weed and soil fertility. Nature 175: 432.

Berner JG, Kling DK, Singh BB (1995). *Striga* research and control: A perspective from Africa. Plant Disease 79: 652-660.

Griffing B (1956). Concept of general and specific combining ability in relation to diallel crossing system. Aust. J. Biol. Sci. 9:463-493.

Gurney AL, Taylor A, Mbwaga A, Scholes JD, Press MC (2002). Do maize cultivars demonstrate tolerance to the parasitic weed *Striga asiatica*? Weed Research 42(4): 299-306(8).

Kim SK (1991). Breeding maize for *Striga* tolerance and the development of a field infestation technique. In Kim SK (ed). Combating *Striga* in Africa. Proc Int. Workshop by IITA ICRI-SAT, and IDRC., Ibadan, Nigeria. 22-24 August, 1988. IITA, Ibadan pp. 96-108.

Kim SK (1994a). Breeding maize for *Striga hermonthica* tolerant open pollinated maize verities in Africa. Pp 263-273. In Menyonga et al. (ed). Progress in food grain research and production in semi-arid Africa. Proc. of the SAFGRAD-Inter-Network Conf.; Niamey, Nigeria, 7-14 March, 1991.

Kim SK (1994b). Genetics of maize tolerance of *Striga hermonthica*. Crop Sci. 34: 900-907.

Kim SK, Adetimirin VO (1997). Effects of *Striga hermonthica* seed inoculums rates on the expression of tolerance and susceptibility of maize hybrids. Crop Sci. 37: 876-894.

Kim SK, Fajesinmi JM, The C, Adepoju A, Kling J, Badu-Apraku B, Versteeg M, Carsky R, Ladoke STO (1998). Development of synthetic maize populations for resistance to *Striga hermonthica*. Plant Breed. Abstract 28: 1628.

Kim SK, Adetimirinn VO, Thé C, Dossou R (2002). Yield losses in maize due to *Striga hermonthica* in West and Central Africa. Int. J. Pest Manag. 48 (3): 211-217.

Lane JA, Child DV, Moore THM, Arnold GM, Bailey JA (1997). Phenotypic characterization of resistance in *Zea diploperennis* to *Striga hermonthica*. Maydica 42: 45-51.

Menkir A, Kling JG, Badu-Apraku B, The C, Ibikunle O (2001). Recent advances in breeding maize for resistance to *Striga hermonthica* (Del.) Benth. Seventh Eastern and Southern Africa Regional Maize Conference pp. 151-155.

Mumera LM, Below FE (1996). Genotypic variation in resistance to *Striga* parasitism of maize. Maydica 41:255-262.

Olakojo SA, Olaoye G (2005). Combining ability for grain yield, agronomic traits and *Striga lutea* tolerance of maize hybrids under artificial *Striga* infestation. Afr. J. Biotechnol. 4(9): 984-988.

Ramaiah KV (1986). Breeding cereal grains for resistance to witchweeds. In Parasitic Weed in Agriculture. I. *Striga*. Musselman LJ (eds), CRC Press, Boca Raton, Florida pp. 227-242.

Ransom JK, Elpee RE, Langston MA (1990). Genetic variability for resistance in *Striga asiatica* in maize. Cereal Research Communication 18: 329-333.

Revilla PA, Butro RA, Malvar K, Orda A (1999). Relationship among kernel weight, ear vigour and growth in maize. Crop Sci. 39: 654-658.

SAS Institute Inc (1999). SAS/STAT user's guide, version 8. SAS Institute Inc. Cary, NC. Sauerborn J (1991). The economic importance of the phytoparasites Orobanche and *Striga*. pp. 137-143. In JK Ransom et al. [eds]. Proceedings of the 5th. International Symposium of Parasitic Weeds. The International Maize and Wheat Improvement Center, Nairobi, Kenya.

Shanghi AK, Agarwal KL, Qadri MI (1983). Combining ability for yield and maturity in early maturing maize under high plant population densities. Indian J. Genet. Plant Breed 43: 123-128.

Shinde VK, Kulkarni N (1982). Genetics of resistance to *Striga asiatica* in sorghum. Proc. of the ICRISAT Working Group Meeting on striga control. pp 134-141. Patancheru. India.

Vasudeva Rao MJ, Chidley VL, House LR (1982). Genetic control of *Striga asiatica* in sorghum. p. 22. *In* RV Vidyabhushanam et al. (eds). ICRISAT-AICSIP (ICAR) Working Group Meeting on *Striga* Control. ICRISAT, Patancheru, India.

Seed germination of java plum (Syzigium cumnii) in three provenances western Kenya

J. L. Okuto and G. Ouma*

Department of Botany, Maseno University, P. O. Box 333, Maseno, Kenya.

Indigenous fruits are very important as sources of food security, balanced households nutrition. Deforestation poses a serious threat to their survival. Domestication of these trees can be successful only if successful propagation techniques are applied. In Kenya Syzigium cuminii has become a very important indigenous fruit tree for local consumption and income generation and it mainly grows in Siaya, Vihiga and Kisumu districts. Studies were conducted in Maseno, Kenya (2004 - 2005) to investigate variation of seed germination of S. cuminii in three districts (provenances) mentioned above. The experimental design used was completely randomized design. The treatments included different watering regimes and light and darkness durations respectively. The results showed that both light and irrigation regimes significantly (P ≤ 0.05) increased the germination of S. cuminii but 24 h darkness inhibited it. The best treatment was 12 h light. S. cuminii seeds are therefore photodormant since they require light to germinate. There was no variation of seed germination among the districts (provenances).

Key words: Light, water, photodormancy, provenance, germination.

INTRODUCTION

Indigenous fruit trees are very important as sources of food security and balanced nutrition in Kenyan households. They also contribute immensely to income for the rural population (Saka et al., 2002; 2004b; Akinnifesi et al., 2004a; Kaaria, 1998; Schoraburg et al., 2002). Indigenous fruit trees are also sources of timber for construction (Campbell et al., 1997). In many African countries, 80% of the indigenous fruit producers/collectors process fruits for home consumption (Kadzere et al., 2001). In South Africa 94% of households use them for making Jam and Juice (Shackleton, 2004). Deforestation poses a serious threat to the survival of indigenous fruit trees. In Kenya Syzigium cuminii has become a very important indigenous fruit for household consumption and income generation (Icraf, 1992). S. cuminii belongs to the family myrtaceae and it is an evergreen tree. Its fruit has one seed which has a low germination rate of 30 - 40% and it is recalcitrant (NAS, 1980). It is also processed locally into jam and juices. Provenance variation of seed germination has been reported in many tree species, including Podocarpus falcatus (Negash, 2003); Uapaca kirkiana (Mwase et al., 2006), Juniperus procera (Mamo et al., 2006; Mkonda et al., 2006) and Cordia africa (Loha et al., 2005). The fruit is mainly grown in the Western Kenya districts of Siaya, Kisumu and Vihiga. The objectives of the present study were to investigate seed germination of S. cuminii fruits from three provenances namely Siaya, Vihiga and Kisumu.

MATERIALS AND METHODS

The fruits were obtained from Siaya, Kisumu and Vihiga districts and seeds extracted in warm water. To determine the seed germination the fruit and pericarp were dried in an oven for 48 h at a temperature of 80°C. Fruits of S cuminii were collected from the three districts (Siaya, Kisumu and Vihiga) and fermented in gunny bags for three days and then hauled by hand (Atangana et al., 2001). The seeds were then cleaned and the fully occupied non-shrunken ones were selected for germination in a glass house at the Kenya Forestry Research Institute centre, Maseno Kenya. Seedbeds were constructed with stones at the bottom to allow

---

*Corresponding author. E-mail: goumaoindo@yahoo.com.

water perculation and sand at the top as germination medium. The soils at Siaya were Luvicphae zones (Anon, 2005) at Vihiga the soils were sandy loam (Anon, 2002), Soils at the provenance were analysed using standard methods (Table 4). In Kisumu eight freshly extracted seeds were sown by spreading them evenly in each seedbed so that none lied on top of the other to avoid damping off disease attack (Gachanga and Ilg 1990). Sand volume equal to the diameter of the seed was used to cover it. Irrigation of the seed bed was done according to the treatment type.

After one week the sand covering the seeds were gently uncovered to check whether germination was evident. The sand was then returned. Germination checks were carried out daily for three weeks and recorded. The experimental design was completely Randomized design (CRD). The treatments were applied as follows:

Light

Three light levels were used for five weeks.

(a) 12 h light, 12 h darkness (L1).
(b) 24 h light, (L2)
(c) 24 h darkness (L3) (control)

Other germination factors e.g. temperature, oxygen etc were kept constant.

Moisture treatments

These treatments were maintained for five weeks.

(a) 0.5 l of water per pot was applied in the morning and in the evening daily (W1) control
(b) 0.5 l of water per pot was applied in the morning daily (W2).
(c) 0.5 l of water per pot was applied in the morning and evening on alternate days (W3).
(d) 0.5 l of water per pot was applied in the morning and evening an alternate days (W4).

Potting medium

Top soil dug 15 cm from the surface of the earth and free from organic matter was taken from randomly sampled sites from every district for use in potting. For the top soils from the districts, chemical analysis was done to ascertain the presence of essential minerals such as nitrogen, phosphorous and potassium (Gachanja and Ilg, 1990) (Table 4).

Transplanting seedlings

Young germinated seedlings from each district were transferred (pricked out) into potted soils from respective districts. Adibbler (a small stick of 20 cm length and 3 cm diameter with a pointed end) was used to remove the seedlings from the seedbed to an empty container (500 ml) capacity with clean water. Seedlings from each district were then transplanted into containers potted with topsoils removed from the respective district randomly. The dibblers were used to open a hole in the potted soils of size reasonable enough to accommodate the rooting system of each seedling. The dibblers were again used to loosen soils around seedlings in the seed bed to improve water percolation within the seedling container and finally to push soil to occupy spaces left in the hole after transferring the seedlings into potted soils making the young seedlings firm enough to withstand the pressure of watering and wind (Gachanja and Ilg, 1990).

Data analyses

Percent germination was recorded for 5 weeks for each treatment and subjected to analysis of variance (ANOVA) and mean separation using the least significant difference L.S.D) method.

RESULTS

Seed germination

In Kisumu district both light duration and watering regimes significantly ($P \leq 0.05$) affected percent germination at all sampling dates. Germination increased with the date of sampling (Tables 2), the interaction between light duration and watering regime was generally significant ($P \leq 0.05$). The best treatment was obtained from the treatment 24 h darkness and water applied morning daily followed by 24 h darkness and water applied morning daily and lastly 24 h light and water applied morning and evening daily. All these observation were made during the first week after sowing. Conversely the situation changed two weeks after sowing when the best germination was from 12 h light and water applied morning, midday and evening and alternate days followed by 12 h light and water applied morning daily and 12 h light and water applied morning and evening daily. Under 24 h darkness, applying water every morning daily was the same as under 12 h light and applying water every morning daily, applying more water under 12 or 24 h darkness (Table 2) Similar trends were obtained three weeks after sowing (Table 2). The lowest germination percent was obtained by the treatments with 24 h darkness and this varied with the amount of water applied apparently worsening when water was applied on alternate days. The best treatment occurred with 12 h light followed by 24 h light depending on the frequency of watering. Watering on alternate days tended to give poor germination at any level of light. Therefore from the results of this study germination increased from 0% in the first week to the fifth week 12 h lighting to peak at 24 h lighting with 93%.

In Siaya district germination of *S cuminii* increased as the date of sampling increased. Light duration and watering regimes significantly ($P \leq 0.05$) affected germination percent. During the first week of sampling the 24 h light treatment had very small germination but it was the highest. The rest of the treatments had negligible or no germination (Table 3). By the second week after sowing germination had increased and highest germination was obtained from the 24 h darkness treatment and water applied every morning daily. This was the same as 12 h light with water applied every morning daily and 12 h light with water applied in the morning, evening or alternate days. Similar results were obtained by 12 h light with water applied morning and evening daily, and 24 h light with water applied in the morning daily and 24 h light and water applied in the morning and evenings on alternate days (Table 3).

The least germination was observed in the treatments where 24 h darkness and watering were applied on alternate days. The above situation changed three weeks after sowing (Table 3) when the 12 h light and 24 h light treatments increased in germination as compared to 24 h darkness treatments. As the frequency of watering and light duration increased, germination increased with an exception of the 24 light treatments with water applied morning, midday and evening on alternate days (Table 3). The 24 h darkness treatments had the lowest germination especially watering on alternate days. Similar trends as above were observed at weeks 4 and 5 after sowing that is germination increased with light duration and irrigation frequency with the least germination at 24 h darkness and watering on alternate days (Table 3).

In Vihiga district light duration and watering regime significantly ($P \leq 0.05$) increased germination of S. cumini. No germination was recorded one week after sowing in all the treatments. Germination percent increased as the date of sampling increased (Table 1). The highest germination was in the second week after sowing and was obtained from the treatment 12 h light with water applied in the morning daily. The rest of the treatments had very low germination. As the days from sowing increased, germination also increased. Germination increased at the 12 and 24 h light treatment with no definite trends. Generally the 12 h light treatment was better than 24 h light treatment at all irrigation levels (Table 1). By three weeks after germination the 12 h light treatment was still had the highest germination followed by the 24 h light treatment at all watering levels. 24 h darkness treatments had the lowest germination (Table 1). Similar trends were obtained four and five weeks after sowing (Table 1) at all watering levels.

These results in germination may also be attributed to the difference in soil types in the provenances as shown in Table 4.

The results of the present study have shown that both light and water are needed to achieve good germination of S. cumini seeds. Water is apparently needed for the early metabolic activities of the seed such as enzyme activation and mobilization of food reserves such as proteins, carbohydrates and lipids for the development of embryo (Hartmann et al., 2001). Early seed germination commences with rapid intake of water. For many seeds without dormancy the availability of water is the early factor limiting germination at a suitable temperature. Water deficit reduces germination percent (Doreen and Gilrary, 1943; Ayers, 1952; Hanks and Thorpe, 1956). The germination levels of some seeds especially those with dormancy are inhibited as moisture levels are decreased. Such seeds apparently contain inhibitors that require leaching (Attwater, 1980).

In all the provenances the degree of germination of S. cumini appeared to be dictated by light duration. Germination was relatively poor in darkness even when it was watered frequently. This shows that S. cuminii is a light sensitive species and needs light for germination but water has to be provided also. Light acts both in dormancy induction and as a mechanism that adapts plants to a specific niche in the environment, often interacting with temperatures (Hartmann et al., 2001). Light appears to break physiological dormancy of S. cumini (Welban et al., 1998).

The germination of S. cumini in the present study contradicts the reported germination rate of 30 - 40% (NAS, 1980). In the present study the germination rate of S cumini was as high as 90%. This was achieved when the seeds were sown when freshly extracted and balancing of watering regimes and lighting levels with an optimum germination period. For example in Kisumu district germination of 93% was achieved when freshly extracted seeds were sown in the seedbed where 500 ml of water was applied in the morning, midday and evening on alternate days for five weeks and lighting applied for 24 h during the germination period. For Vihiga and Siaya district seeds (in the same conditions as that for Kisumu district above) were shown in the beds with the same lighting and watering conditions as those of Kisumu above and germination percent in each district was 92%.

In both Kisumu and Siaya districts the 24 h darkness treatment had the best germination at the second week of sampling only but not any other date of sampling only but not any other date of sampling. This was an isolated case and is difficult to explain. In the present study the germination rate and trends were generally similar in all the provenances with the 24 h light being superior to all treatments and adding water to the light treatments increased germination. S. cuminii seeds are therefore photodormant. Similar results have been reported in Juniperus procera by Mamo et al., 2006. Lack of water or enough water in the 24 h darkness treatment reduced germination further. Water and light are therefore essential for the germination of S. cuminii. Similar results on effects of light on germination have been reported by Mamo et al., 2006, working on J. procera. The results of the present study are at variance with those of Loha et al. (2006), Mamo et al. (2006) who reported provenance variation in seed germination of some indigenous tree species such as Procera and Cardia africana, Podocarpus falcatus (Negash, 2003) Strychnos cocculoides (Mkonda et al., 2005). This could be explained by the different germination requirements between S. cuminii and the above mentioned tree species.

Conclusion

The conclusions to be drawn from the present study are:

- Varying levels of light and watering regimes affect the

Table 1. (a) Effects of lighting levels on germination percent of *S. cumini* trees seeds collected from Vihiga district taken four weeks after sowing. Means with the same letter are not significantly different (NS).

Treatment	% Germination
Lighting	
12 h light	68.748 a
24 h light	68.602 a
24 h darkness	28.198 b
LSD	11.799

Table 1. (b) Effects of watering regimes on germination percent of *S. cumini* tree seeds collected from Vihiga district taken four weeks after sowing. Means with the same letter are significantly different (NS).

Treatment	% Germination
500 ml of water	
Applied morning and evening daily	57.935a
Applied morning daily	59.522a
Applied morning, midday and evening on alternate days	45.436b
Applied morning and evening on alternate days	57.837a
LSD	13.624

Table 1. (c) Effect of interaction between lighting levels and watering regimes on mean germination percent of *S. cumini* trees seeds collected from Vihiga district.

Treatment		Mean germination % recorded after sowing		
		Week 2	Week 3	Week 4
Lighting	**500 ml water**			
	Applied morning and evening	14.288	35.712	80.950
12 h	Applied morning daily	15.478	38.095	78.570
	Applied morning, midday and evening on alternate days	14.288	36.905	80.950
	Applied morning and evening on alternate days	7.140	35.715	86.905
Mean		12.799	36.607	81.844
	Applied morning and evening daily	8.330	32.143	84.525
24 h	Applied morning daily	3.570	58.335	92.860
	Applied morning, midday and evening on alternate days	0.000	36.905	86.805
	Applied morning and evening on alternate days	4.165	37.798	86.881
Mean		4.165	37.798	86.881
Darkness				
	Applied morning and evening daily	9.520	21.430	48.805
24 h	Applied morning daily	11.903	16.668	51.190
	Applied morning, midday and evening on alternate days	0.000	1.190	14.285
	Applied morning and evening on alternate days	1.190	8.335	52.380
Mean		5.653	11.906	41.665
G/Mean		7.539	28.770	70.13
Standard error		0.962	3.462	2.582

Table 1. (d) The effects of lighting levels and watering regimes on mean germination percent of seeds collected from *S. cuminii* trees from Vihiga district.

Treatment		Mean germination % recorded after sowing
		Week 2
Lighting	**500 ml of water**	
	Applied morning and evening daily	14.3
12 h	Applied morning and daily	15.5
	Applied morning, mid day and evening on alternate days.	14.3
	Applied morning and evening on alternate days	7.1
	Mean	12.8
	Applied morning and evening daily	4.8
24 h	Applied morning daily	8.3
	Applied morning, mid day and evening on alternate days.	3.6
	Applied morning and evening on alternate days	0.0
	Mean	4.2
Darkness		
	Applied morning and evening daily	9.5
24 h	Applied morning daily	11.9
	Applied morning, mid day and evening on alternate days.	0.0
	Applied morning and evening on alternate days	1.2
Mean		5.6
Grand mean		7.5
Standard error		0.962

Table 1. (e) The effects of lighting levels on germination percent of *S. cuminii* trees seeds collected from Vihiga district taken on the third, fourth and fifth weeks after sowing. Means with the same letter are not significantly different ($P > 0.05$).

Lighting time	Mean % germination recorded weeks after sowing		
	3rd	4th	5th
12 h	36 a	68 a	81 a
24 h	37 a	68 a	86 a
24 h darkness	11 b	28 b	41 b

Table 1. (f) The effects of watering regimes on germination percent of *S. cuminii* tree seeds collected from Vihiga provenance taken four weeks after sowing. Means with the same letter are not significantly different ($P > 0.05$).

Treatment	Mean germination%
500 ml of water	
Applied morning and evening daily	58.0 a
Applied morning daily	59.5 a
Applied morning, midday and evening on alternate days	45.4 b
Applied morning and evening on alternate days	57.8 a
L.S.D.$_{(0.05)}$	13.6

Table 2(a). Effect of interaction between lighting levels and watering regimes on mean germination percent of *S. cumini* trees seeds collected from Kisumu district.

Treatment		Mean germination % recorded after sowing				
		Week 1	Week 2	Week 3	Week 4	Week 5
Lighting	**500 ml water**					
12 h	Applied morning and evening	0.000	16.668	51.190	79.763	81.548
	Applied morning daily	0.000	14.288	44.348	75.000	80.950
	Applied morning, midday and evening on alternate days	0.000	20.240	42.860	76.190	84.525
	Applied morning and evening on alternate days	0.000	11.903	48.810	77.975	84.523
Mean		0.000	14.484	47.472	77.480	83.432
24 h	Applied morning and evening daily	0.893	4.760	41.665	78.570	86.900
	Applied morning daily	0.000	9.520	45.235	75.000	82.140
	Applied morning, midday and evening on alternate days	0.000	8.330	57.499	92.858	93.450
	Applied morning and evening on alternate days	0.000	1.190	59.525	79.166	86.903
Mean		0.000	7.933	50.981	81.400	87.348
Darkness						
	Applied morning and evening daily	1.190	10.710	25.000	40.476	79.760
24 h	Applied morning daily	2.698	14.288	32.143	48.810	53.570
	Applied morning, midday and evening on alternate days	0.000	1.190	2.380	13.393	19.050
	Applied morning and evening on alternate days	1.488	1.190	17.860	32.735	55.953
Mean		1.339	6.845	19.346	33.854	52.083
G/Mean		0.446	9.754	39.267	64.245	77.288
Standard error		0.386	1.866	4.044	5.358	2.521

Table 2(b). Effect of interaction between lighting levels and watering regimes on mean germination percent of *S. cumini* trees seeds collected from Kisumu district.

Treatment		Mean germination % recorded after sowing				
		Week 1	Week 2	Week 3	Week 4	Week 5
Lighting	**500 ml water**					
12 h	Applied morning and evening	0.000	16.668	51.190	79.763	81.548
	Applied morning daily	0.000	14.288	44.348	75.000	80.950
	Applied morning, midday and evening on alternate days	0.000	20.240	42.860	76.190	84.525
	Applied morning and evening on alternate days	0.000	11.903	48.810	77.975	84.523
Mean		0.000	14.484	47.472	77.480	83.432
24 h	Applied morning and evening daily	0.893	4.760	41.665	78.570	86.900
	Applied morning daily	0.000	9.520	45.235	75.000	82.140
	Applied morning, midday and evening on alternate days	0.000	8.330	57.499	92.858	93.450

Table 2(b). Contd.

	Applied morning and evening on alternate days	0.000	1.190	59.525	79.166	86.903
Mean		0.000	7.933	50.981	81.400	87.348
Darkness						
	Applied morning and evening daily	1.190	10.710	25.000	40.476	79.760
24 h	Applied morning daily	2.678	14.288	32.143	48.810	53.570
	Applied morning, midday and evening on alternate days	0.000	1.190	2.380	13.393	19.050
	Applied morning and evening on alternate days	1.488	1.190	17.860	32.735	55.953
Mean		1.339	6.845	19.346	33.854	52.083
G/Mean		0.446	9.754	39.267	64.245	77.288
Standard error		0.386	1.866	4.044	5.358	2.521

Table 2(c). The effect of lighting levels and watering regimes on mean percent germination of *S. cuminii* trees seeds collected from Kisumu district.

Treatment		Mean germination % recorded after sowing				
		Week 1	Week 2	Week 3	Week 4	Week 5
Lighting						
12 h	**500 ml of water**					
	Applied morning and evening	0.0	16.6	51.1	79.7	81.5
	Applied morning daily	0.0	14.2	44.3	75.0	80.9
	Applied morning, mid day, and evening on alternate days	0.0	20.2	42.8	76.1	84.5
	Applied morning and evening on alternate days	0.0	11.9	48.8	77.9	84.5
Mean		0.0	14.4	47.4	77.4	83.4
24 h	Applied morning and evening daily	0.8	4.7	41.6	78.5	86.9
	Applied morning daily	0.0	9.5	45.2	75.0	82.1
	Applied morning, mid day, and evening on alternate days	0.0	8.3	57.4	92.8	93.4
	Applied morning and evening on alternate days	0.0	1.1	59.5	79.1	86.9
Mean		0.0	7.9	50.9	81.4	87.3
Darkness						
	Applied morning and evening daily	1.1	10.7	25.0	40.4	79.760
24 h	Applied morning daily	2.6	14.2	32.1	48.8	53.5
	Applied morning, midday and evening on alternate days	0.0	1.1	2.3	13.3	19.0
	Applied morning and evening on alternate days	1.4	1.1	17.8	32.7	55.9
Mean		1.3	6.8	19.3	33.8	52.0
G/Mean		0.4	9.7	39.2	64.2	77.2
Standard error		0.3	1.8	4.0	5.3	2.5

Table 3(a). Effect of interaction between lighting levels and watering regimes on mean germination percent of S. cumini trees seeds collected from Siaya district.

Treatment		Mean germination % recorded after sowing				
		Week 1	Week 2	Week 3	Week 4	Week 5
Lighting	**500 ml water**					
12 h	Applied morning and evening	0.000	16.668	51.190	79.763	81.548
	Applied morning daily	0.000	14.288	44.348	75.000	80.950
	Applied morning, midday and evening on alternate days	0.000	20.240	42.860	76.190	84.525
	Applied morning and evening on alternate days	0.298	11.903	48.810	77.975	84.523
Mean		0.075	14.484	47.472	77.480	83.432
24 h	Applied morning and evening daily	0.893	4.760	41.665	78.570	86.900
	Applied morning daily	0.000	9.520	45.235	75.000	82.140
	Applied morning, midday and evening on alternate days	0.000	8.330	57.499	92.858	93.450
	Applied morning and evening on alternate days	0.000	1.190	59.525	79.166	86.903
Mean		0.223	7.933	50.981	81.400	87.348
Darkness						
24 h	Applied morning and evening daily	0.000	10.710	25.000	40.476	79.760
	Applied morning daily	0.900	14.288	32.143	48.810	53.570
	Applied morning, midday and evening on alternate days	0.000	1.190	2.380	13.393	19.050
	Applied morning and evening on alternate days	1.488	1.190	17.860	32.735	55.953
Mean		0.225	6.845	19.346	33.854	52.083
G/Mean		0.174	9.754	39.267	64.245	77.288
Standard error		0.239	1.866	4.044	5.358	2.521

Table 3(b). The effect of lighting levels and watering regimes on mean germination percent of S. cuminii trees seeds collected from Siaya district.

Treatment		Mean germination % recorded after sowing				
		Week 1	Week 2	Week 3	Week 4	Week 5
Lighting	**500 ml of water**					
12 h	Applied morning and evening	0.0	8.3	40.4	79.7	84.5
	Applied morning daily	0.0	10.7	48.8	72.6	82.1
	Applied morning, mid day, and evening on alternate days	0.0	10.7	41.6	62.4	79.7
	Applied morning and evening and alternate days	0.2	5.9	42.8	76.1	84.5
Mean		0.0	8.9	43.4	72.7	82.7
24 h	Applied morning and evening daily	0.8	4.4	32.1	79.1	83.3

Table 3(b). Contd.

Treatment		Mean germination % recorded after sowing				
		Week 1	Week 2	Week 3	Week 4	Week 5
	Applied morning daily	0.0	8.3	44.0	72.6	82.1
	Applied morning, mid day, and evening on alternate days	0.0	6.5	58.3	92.8	92.8
	Applied morning and evening on alternate days	0.0	1.1	48.8	79.7	85.7
	Mean	02	5.1	45.8	81.1	86.0
Darkness						
24h	Applied morning and evening daily	0.0	8.3	23.8	33.3	48.8
	Applied morning daily	0.9	10.7	30.9	48.8	52.3
	Applied morning, midday and evening on alternate days	0.0	0.0	1.1	8.3	24.4
	Mean	0.2	4.7	13.9	22.6	31.3
	Grand mean	0.1	4.7	25.8	58.8	66.7
Standard error		0.2	1.0	2.6	2.1	3.3

Table 4. Soil analysis report.

District	Exchange %K in me/100 gm of soil	Total P	% N
Siaya	1.43	0.53	0.164
Vihiga	0.53	0.059	0.1640
Kisumu	2.21	0.05	0.118

germination of S. cumini.
- Light is needed for the germination of S. cuminii so it has photodormancy.
- The best treatment for germination is 12 h light and the least is 24 h darkness.
- Germination rate of S. cumini is over 92% not as reported on 30 - 40%.
- Increasing light and water increase the germination of S. cumini.
- There was no geographic variation in seed germination of S. cuminii.

REFERENCE

Akinnifesi FK, Kwesiga FR, Mhango J, Mkonda, A Swai R (2004a). Priority Miombo. Indigenous fruit trees as a promising livelihood option for smallholder farmers in southern Africa. Acta Hort., 632: 15 - 30.
Att water BR (1980). Germination, Dormancy and morphology of the seeds of herbaceous ornamentals. Seed Sci. Tech., 8: 523 - 573.
Campbel B, Lukert M, Scoones I (1997). Local Level evaluation of Savanna land resources. A study in Zimbabwe, Econ. Bot., 51: 57 - 77.
Doreen LD, MacGillary JH (1943). Germination of vegetable seed as affected germination of vegetable seed as affected by different soil conditions, Plant Physiol., 18: 524 -529.
Hartmann HT, Kester PE, Davies FT, Geneva RI (2001). Plant Propagation and Practices. Prentice Hall Publishers, New York Pp. 450 - 490.
Gachanja SP, Ilg P (1990). Fruit tree nurseries, Government of Kenya, Ministry of Agriculture, Nairobi, Pp. 8 - 9,16.
Kadzere L, Hove L, Gatsi T, Masarimbari MT, Makaya PR (2001). Current practices on postharvest landling and traditional processing of indigenous fruits in Zimbawe, Finnal Technical Report to Department of Agricultural Research and Technical services, Zimbabwe P. 60.
Kaaria SW (1998). The Economic potential of wild fruits in Malawi, M.Sc. Thesis, University of Minnesota, U.S.A. Pp. 173.
Loha AM, Tigab D, Teketay K, Lundkvist J , Fries A (2006). Provenance variation in seed Morphometric Traits, Germination and Seedling growth of cordia africana L, New Forests, 32(1): 71 - 86.
Mkonda A, Lungu S, Maghembe JA, Mafongoya PL (2004). Agroforestry Syst., 58(1): 25-31.
Mwase WF, Bjarnstad P, Ntupanyama, YM (2006). Phenotypic variation in fruit and seed of Uapaca Kirkiana provenance found in Malawi Southern Afr. Forestry J. 208(1): 15- 21.
Mamo NM, Miheretu M. Fekadu T, Mulualem T, Tekelalya D 2006). For. Ecol. Manag. 225(1-3): 320- 327.
Saka JD, Mwendo-phiri E, Akinnifesi FK (2002). Community processing and nutritive value of some Miombo Indeginous

(fruits in central and southern Malawi, in Kwesiga F, Ayuk E,and Agumya A (eds) Proc. 14th Southern African Regional Review and Planning Workshop 3 – 7 September, Harare, Zimbabwe, Saka JK, Swai K, Pp. 164 -169.

Mkonda R, Schora burg A, Kwesiga FF, Akinnifesi FK (2004). Processing and utilization of Indeginous fruits of the Miombo in southern Africa Agroforestry Impacts Livelihoods in Southern Africa; putting Research into Practice in Rao MR., Kwesiga FR. (eds) Proc. Regional Agroforestry conference on Agroforestry Impact on livelihoods in Southern Africa; Putting Research into Practice Pp. 305 - 306.

Negash J (2003). Insitu fertility decline and provenance difference in the East Africa yellow wood (*Podocarpus falcatus*) measured through *in vitro* seed germination. Forest Ecology and management 174(1-3): 127-138.

Shackleton CM (2004). Use and selection of *Sclerocarya birrea* (Marula) in the Bushbackridge low veld, S. Africa in Rao MR and Kwesiga FR. (eds) Proc. Regional Agroforestry Conference on Agroforestry Impacts on Livelihoods in Southern Africa; putting Research into Practice, Nairobi, Kenya. Pp. 72 – 92.

Ayers AD (1952). Seed germination, dormancy, and germination of the seeds of herbaceous ornamentals plants. Seed Sci. Tech., 16: 1 - 113.

Hanks RS, Thorpe KC (1956). Seedling emergence of wheat as related to soil moisture, bulk density, oxygen diffusion rate, crust strength. Proc. Soil. Sci. Am. 20: 307 - 310.

Icraf(1992). A selection of useful trees and shrubs for Agroforestry, International Centre for Research in Agroforestry (ICRAF), Pp. 182.

Schoraburg A, Mhango J, Akinnifesi FK (2002). Marketing of *U. Kirkiana* and *Z. mountania* fruits and their protential for processing by rural communities in southern Malawi in Kwesiga, F, Ayuk E. and Agumya, A. Proc. 14th Southern African Regional Review Planning Workshop 3 - 7 September 2001, Harare, Zimbabwe. Pp. 169 - 176.

Welban GE, Bradford KJ, Booth DT (1998). Biophysical, physiological and biochemical process regulating seed germination, seed set Tec. 16: 1 - 113.

Structural features of a cytoplasmic male sterility source from *Helianthus resinosus*, CMS RES1

F. Ardila[1,2], M. M. Echeverría [1,3]*, R. Rios[2] and R. H. Rodríguez[3]

[1]Ex Aequo asbl, Rue Locquenghien, 41 B-1000 Brussels, Belgium.
[2]Instituto de Genética CICV y A, CNIA, INTA. C.C. 25 (1712) Castelar, Buenos Aires, Argentina.
[3]Unidad Integrada: Facultad de Ciencias Agrarias, Universidad Nacional de Mar del Plata - Estación Experimental INTA Balcarce. C.C. 276 (7620) Balcarce, Buenos Aires, Argentina.

Comparisons in the structural organization of mitochondrial DNA between a cytoplasmic male sterility source from *Helianthus resinosus* (CMS RES1) and HA89 fertile line have been performed surrounding *cox*III and *atp*9 genes. Differences have also been recognized by PCR analyses between HA89 (CMS RES1) and either HA89 (CMS PET1) or HA89 (CMS MAX1), indicating that the molecular basis of their sterility mechanism is different. To be specific, RES1 lacks *orf*H522, which is present in the mitochondrial DNA of PET1 and MAX1.

Key words: *Helianthus annuus*, *Helianthus resinosus*, sunflower, cytoplasmic male sterility, CMS, *cox*III, *atp*9.

INTRODUCTION

Hybrid breeding in sunflower is based on a single source of cytoplasmic male sterility (CMS), the so-called PET1, which was originated from an interspecific cross between *Helianthus petiolaris* and *H. annuus* (Leclercq, 1969). However, the use of a single CMS source implies a potential risk as a result of the vulnerability of such a narrow genetic basis (Haevekes et al., 1991). In sunflower, up to date 72 CMS sources are available and could be potentially used for hybrid breeding (Serieys and Christov, 2005). However, most of them lack detailed information about structural and functional characteristics. Therefore, thorough investigations on the molecular level of the available CMS are advisable to identify cytoplasms different in respect to the male sterility mechanism and the organization of the mitochondrial DNA (mtDNA).

Mapping studies of mtDNAs from isonuclear fertile and CMS PET1 lines have revealed that detectable alterations are restricted to an mtDNA region of about 16 kb, between *atp*A and *cob* genes. This region includes an 11 kb inversion and a 5 kb insertion, creating an open reading frame (*orf*H522) immediately downstream of the *atp*A gene (Sicullela and Palmer, 1988). Nine additional CMS sources also have the same organization of the *atp*A and CMS mechanism as PET1 (Horn and Friedt, 1999).

We have agronomically described a CMS source from *H. resinosus*, CMS RES1. This cytoplasm is the result of controlled crosses between the perennial species *H. resinosus* (2n = 6x = 102) and cultivated sunflower *H. annuus* (2n = 2x = 34) followed by several backcrosses using *H. annuus* as a recurrent parent, in order to restore the diploid level. This cytoplasm is stable in different environments, the female-fertility is not affected and meiosis is normal. The absence of pollen is caused by postmeiotic alterations and the factors which control the pollen fertility restoration have not been found in diploid germplasm (Echeverria et al., 2003).

In this communication, we report structural features from its mtDNA that show architectural differences in the region around *atp*6 – *cox*III genes from some fertile inbred lines, such as HA89, RHA271, and RHA801 compared with male – sterile lines, CMS RES1, as well as with HA89 (CMS PET1) and (CMS MAX1).

MATERIALS AND METHODS

Plant materials

CMS sources

The fertile inbred lines HA89 (B), RHA271 (B), RHA801 (B), and the male-sterile isoplasmic lines: HA89 (CMS MAX1), HA89 (CMS

---

*Corresponding author. E-mail: mecheverria@balcarce.inta.gov.ar.

PET1), HA89 (CMS RES1), RHA271 (CMSRES1), RHA801 (CMSRES1) have been used. The inbred lines RHA271 (B) and RHA 801 (B) carry the CMS PET1 and the gene of pollen fertility restoration for this cytoplasm.

DNA isolation

Total DNA (tDNA) was obtained from leaves as indicated by Saghai-Maroof et al. (1984).

PCR analyses

Universal mitochondrial primers

Primers coxIII5´/coxIII3´ were used for the amplification of coxIII gene (Iglesias, 1994). For amplifying the region rpS14 – cob, primers rpS14 and cob were used and for nad gene the pairs nad4exon1/nad4exon2 and nad4exon2/nad4exon4, were used according to Demesure et al. (1995).

Amplicons obtained by the utilization of primer pairs coxIII 5´/3´, rpS14/cob and nad4 exon2/4 were digested by using HaeIII and HpaII (Promega) in order to search for structural differences between lines.

Specific primers

Primers orfH522F/orfH522R for amplifying the sunflower specific region, were used according to Rambaud et al. (1997). A three-primer strategy, using primers atpAF, orfH522R and orfH873R homologous to portions of mitochondrial atpA, orfH522 and orfH873, respectively, was applied as described by Rieseberg et al. (1994).

Southern hybridization

Mitochondrial probes

Probes for atpA (700 bp), coxIII (1.1 kb) and atp9 (2.2 kb) were obtained by PCR amplification using pUC/M13 primers (Promega) and pUC plasmids containing the genes atpA (Laver et al., 1991), coxIII and atp9 (Dewey et al., 1985). The probe for orfH522 was obtained by PCR amplification from HA89 (CMS PET1) total DNA (tDNA) using sequence specific primers orfH522F/orfH522R (Rambaud et al., 1997). All the PCR-amplified fragments were labeled with alkaline phosphatase using the AlkPhos Direct Labelling system (Amersham).

Total DNA was digested using the restriction enzymes BglII, EcoRI, HindIII and SalI (Promega), according to the manufacturer's manuals.

Southern blot

Digested tDNAs (2µg) were separated on a 1% agarose gel and blotted onto positively charged nylon membranes (Hybond-N+, Amersham). Hybridization was performed according to Sambrook et al. (1989). After hybridization, the specific fragments were detected by chemiluminescence using the AlkPhos Detecting system (Amersham).

RESULTS

PCR analyses

Universal mitochondrial primers

No differences in amplified fragment sizes were observed between HA89 fertile, CMS PET1, and CMS RES1, from the genes coxIII (700 bp product), rpS14-cob (1000 bp product), and nad4 (2 kb for the stretch between exon 1 and 2, and 600 bp from exon 2 to 4). Restriction analyses, using HaeIII and HpaII on PCR products obtained by using coxIII, nad4 (exon 2/4) and rpS14/cob primer pairs, did not detect any difference among the amplified products from HA89 fertile, CMS PET1 and CMS RES1 lines (not shown).

Specific primers

When PCR analyses were performed using orfH522 primer pairs, no amplification was detected when tDNA of HA89 (B) was used as template, and fragments of 344 bp were observed using tDNA from HA89 (CMS PET1) and HA89 (CMS MAX1), as expected. When the same analysis was carried out using tDNA from HA89 (CMS RES1) no amplification was obtained. These results were confirmed using tDNA from the inbred lines RHA271and RHA801 fertile lines and the isoplasmic form (CMS RES1) (data not shown). These fertile lines show the fragment of 344 bp, because they carry the CMS PET1. Similarly, when using the three-primer strategy (Figure 1A), the expected fragments of 1450, 840 and 790 bp, were obtained when the technique was applied on HA89 (CMS PET1), HA89 (B) and HA89 (CMS MAX1) tDNAs, respectively (Figure 1A, Lanes 1, 2 and 4) (Rieseberg et al., 1994; Hahn and Friedt, 1994 and GenBank Acc# X55963). The fertile inbred lines RHA271 and RHA801 exhibited the fragment of 1450 bp, (Figure 1B, Lanes 2 and 3) like CMS PET1. As on the fertile line HA89 (B), (Figure 1A, Lane 2) fragments of 840 bp were amplified from HA89 (CMS RES1) (Figure 1A, Lane 3), RHA271 (CMSRES1) and RHA801 (CMSRES1) tDNA (Figure 1B, Lanes 5 and 6).

Southern hybridization

Southern blot analyses

Hybridization with orfH522 and atpA probes on tDNA samples digested with BglII, SalI, HindIII or EcoRI showed no differences between fertile and CMS (RES1) lines (not shown). Similarly, the same hybridization pattern was observed on these lines when digested with HindIII and by using coxIII and atp9 probes (not shown). However, we could detect structural differences by using coxIII probe when both materials were digested with BglII or HindIII and by using atp9 probe on EcoRI digested tDNAs. Figure 2A shows that when using coxIII probe on a BglII-digested DNA, a band of 8 kb was detected in the fertile line HA89 (Figure 2A, Lane 1) while a 5.5 kb band hybridized in the CMS RES1 (Figure 2A, Lane 2). In addition, the same analysis on

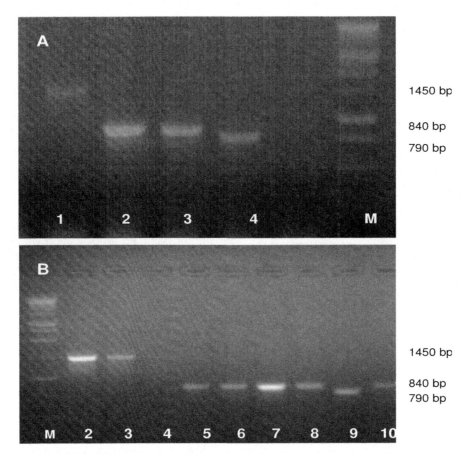

Figure 1. PCR analyses using the three-primer strategy. A: on tDNA from HA89 CMS PET1 (1), HA89 (B) (2), HA89 CMS RES1 (3), and HA89 CMS MAX1 (4). M: 1 Kb DNA ladder. B: on tDNA from RHA271 fertile (B) (2) and male-sterile (CMSRES1) (5); RHA801 fertile (B) (3) and male-sterile (CMSRES1) (6); and HA89 fertile (B) (7) and male-sterile isoplasmic lines (CMS RES1 (8), MAX1 (9) and PET2 (10). M: 1 Kb DNA ladder.

HindIII-digested tDNAs, showed a band of 7 kb in the fertile line HA 89 (Figure 2A, Lane 3) but exhibited a 5 kb fragment in the case of CMS RES1 (Figure 2A, Lane 4). Finally, when a fragment corresponding to *atp9* gene was employed as probe on an *Eco R*I-digested tDNAs, the results were that, while a 6 kb fragment was recognized in both lines (Figure 2B, Lanes 5 and 6), a fragment of 18.5 kb detected in HA89 (B) (Figure 2B, Lane 5) was not present in CMS RES1 and an 11 kb fragment was hybridized on this material (Figure 2B, Lane 6).

DISCUSSION

Despite their apparent unrelated nature, most CMS associated mitochondrial genes share some characteristics. A common feature of CMS-associated genes is their physical association and co-transcription with essential mitochondrial genes. CMS genes appear as the products of the recombination activity of mitochondrial genomes in plants. Investigations in a variety of plant species including maize, *Petunia*, sunflower, *Brassica* and bean have shown that different alterations of the mitochondrial genome are associated with the CMS phenotype (Budar and Pelletier, 2001). A hypothesis suggests that CMS gene products somehow interfere with the normal physiology of mitochondria (possibly via the portions of normal mitochondrial polypeptides included in their sequences), leading to less efficient respiration and/or ATP production, impairing pollen production, which is an energy demanding developmental program (Budar et al., 2003).

Southern blot analyses showed that several CMS sources of the genus *Helianthus* presented differences in the hybridization patterns of the regions of *cox* and *atp* gene families (Horn, 2002). Structural alterations around these genes are also present in the CMS lines in other species like *Petunia* (Young and Hanson, 1987), *Brassica napus* (Dieterich et al., 2003), carrot (Szklarczyk et al., 2000), bean (Chase, 1994), *Lolium perenne* (Kiang and Kavanagh, 1996), and rice (Eckardt, 2006). Schnable and Wise (1998) reported differences between the male-sterile and male-fertile mitochondrial genomes that are confined to a rearranged region around the *atp6, atp9,* and *atpA*

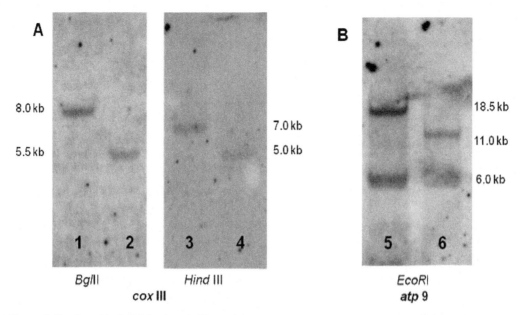

Figure 2. Southern blot hybridization. A: The probe was a fragment from *cox*III gene. tDNA from HA89 (B) (1, and 3) and from HA89 CMS RES1 (2, and 4), were digested with *Bgl*II (1, and 2) or *Hind*III (3, and 4). B: The probe was a fragment of *atp9* gene. tDNA from HA89 (B) (5) and from HA89 CMS RES1 (6) were digested with *Eco*RI.

genes. Specifically, the central role of the abnormal transcripts of *atp9* linked to male sterility was reported in *Brassica napus* (Dieterich et al., 2003), *Petunia* (Hanson et al., 1996), and sorghum (Tang et al., 1996). In sunflower, Spassova et al. (1994) detected this *locus* in the cytotype CMS3, then De la Canal et al. (2001) explained the male sterility condition in CMS PEF1 by the insertion of 500 bp in the 3´*atp9* gene.

We have shown here some structural features of CMS RES1 mtDNA. Amplicon sizes obtained through PCR amplification using primers *orf*H522F and *orf*H522R on CMS PET1 and CMS MAX1 tDNA were the same as those of Rambaud et al. (1997). When the same analytical strategy was applied on CMS RES1 no amplification was observed. These results indicate that *orf*H522 is not involved in the male sterility condition in RES1 and that CMS RES1 has a different architectural organization than CMS PET1 and MAX1. When using the three primers system, the results pointed out to structural similarities between HA89 (B) and the three lines CMS RES1 in the 3´*atp*A - *orf*H873 region and, importantly, confirmed the differences that CMS RES1 has, in this region, with CMS PET1 and CMS MAX1 (Figure 1). In addition, CMS RES1 showed no differences associated with *rp*S14, *cob*, and *nad4* regions with the fertile line by PCR criteria.

On the other hand, Southern blot analyses showed that CMS RES1 also presents some differences in the mtDNA architecture compared to the fertile line. We detected hybridization differences between fertile HA89 (B) and CMS RES1 lines in areas surrounding *atp9* and *cox*III genes (Figure 2). Therefore, these areas could locate structural differences responsible for the sterile phenotype RES1.

Studies are underway to further describe the molecular nature of the CMS RES1 and determine the sequences of mtDNA in the areas surrounding the *loci atp9* and *cox*III to assess the homologies and differences in these areas with other CMS sources.

ACKNOWLEDGMENTS

This work was conducted in partial fulfilment of M.M. Echeverria's Doctorate and was supported by Instituto Nacional de Tecnología Agropecuaria (INTA, Argentina) and Universidad Nacional de Mar del Plata (UNMdP, Argentina).

REFERENCES

Budar F, Pelletier G (2001). Male sterility in plants: occurrence, determinism, significance and use. C. R. Acad. Sci. III 324: 543-550.

Budar F, Touzet P, De Paepe R (2003). The nucleo-mitochondrial conflict in cytoplasmic male sterilities revisited. Genetica 117: 3-16.

Chase CD (1994). Expression of CMS-unique and flanking mitochondrial DNA sequences in *Phaseolus vulgaris* L. Curr. Genet. 25: 245-251.

Demesure B, Sodzi N, Petit RJ (1995). A set of universal primers for amplification of polymorphic non-coding regions of mitochondrial and chloroplast DNA in plants. Mole. Ecol. 4: 192-131.

De la Canal L, Crouzillat D, Quetier F, Ledoigt G (2001). A transcriptional alteration on the atp9 gene is associated with a sunflower male-sterile cytoplasm. Theor. Appl. Genet. 102: 1185-1189.

Dewey R, Schuster A, Levings C, Timothy D (1985). Nucleotide sequence of F_0-ATPase proteolipid (subunit 9) gene of maize mitochondria. Proc. Natl. Acad. Sci. USA, 82: 1015-1019.

Dieterich J, Braun H, Schmitz U (2003). Alloplasmic male sterility in *Brassica napus* (CMS "Tournefortii-Stiewe") is associated with a special gene arrangement around a novel atp9 gene. Mole.. Genet. Genomics. 269: 723-731.

Eckardt NA (2006). Cytoplasmic male sterility and fertility restoration. Plant Cell 18, 515-517.

Echeverria MM, Salaberry MT, Rodríguez RH (2003). Characterization for agronomic use of cytoplasmic male-sterility in sunflower (*Helianthus annuus* L.) introduced from *H. resinosus* Small. Plant Breed, 122: 357-361.

Haevekes FWJ, Miller JF, Jan CC (1991). Diversity among sources of cytoplasmic male sterility in sunflower (*Helianthus annuus* L.). Euphytica, 5: 125-129.

Hahn V, Friedt W (1994). Molecular analysis of the cms-inducing MAX1 cytoplasm in sunflower. Theor. Appl. Genet. 89: 379-385.

Hanson MR, Sutton CA, Lu B (1996). Plant organelle gene expression: altered by RNA editing. Trends Plant Sci. 1: 57-64.

Horn R (2002). Molecular diversity of male sterilty inducing and male-fertile cytoplasms in the genus *Helianthus*. Theor. Appl. Genet. 104: 562-570.

Horn R, Friedt W (1999). CMS sources in sunflower: different origin but same mechanism? Theor. Appl. Genet. 98: 195-201.

Iglesias V (1994). Genetic transformation studies in wheat using particle bombardment. Swiss Federal Institute of Technology, Zurich. Dis # 10628.

Kiang A, Kavanagh T (1996). Cytoplasmic male sterility (CMS) in *Lolium perenne* L, 2. The mitochondrial genome of a CMS line is rearranged and contains a chimaeric atp9 gene. Theor. Appl. Genet. 92: 308-315.

Laver H, Reynolds S, Moneger F, Leaver C (1991). Mitochondrial genome organization and expression associated with cytoplasmic male sterility in sunflower (Helianthus annuus). Plant J. 1: 185-193.

Leclercq P (1969). Une stérilité mâle cytoplasmique chez le tournesol. Ann. Amélior. Plantes 19: 99-106.

Rambaud C, Bellamy A, Dubreucq A, Bourquin J, Vasseur J (1997). Molecular analysis of the fourth progeny of plants derived from a cytoplasmic male sterile chicory cybrid. Plant Breed. 116: 481-486.

Rieseberg LH, Van Fossen C, Arias D, Carter RL (1994). Cytoplasmic male sterility in sunflower: Origin, inheritance, and frequency in natural populations. J. Heredity 85: 233-238.

Saghai-Maroof M, Soliman K, Jorgensen R, Allard R (1984). Ribosomal DNA spacer length polymorphisms in barley: Mendelian inheritance, chromosomal location, and population dynamics. Proc. Natl. Acad. Sci. USA 78: 7463-7467.

Sambrook J, Fritsch EF, Maniatis T (1989). Molecular cloning: A laboratory manual. 2nd ed. Vol 2. Cold spring Harbor, New York, USA, pp. 9.31-9.57.

Schnable PS, Wise RP (1998). The molecular basis of cytoplasmic male sterility and male fertility restoration. Trends Plant Sci. 3: 175-180.

Serieys H, Christov M (2005). Progress Report 1999-2004 activities of the FAO working group: Identification, study and utilization in breeding programs of new CMS sources. European Cooperative Research Network on Sunflower. Proc. X FAO Consultation Meeting, Novi Sad, Serbia and Montenegro, July 17-20, 2005. FAO, Rome, Italy.

Sicullela L, Palmer J (1988). Physical and gene organization of mitochondrial DNA in fertile and male sterile sunflower, CMS-associated alterations in structure and transcription of the ATP A gene. Nucl. Acids Res. 16: 3787-3799.

Spassova M, Moneger F, Leaver CJ, Petrov P, Atanassov A, Nijkamp P (1994). Characterization and expression of the mitochondrial genome of a new type of cytoplasmic male-sterile sunflower. Plant Molec. Bio. 26: 1819-1831.

Szklarczyk M, Oczkowski M, August VH, Borner T, Linke B, Michalik B (2000). Organization and expression of mitochondrial atp9 genes from CMS and fertile carrots. Theor. Appl. Genet. 100: 263-270.

Tang HD, Pring DR, Shaw LC, Salazar RA, Muza FR, Yan B, Schertz KF (1996). Transcript processing internal to a mitochondrial open reading frame is correlated with fertility restoration in male-sterile sorghum. Plant J. 10: 123-133.

Young E, Hanson M (1987). A fused mitochondrial gene associated with cytoplasmic male sterility is developmentally regulated. Cell 50: 41-49.

Behavioural pattern of upland rice agronomic parameters to variable water supply in Nigeria

Christopher O. Akinbile

Department of Agricultural Engineering, Federal University of Technology, Akure, Nigeria.
E-mail: coakinbile@futa.edu.ng, cakinbile@yahoo.com.

The response of rice crop to water application is of paramount importance to researchers as it helps in determining its agronomic development. The study was aimed at establishing the agronomic responses of rice crop to differential water supply. A two-year dry season experiment was conducted at the research farm of International Institute of Tropical Agriculture, IITA Ibadan, Nigeria. Two upland rice varieties (NERICA 2 and NERICA 4) were planted on a 5 × 5 m plot in a randomized complete block design with four treatments based on different water application. Agronomic parameters such as plant height, root depth, canopy shading, leaf area index, panicle and tiller configuration, biomass and grain yield in relation to crop water use were obtained and the results were subjected to statistical analysis. Maximum plant height (89.0 and 100.3 cm), maximum root depth (22.1 and 23.8 cm), panicle diameter (3.9 and 4.5 cm), panicle length (26.1 and 25.7 cm), leaf area index (LAI) 3.27 and 3.95, canopy shading (CS) 0.22 and 0.99 were obtained for both NERICA 2 and NERICA 4, respectively. Leaf width (1.3 and 1.4 cm), total tillers (14 and 12) and leaf length (36.9 and 38 cm) were also observed for the two varieties respectively. The highest total grain and biomass yields of 1.94 and 1.95 t/ha were observed in treatment A for NERICA 2 while the least values of 0.29 and 1.09 t/ha were observed in treatment D. As for NERICA 4, the highest values (1.90 and 2.27 t/ha) were from A and the least (0.38 and 2.29 t/ha) in D. The result of ANOVA showed significant differences in biomass and grain yield, LAI, CS, plant height and root depth among treatments ($P > 0.05$). Specific behavioural pattern to differential water application was observed in the development of agronomic parameters monitored during the experiment. This was indicative that water is a major yield - influencing factor in rice production.

Key words: Upland rice, LAI, CS, agronomic parameters, water supply.

INTRODUCTION

Rice (*Oryza sativa L*) constitutes one of the most important staple foods of over half of the world's population. Globally, it ranks third after wheat and maize in terms of production (Bandyopadhay and Roy, 1992). In Nigeria, rice is the sixth major crop in cultivated land area after sorghum, millet, cowpea, cassava and yam (Dauda and Dzivama, 2004; Olaleye et al., 2004). It is the only crop grown nationwide and in all agro ecological zones from Sahel to the coastal swamps. Rice could be cultivated in about 4.6 - 4.9 million ha of land in Nigeria, but the actual area under cultivation is only 1 million ha representing 22% of the total potential available area (Kehinde, 1997). Before the oil boom of the 1970s, Nigeria had been largely self sufficient in rice production with negligible imports to take care of the taste of small European population in the country. The resultant buoyant foreign exchange earnings of the country from the oil boom of 1970 - 1980 raised the general standard of living and taste, which resulted in massive importation of all kinds of manufactured goods and commodities, including rice. Local rice production was no longer encouraged and therefore national self-sufficiency declined from over 99% from 1961 to 1973 to about 23% in 1984 (Akintola, 2000). Rice importation rose from 7,000 tons in the 1960s to 657,000 tons in the 1990s (WARDA, 2003). Nigeria is the World's second largest rice importer, spending over US$300 million on rice imports annually. It imported 1.7 and 1.5 million tons in 2001 and 2002 respectively, (WARDA, 2003). This created a serious drain on Nigeria's foreign exchange reserve and also raised a big question: Why should the country continue to spend that much on rice imports when it has the capacity to become self sufficient in rice production? The answer could lie in increasing productivity using irrigation. Water is essential for rice cultivation and its supply in adequate quantity is one of the most important factors in rice production. In Asia and other parts of the world, rice crop suffers either from too little water (drought) or too much (flooding or submergence). Most studies on constraints to high rice yield shows that water is the main factor for yield gaps and yield variability from experiment stations to farm (Papademetriou, 2001). Irrigated agriculture is the dominant use of water, accounting for about 80% of global and 86% of developing countries water consumption as at 1995 (Rosegrant et al., 2002). By 2025, global population will likely increase to 7.9 billion, more than 80% of whom will live in developing countries and 58% in rapidly growing urban areas (IWMI, 2000). About 250 million ha, representing 17% of global agricultural land, is irrigated worldwide today, nearly five times more than at the beginning of the 20th century. This contributes about 40% of the global production of cereal crops. Irrigated rice was responsible for about 75% of the world's total rice production. Irregular water application often leads to a high amount of surface runoff, seepage and percolation which accounts for about 50 - 80% of the total water input into the field (Guerra et al., 1998). Therefore, the water crisis being experienced today is not about having too little water to satisfy our needs especially in agriculture but a crisis of proper management (Akinbile, 2009). The objective of this study therefore, is to investigate the effect of this important factor of production on the agronomy of rice crop under variable supply in Nigeria.

MATERIALS AND METHODS

The study was carried out at the farmyard of the International Institute of Topical Agriculture (IITA) Ibadan, the Oyo State capital, Nigeria. It is located between latitude 3° 54'E and 7° 30' N, at elevation of 200 m above the mean sea level. It has an annual rainfall range of between 1300 and 2000 mm while its rainfall distribution pattern is bimodal. The annual mean temperature is 27.2°C during dry season and 25.6°C during the rainy season. The soil class is *Oxic paleustaff* which belongs to Egbeda Series and is described as Alfisol (Apomu Sandy loam). The vegetation is humid rain forest with an average relative humidity of between 56 and 59% during the dry season and 51 - 82% during the wet season (IITA, 2002).

Field experiment were conducted for two dry seasons to ascertain the crop's water use under irrigated conditions, between November, 2005 and March, 2006 and November, 2006 to March, 2007. The experimental design was a Randomized Complete Block Design (RCBD) with four treatments. NERICA 2 and 4 were planted on all the plots and irrigation water was delivered through an overhead sprinkler systems. There were four treatments based on the level of irrigation water application. Plot A (first treatment) received water seven times continuously in one week (100%ET) and plot B (second treatment) received water six times a week (75%ET). The third treatment (plot C) received water five times a week (50%ET) and the fourth treatment (plot D) received water four times a week (25%ET). A controlled experiment to monitor the behaviour of rice on the field was carried in a lysimeter situated in a screen house located 50 m away from the field (Akinbile, 2009). Weekly measurements of plant height of rice were made using the measuring rule from two weeks after planting (that is from emergence) to maturity stage (that is fifteen weeks after planting) in order to monitor the crop growth response to the variability of water supplied. Canopy Analyzer was used in measuring Leaf Area Index (LAI) and Canopy Shading (CS) non-destructively. LAI is simply referred to as foliage orientation or density while CS is the amount of foliage present per plot. Other agronomic parameters determined include, the grain yield, grain size, leaf length and width, number of leaves panicle length, tillering ability, root depth, flowering and maturity days using convectional equipment such as weighing balance, measuring rule and vernier caliper. For grain yield, a 5 m by 5 m sub area was harvested per plot with the aid of sickles. The field weight of grain yield was corrected to 12% moisture content before storage. Results obtained during field experimentation were subjected to statistical analysis using SAS 9.1 version.

RESULTS AND DISCUSSION

Responses with respect to root depth

The root depths of the crop throughout the entire growing season are as shown in Figure 1. The maximum root depths of 22.6 and 23.8 cm were recorded in Plot A in the 2 varieties (N2 and N4) in both the field and controlled experiment during the first trial while the lowest was recorded in Plot C in controlled (17.3 cm) and predictably in plot D, (18.1 cm) in field observations respectively. The variation in the lowest root depth may be due to differential water application towards the end of mid season stage as a result of deficit irrigation. In the observations during the first trial in 2006, the maximum and minimum root depths were in Plots A and D respectively and thus agreed with Lafitte et al. (2004 and 2007). This is a clear indication that the length of roots of rice has a direct bearing on the water application and use of water in all the treatments considered.

Reponses with respect to plant height

The heights of rice during the vegetative, ripening and maturity stages for the four treatments are given in Figure 2 and detailed graphs of mean heights and the standard deviation of each of the treatments are given in Figure 3. NERICA 2 variety had a maximum plant height of 89 cm while NERICA 4 had a height of over 100 cm (Figure 2). This is similar in values in the second experimental trial indicating reliability in the methods adopted and quantity of water applied. This agreed with the findings of Becker and Johnson (2001) and Fujii et al. (2005) which in separate instances confirmed the observations of WARDA (2006) that N2 is shorter than N4. The steady and consistent rise in the crop's height may be attributed to the quantity of irrigation water applied, which is a reflection of the glaring differences in the parameters for the different plots. There is a difference in the plant height at ripening (72 DAP) and maturity (93 DAP) stages

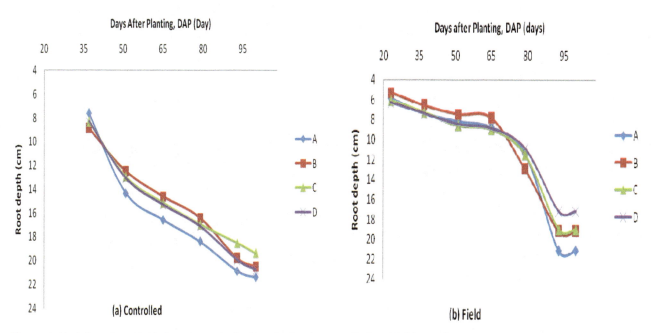

Figure 1. Variations in root depths versus days after planting in controlled and field experiments.

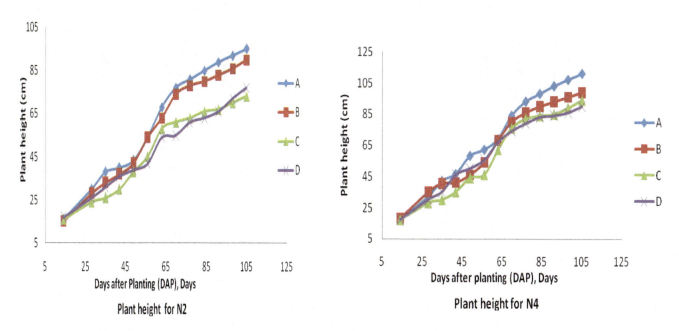

Figure 2. Plant height versus days after planting (DAP) for N2 and N4 on all the plots.

among treatments at 5% level of significance. It was evident that differences in plant height among field treatments were obvious as the controlled treatments especially at 37 DAP. At vegetative stage, the monitored soil effect was more as crop emergence was mostly soil and water dependent. At 72 DAP (ripening stage), the combined effect of soil and weather was more hence the little variation in the plant height (particularly in field experiment). At 92 DAP; the weather effect was more evident hence the pronounced variation in the mean plant height and deviation.

Responses with respect to Leaf Area Index (LAI)

Figure 4 showed the correlation between LAI and DAP in the treatment plots. The values of LAI were found to be highest in treatment A and lowest in treatment D. In the

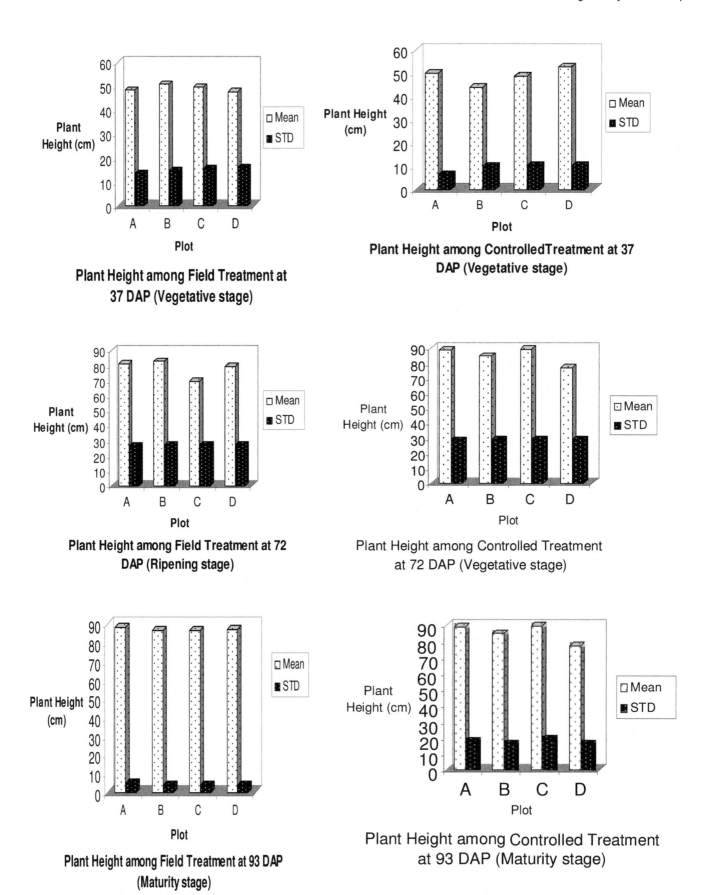

Figure 3. Plant height among treatments in field and controlled experiment at different stages of growth.

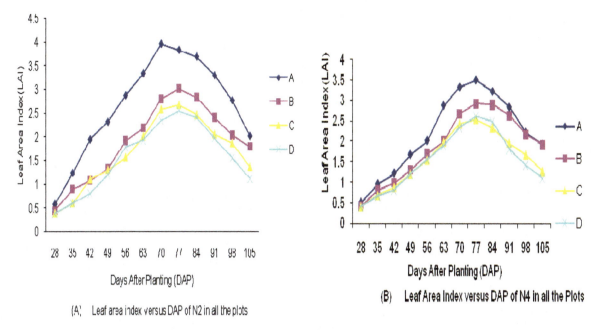

Figure 4. Leaf area index (LAI) versus days after planting (DAP) of N2 and N4 in all the plots during the second experimental trial.

first trial, LAI was found to be 1.78 in N2 variety but in the second experiment, N2 has a LAI of 3.27 while N4 was measured to be 3.95 (Figure 4). These readings were observed during the mid season/ripening stage in both N2 and N4 varieties (65 to 85 DAP). However, the values were nearly equal particularly at the maturity stage. This behaviour followed the water distribution pattern which is a function of %ET of water applied, since any effect on LAI would definitely affect yield. The data confirmed that the variable pattern in water applied affected the LAI and subsequently yields in the treatment plots. The observation was also an indication of increased water application resulting in decreased crop water use since water stress had been eliminated hence the leaves orientation during the ripening stage. This also agreed with the findings of Nwadukwe and Chude (1998) and Lafitte et al. (2004, 2007) who stressed that the leaf area orientation is a function of water application. At 5% level of significance, the difference in values of LAI between the treatments was significant

Responses with respect to canopy shading (CS)

Figure 5 showed Canopy shading (CS) and its variation with days after planting (DAP) for the second trials. From it, the highest CS (0.2) was during the heading, booting, flowering and milky phases of the ripening stage that is between 60 - 79 DAP. By 105 DAP; CS has dropped to between 0.8 and 1.0 among the treatments. This was due to the fact that at maturity stage, leaves colouration has changed from lush green to brown and the canopy had collapsed in readiness for grain harvesting. This behaviour was similar to LAI as the highest CS was observed during the ripening stage; the same time LAI was maximum in all the plots. Similar observations were recorded for the first trial and in both varieties. LAI had maximum values (3.95 for N4 and 3.27 for N2) during ripening stages (65 - 85 DAP) in both varieties. This implies that increased water application increases LAI as well as CS indicating the need for irrigation scheduling at a certain stage of crop growth. It must be noted that irrigation water was increased by 100% (full ET) during the mid season/ripening stage in all the treatments and trials. This was to cater for increased metabolic activities of the crop at this stage. Similar trends in behaviour was observed and recorded for LAI vs DAP and CS vs DAP in the first trials.

Responses with respect to measured post – harvest parameters

Panicle length, panicle diameter, total tiller, grain length, width and diameter and 100 grain weight were all measured after harvesting for both experiments. The results presented in Tables 1 and 2 conform to the standard frequently quoted by the Africa Rice Institute (ARI) and West Africa Rice Development Authority (WARDA) for upland NERICA 2 and 4 rice varieties. The maximum plant height and leaf length were 89 and 37 cm, respectively. Panicle length and diameter were 26 and 3.9 cm respectively. Leaves number, width and total tillers were 11, 1.4 cm and 14 respectively. Similar

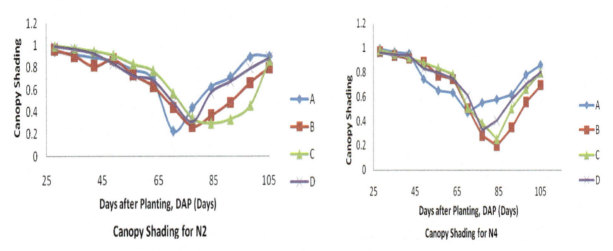

Figure 5. Canopy shading (CS) versus days after planting (DAP) for N2 and N4 in all the plots.

Table 1. Results of measured plant parameters (N2) after harvest from field experiment LSD (P > 5%).

Plots	Plant height (cm)	Root depth (cm)	No. of leaves	No. of tillers	Leaf length (cm)	Leaf width (cm)	Panicle diameter (cm)	Panicle length (mm)
A	88.8	22.6	11	15	36.89	1.44	3.92	26.08
B	85.6	19.2	9	13	35.94	1.30	3.76	25.50
C	88.8	19.1	11	8	32.30	1.28	4.50	25.60
D	76.4	17.2	8	10	29.46	1.24	3.34	23.84

Table 2. Results of measured plant parameters (N4) after harvest from field experiment.

Plots	Plant height (cm)	Root depth (cm)	No of leaves	No of tillers	Leaf length (cm)	Leaf width (cm)	Panicle diameter (cm)	Panicle length (mm)
A	100.33	23.80	12	15	37.98	1.28	4.55	24.65
B	87.33	21.51	11	12	34.97	1.43	4.10	25.97
C	86.63	20.38	12	12	30.33	1.32	3.48	24.48
D	85.75	19.78	10	12	29.25	1.35	4.03	25.75

LSD (P > 5%).

observations were recorded in Table 2 for N4 variety; maximum plant height and leaf length were 100.3 and 38 cm respectively. Panicle length and diameter were 25 and 4.5 cm respectively while leaves number, width and total tillers were 12, 1.3 cm and 12 respectively. These were indicative of definite behaviour to water application at different stages of crop development. Further increase in water application may not result in any pronounced change(s) in these parameters.

Grain and biomass yield

The values of grain yield, biomass and total yield in t/ha are given for the first trial in Table 3 and second trial in Table 4. These parameters were highest in treatment A and lowest in treatment D. The variation in the yield per treatment was due to the quantity of water received and days assigned for irrigation per plot which varied tremendously. The crop in treatment A received water, 7 days a week while treatment D received water 4 days a week. Increased irrigation water application resulted in increase in soil moisture availability for crop use since no water stress was allowed on the crop.

This is a clear indication that increased irrigation water applied does not increase crop water use but decreased water application increases crop water use. Similar observations were recorded for irrigation water applied and total ET for rice in plots C and D. This trend also affected all agronomic parameters including biomass and

Table 3. Grain yield, biomass and total yield of rice in the 2005/2006 in N2 experiment.

Plot	Grain yield (t/ha)	Biomass yield (t/ha)	Total yield (t/ha)
A	1.36	1.84	3.2
B	0.81	2.39	3.2
C	0.30	2.30	2.6
D	0.16	1.64	1.8

Table 4. Grain yield, biomass and total yield for N2 and N4 for the 2006/2007 experiment.

Treatment plots	Rice type	Grain yield (t/ha)	Biomass yield (t/ha)	Total yield (t/ha)
A	N2	1.94	1.95	3.89
	N4	1.90	2.27	4.17
B	N2	1.25	2.70	3.95
	N4	1.43	2.55	3.98
C	N2	0.66	2.15	2.81
	N4	0.91	2.22	3.13
D	N2	0.29	1.09	1.38
	N4	0.38	2.29	2.67

Table 5. Number of whiteheads in the N2 Plots at 78 DAP in the 2005/2006 trial.

Plot	Whiteheads	% Composition per plot
A	47	7.4
B	81	15.7
C	94	16.1
D	98	16.6

grain yield as there were noticeable reduction in all the plots as water use increases (Tables 3 and 4). The grain yield was highest in plot A (1.36 and 1.94 t/ha) in the two trials and steady but gradual decline was observed in all other plots (B and C). In plot D, the grain yield was minimum (0.16 and 0.29 t/ha) indicating that water has a yield-limiting influence on the rice crop. One major factor that limit the yield was the emergence of whiteheads on all the four plots during milky and flowering stages of ripening (78 DAP) in the first trial. Table 5 showed variation of the whiteheads, an indication of water deficit at that stage. The lowest was in Plot A, 7.4% while the highest expectedly was found in plot D 16.6%. This may also significantly affect the final outcome of the grain yield. This was the resultant effect of the introduction of temporal deficit irrigation towards the later part of ripening stage. The emergence of whiteheads was a clear indication that deficit irrigation during mid season stage of crop development was not healthy for an optimum growth. However, this does not imply that deficit irrigation was not possible during rice growing season but its introduction during the mid season/ripening stage will greatly affect the crop development and yield. The findings of Becker and Johnson (2001), Lafitte and Courtois (2002), Lafitte et al. (2007) agreed with the observation in yield variation as a result of differential irrigation among the treatments.

CONCLUSION AND RECOMMENDATIONS

A two year field experiment to evaluate differential water application of upland NERICA rice as it relates to increased productivity was carried out at the farmyard of the International Institute of Tropical Agriculture, (IITA) Ibadan, Nigeria. The choice of upland rice was informed by the fact that the irrigated upland ecology has very high potential for rice production but contributes between 10 and 15% to national production. Adopting strict water conservation measures will lead crop failure at a certain stage, indicating that the effect of water stress leads to corresponding increase in water use at certain stages such as midseason/ripening stage of crop growth and development. The emergence of whiteheads was an indication of shortfalls in water requirements at this stage. There were corresponding responses of all agronomic parameters of the rice crop observed to changes in water application, indicating the dominant effect of water to growth and development.

The recommendations are:

1. Interspecific cross hybridization of the very good traits of other local rice varieties such as Ofada, Igbimo, and Aroso with NERICA cultivars for replication and multiplication is encouraged. This is to complement farmers' efforts with their increased yield production while maintaining some of its very good African traits.

2. Research should be conducted into using modern technologies for bird scaring to reduce considerable yields of rice being lost annually. This is to ensure reliable yield data. The age long, primitive method of human bird scaring is effective only on small fields. However, the use of nets as temporary measures to prevent rice invasion by birds (small or medium fields) is suggested. Similarly, chicken wire mesh should be placed round the field to prevent rodents and grass cutters invasion. This is useful where human efforts (bird scarer) may not be enough to prevent birds from attacking rice fields.

3. Application of Alternate Wetting and Drying (AWD) technique to upland rice cultivar in Nigeria should be carried out. This is to investigate its effect on the yield and other agronomic parameters while considering water saving as one of the mitigating strategies against impact of climate change on food and agriculture productivity.

ACKNOWLEDGEMENTS

The author is grateful for the support of Africa Rice Centre (WARDA), Ibadan, Nigeria for providing support and other logistics for taking measurements. The study was funded by WARDA.

REFERENCES

Akinbile CO (2009). Crop Yield Responses of Upland Rice to Differential Water Application under Sprinkler Irrigation System, Unpublished Ph.D Thesis, Department of Agricultural and Environmental Engineering, University of Ibadan 21: 195.

Akintola AA (2000). Increased rice production and processing in Nigeria: The way forward, in: Niger. Agric. Eng., 1(1): 10-17. SNAAP Press Limited, Enugu, Nigeria

Bandyopadhay S, Roy CN (1992). Rice Processing Technology, IBTT Publishing Co. PVT. Ltd, 66 Japath, New Delhi, India pp. 23-29.

Becker M, Johnson DE (2001).Cropping intensity effects on upland rice yield and sustainability in West Africa Nutrient Cycling in Agroecosystems. Kluwer Academic Publishers, the Netherlands 59: 107-117.

Dauda SM, Dzivama AU (2004). Comparative performance of a locally developed rice thresher with an imported Votex Rice Fan. Proceedings of the 5th International Conference of the Nigerian Institution of Agricultural Engineers, Ilorin 26: 29-32.

Fujii M, Andoh C, Ishihara S (2005). Drought resistance of NERICA (New Rice for Africa) compared with Oryza sativa L. and millet evaluated by stomata conductance and soil water content. Proceedings of the 4th International Crop Science Congress, New Delhi, India pp. 331-338.

Guerra LC, Bhuiyan SI, Tuong TR, Barker R (1998). Producing more Rice with less Water, SWIM, Paper 5, Colombo, Sri Lanka, International Water Management Institute (IWMI) p. 21.

International Institute of Tropical Agriculture (IITA) (2002). Annual Report. http://:www.iita.org/annual_report2002.aspx.

International Water Management Institute (IWMI) (2000). Research contributions to the World Water vision, Colombo Sri Lanka, IWMI. pp. 450-452.

Kehinde JK (1997). Advances in Rice Production. In Rice Production and Processing Workshop, National Cereals Research Institute, Badeggi, Niger State 1-14.

Lafitte HR, Yongsheng G, Yan S, Li ZK (2007). Whole plant responses, key processes and adaptation to drought stress: the case of rice, J. Exp. Botany 58(2): 169-175.

Lafitte HR, Price AH, Courtois B (2004). Yield response to water deficit in an upland rice mapping population: associations among traits and genetic markers. Theo. Appl. Genet. 109: 1237-1246.

Lafitte HR, Courtois B (2002). Interpreting cultivar X environment interactions for yield in upland rice: assigning value to drought-adaptive traits, Crop Sci. 42: 1409-1420.

Nwadukwe PO, Chude VO (1998). Manipulation of the irrigation schedule of rice (Oryza sativa L.) as a means of maximizing water use efficiency and irrigation efficiency in the semi-arid tropics, J. Arid Environ. 40(3): 331-339.

Olaleye AO, Osiname OA, Cofie O (2004). Soil and water management for rice production in Nigeria, Proceedings of the West Africa International Conference on Hunger without frontiers, West African Society for Agricultural Engineering (WASAE), 20-24 September, Kumasi, Ghana pp. 259-265.

Papademetriou MK (2001). Issues and perspectives of rice production in Asia and the pacific Region: In Yield Gap and productivity Decline in Rice production consultation held in Rome 5-7: 39-67.

Rosegrant WM, Cai X, Cline SA (2002). World water and food to 2005: Dealing with Scarcity, International Food Policy Research Institute (IFPRI), Washington DC pp. 3-12.

WARDA (2006). Africa Rice Center (WARDA) Annual Report 2005-2006, Providing what's needed, Cotonou, Benin p. 52.

WARDA (2003). Strategy for Rice Sector Revitalization in Nigeria, Draft for Discussion at Technical Workshop 20-21 August, 2003 IITA, Ibadan, Nigeria. pp. 23-25.

WARDA (2006). Rice Trends in sub-Saharan Africa, Second edition, WARDA Bouake p. 10.

Assessment of dissimilar gamma irradiations on barley (*Hordeum vulgare* spp.)

S. Sarduie-Nasab[2], G. R Sharifi-Sirchi[1*] and M. H. Torabi-Sirchi[3]

[1]Horticulture and Dates Research Institute, College of Agriculture, Shahid Bahonar University of Kerman, Kerman, Iran.
[2]Graduate student of Agronomy and Plant Breeding Department, Shahid Bahonar University of Kerman, Kerman, Iran.
[3]Department of Agriculture Technology, Faculty of Agriculture, University Putra Malaysia, 43400 Serdang, Selangor Darul Ehsan, Malaysia.

Hordeum vulgare (barley), is an important agricultural crop for food, feed and also has been used virtually worldwide as a model plant for biological research. It is a diploid crop with a low chromosome number (2n = 14) and targeted as a proper crop for intense research on mutagenesis, mutagens and mutants (Khattak and Klopfenstein, 1989). Recently, heavy-ion beams have been used as novel and efficient mutagens in plant breeding. Many plant variations were induced by irradiation and many novel experimental materials and practical cultivars were generated. The present research aimed to estimate the effect of different gamma radiation on germination and emergence indexes of Barley variety Nosrat. Barley seeds were treated with 4 different gamma ray doses (0, 200, 700, 1200 Gy) from 3 angels. Results of this study expressed that high gamma ray doses decreased emergence index compare with control treatment and also, radiation has inhibitory effect on stem height and width. Base on experimental and green house studies doses of 700 and 1200 Gy strongly decline plant growth and treated plant seeds with these radiation doses did not emerge in the field. Therefore we suggest that 200 gamma ray dose was the best treatment to screen mutant Barley.

Key words: Barley, emergence index, gamma ray, germination index, *Hordeum vulgare*. mutation.

INTRODUCTION

Barley (*Hordeum vulgare* spp.) belonging to family of "*Poaceae*" order of *poales* genus of *Hordeum* it has cultured from 7000 B.C. till yet (Khattak and Klopfenstein, 1989). This annual plant is the second strategic agricultural crop subsequent of wheat that is utilized for feeding and brewery purpose moreover, defines 91 million ha of agricultural world farm lands to itself. Nosrat variety cultivated as one of most feed crops in temperate area of Iran, it has some important characters like: high yield, resistant to drought stressed and cool environments. Nosrat variety breeds from crossing of Karon and Kavir that these are from commercial varieties of Iran. Due to artificially induced mutation, not only a broader variation of desirable traits was obtained but also, in the results of breeding, many mutation varieties were released. Mutation technology for development of new hybrid varieties tends to make new plants with high better quality and quantity characters, resistant against biotic and abiotic stresses (ASTM, 2007). It is well established that seedling growth reveals a significant influence on nutrient value of the plant part/seed, which is consumed as food (Adilson et al., 2002; Satter et al., 1990). The quality or nutritional value of any consumable plant part, including seeds, depends on its basic constituents, including proteins, carbohydrates, minerals and vitamins (Al-Kaisey et al., 2003; Khattak and Klopfenstein, 1989). Mutation breeding with molecular techniques and *in vitro* methods are suitable techniques to improve food industries (Selim and Banna, 2001). From the end of 1930 decade, ionization radiation to create mutations on agro plants is carried out. Physical parameters like electromagnetic radiation with wave length between 1 - 10 angstrom (gamma ray) could cause genetic changes with permanent differentiations on plant genome. Gamma ray with ionization molecules,

---

*Corresponding author. E-mail: sharifisirchi@yahoo.com, sharifi-sirchi@mail.uk.ac.ir.

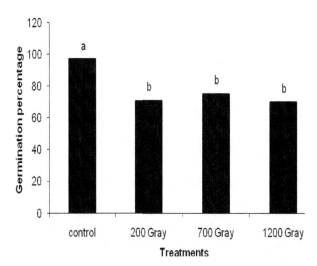

Figure 1. Effect of four different gamma radiations on percentage of seed germination after one week of culture. Different abbreviation letters showed significant different at α = 0.01

specially the water of DNA around cause to make free radicals that these free radicals attacked to DNA molecule and cause affect differentiation on one alkali but at most of cases it cause to breaking one or two chains of DNA (Hagberg and Persson, 1968; Jyoti et al., 2009). Almost 89% of mutant varieties developed by using of physical mutagens and about 60% of them were created by applying gamma ray mutants at barley caused to product high yield, resistant to mildew, strong stem, high protein and skinless seeds (Ananthaswamy et al., 1971). Great of high yield and dwarf barley mutant like, Diamant, Golden Promise cultivars had positive effect on beer brewing industry at Europe (Lundqvist and Franckowiak., 1997; Selim and Banna, 2001). The present study was conducted to evaluate the effects of gamma ray on germination index, emergence index and some phonotypic characters on barley plant variety Nosrat. Another aim of this study was defining the best Gamma ray doses to barley variety Nosrat.

MATERIALS AND METHODS

This research was carried out on 1388 at Agriculture College of Shahid Bahonar University of Kerman-Iran. Nosrat variety seeds were irradiated with Gamma rays 200, 700 and 1200 Gy at nuclear organization of Tehran-Iran. For radiation purpose, 1.0 kg seed cylinder (12 cm diameter x 14 cm height) that it considered for maintained uniformity of purposed dosages were applied and each dosage 3 times from all angels were radiated. The radiation dosages were measured with a Fricke Dosimeter (ASTM, 2007).This research was divided in two parts, experimental and Green house studies

Laboratory experiment

Seeds from each treatment were placed on one layer of 10 cm diameter Whitman No. 1 filter paper containing 10 ml of distilled water at non stress conditions at 25°C. Seeds were scored as germinated when a seed coat was broken and 2 mm of radical was emerged. Germination of individual seeds was measured for 24 h interval until no further germination exposure. Also other traits like, stem width, stem length, number of auxiliary roots and plant height were measured. Petri-dishes were arranged based on Completely Randomized Design (CRD) with three replications.

Green house experiment

Green house studies were carried out during 2008. Pots were filled with sandy loam soil with pH 7.4 that is general pH of Kerman Province soils where this study was performed. The field capacity (FC) of the soil was measured as 15% in soil science laboratory. Drainage holes (0.5 mm diameter) were made at the bottom of each pot. Then pots were arranged based on Completely Randomized Design (CRD) with three replications in the green house under controlled condition (25/15°C day/night). Hundred seeds from the same treatments that utilized in laboratory experiment were sown approximately 4.0 cm deep in 4 L pots (18 cm diameter x 30 cm height). All pots were weighed daily and irrigated twice per day with water to maintain soil moisture at field capacity (FC). Plants within each pot counted and scored as emergence percentage after 7 days of cultivation.

RESULTS

In laboratory experiment, Variance analysis of germination percentage of treatments showed significant differences between them. Non irradiated seeds had highest germination percentage (97.3%) which came to class a basis on Duncan's New Multiple Range Test (DNMRT) however, irradiated barley seed had lower germination percentage and came to class b (Figure 1). Results of analysis of variance for root number trait at Petri dish showed significant difference between

Table 1. Effect of four different gamma radiations on barley seeds at Petri dish and pot.

Shoot length (pot)	Stem diameter (Petri dish)	Torque diameter (Petri dish)	Root Number (Petri dish)	Treatment
20.20 a	0.83 a	0.93 a	5 a	Control
19.78 a	0.77 a	0.88 a	5 a	200 Gray
10.36 b	0.73 a	0.87 a	4 ab	700 Gray
-	0.70 a	0.75 a	3 b	1200 Gray

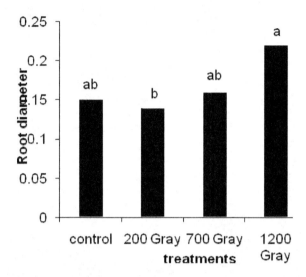

Figure 2. Effect of four different gamma radiations on root diameter after one week of culture. Different abbreviation letters showed significant different at α = 0.01

treatments (control, 200, 700, 1200) at 1% level of signification (Table 1). Base on data analysis, root number of treated seeds had a reverse relationship with dosage of Gamma radiation. So that in treated seeds with 1200 Gy Gamma ray number of root decreases to 3. Analysis of variance for stems and torque width traits showed no significant differences between treatments and control at 1% level of signification (Table 1). Statistic analysis for root diameter trait showed significant difference between control and irradiated plants at 1% level, so that irradiated plant with 1200 Gy had highest root diameter (0.22 mm) which came to class a (Figure 2). Analysis of shoot length trait showed significant difference between the treatments and non irradiated plants had highest shoot length (4.95 cm class a) but in irradiated plant, shoot length decrease with increase of gamma ray dosage, thus shoot length at 1200 Gy decrease to 1.15 cm and came to class d (Figures 3 and 5). In green house experiment, analysis of obtained data from shoot length trait showed that there was significant difference between treatments on irradiated plants and irradiated plants with 200 Gy which had highest shoot length (20.20 and 19.78 cm length) respectively which basis on Duncan's New Multiple Range Test (DNMRT) these two came to class a and irradiated plants with 700 Gy had lower length (10.36 cm)

as class b (Table 1). However, Irradiated plant with 1200 Gy died after short period of growth (1.0 cm) in the green house. Variance analysis of emergence percentage; data showed significant difference between treatments at 1% level of signification. None irradiated plant with 81% and irradiated plant with 56% had highest (class a) and lowest (class d) emergence percentage respectively (Figure 4).

DISCUSSIONS

In the present study observed for the first time, the effects of different gamma radiations on the growth of Iranian variety of barley (Nosrat). Results of this experiment revealed that with increasing gamma ray dosage, the percentage of germination and emergence, plant height, root number, root length, torque and stem width decreased (Ananthaswamy et al., 1971). The highest implement gamma ray dosage 1200 Gy had negative and hazardous effects on barley morphology and growth compared to control. Base on previous researcher reports, the total protein and carbohydrate contents decrease with increasingly higher dosages of gamma irradiation due to higher metabolic activities and hydrolyzing enzyme activity in germinating seed

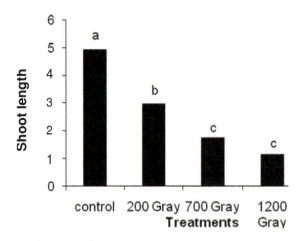

Figure 3. Effect of four different gamma radiations on shoot length after one week of culture. Different abbreviated letters showed significant different at α = 0.01.

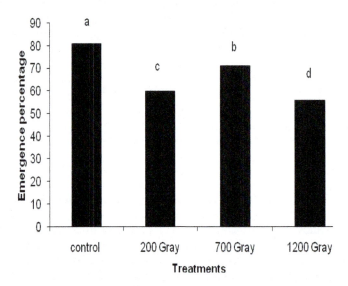

Figure 4. Effect of four different gamma radiations on percentage of seed emergence after one week of culture. Different abbreviation letters showed significant different at α = 0.01.

Figure 5. Effect of four different gamma radiations on barley seedlings germination after one week of culture (Bar=1.4cm). (A) 1200 (Gy) (B) 700 (Gy) (C) 200 (Gy) (D) control.

(Adilson et al., 2002; Maity et al., 2004). The gamma ray breaks the seed protein and produces more amino acid (Adilson et al., 2002; Maity et al., 2004; Tipples and Norris, 1963).

This process may also cause inhibition of protein synthesis. The total protein and carbohydrate was decreased with increasing gamma ray dosage at wheat and rice plants too (Hagberg and Persson, 1968; Inoue et al., 1975). The total DNA content and bonds intension decreased with increasing gamma ray dosage that is anticipated main effect of gamma irradiation on viability of Barley seed (Jyoti et al., 2009).

Bilge and Ersoy 1972, defined that low dosage of gamma ray such as 20 Gy, resulted to increase plant height at barley, furthermore high dosages such as 40, 80 and 120 Gy decreased plant height (Hagberg and Persson, 1968) Also, the results of this study were consistent with Hell fundings (1974) on *Phaseolus valgari* whom observed that under treatment seeds with high gamma ray had low seed germination (Hell and Silveria, 1974).

Conclusion

Base on present study results authors suggest that gamma ray with 200 Gy dosage is the pre-eminent applied gamma ray dosage among all dissimilar dosages used, in this case is advisable for further researcher studies which looking for dominant and recessive mutation on barley Nosrat variety plant seed, to have mutant plants and mutant screening for biotic and abiotic stresses.

ACKNOWLEDGMENTS

Authors would like to dedicate best gratuities to Associate Prof Dr. Gholam Reza Sharifi Sirchi for being abundantly helpful, offered invaluable assistance, support and guidance. This research project would not have been possible without his support. Authors would strongly like to express their sincere appreciation to Madam Sepideh Ghotbzadeh for her endless supports too. Also they like to convey thanks to the Ministry of Science Iran and Faculty of Agriculture, University Shahid Bahonar Kerman-Iran for providing the financial means and laboratory facilities.

REFERENCES

Adilson CB, Maria TLF, Ana Lćia CHV, Henry D, Valter A (2002). Identification of irradiated wheat by germination test, DNA comet assay and electron spin resonance. Radiat. Phys. Chem. 63: 423- 426.

Al-Kaisey MT, Alwan AKH, Mohammad MH, Saeed AH (2003). Effect of Gamma irradiation on anti nutritional factors in broad bean. Radiat. Phys. Chem. 67: 493-496.

Ananthaswamy HN, Vakil UK, Sreenivasan A (1971). Biochemical and physiological changes in gamma-irradiated wheat during germination. Radiat. Bot. 11: 1-12.

ASTM (2007). Standard practice for using the Fricke reference-standard dosimetry system. E1026–04e1, annual book of ASTM standard PA, USA: Americ. Soc. Test. Mat. 12: 2.

Hagberg A, Persson G (1968). Induced mutations in barley breeding, Heredit. 59: 396-412.

Hell KG, Silveria MAV (1974). Imbibation and germination of gamma irradiation *phaseolus vulgaris* seed. Field Crop Abst. 38: 300.

Inoue M, Hasegawa H, Hori S (1975). Physiological and biochemical changes in gamma irradiated rice. Radiat. Bot. 15: 387-395.

Jyoti PM, Sukalyan CSK, Subrata P, Jiin-Shuh J, Alok C, Anindita C, Subhas CS (2009). Effects of gamma irradiation on edible seed protein, amino acids and genomic DNA during sterilization. Food Chem. 114: 1237-1244.

Khattak AB, Klopfenstein CF (1989). Effect of gamma irradiation on the nutritional quality of grains and legumes. II. Changes in amino acid profiles and available lysine. Cer. Chem. 66: 171-172.

Lundqvist U, Franckowiak JD (1997). Descriptions of Barley Genetic Stocks for 1997. Barley Genetics New. Let. 28: 26-54.

Maity JP, Chakraborty A, Saha A, Santra SC, Chanda S (2004). Radiation induced effects on some common storage edible seeds in India infested with surface microflora. Radiat. Phys. Chem. 71: 1065–1072.

Satter A, Neelofar Z, Akhtar MA (1990). Irradiation and germination effects on phytate, protein and amino acids of soybean. Pla. Fo. Hum. Nut. 40: 185-195.

Selim AH, Banna EN (2001). Ionizing irradiation effects on germination, growth, some physiological and biochemical aspects and yield of pea (*Pisumsativum* L) plant. In: GEAR symposium on environmental pollution in Egypt.Cairo.Egypt.

Tipples KH, Norris FW (1963). Some effects of gamma-irradiation on barley and its malting properties. J. Sci. Fo. Agric. 14: 646-654.

Grain yield and yield components of maize (*Zea mays* L.) as affected by crude oil in soil

O. M. Agbogidi

Faculty of Agriculture, Delta State University, Asaba Campus, Asaba, Nigeria.
E-mail: omagbogidi@yahoo.com.

A field experiment on the effect of crude oil levels: 0.0 (control), 5.2, 10.4, 20.8 and 41.6 ml applied at different stages of growth on maize yield and yield attributes, with a view to making appropriate recommendation to maize growers in the oil producing areas of the Niger Delta, was conducted in Ozoro, Delta State during the 2003 and 2004 cropping seasons. Seven maize varieties: Composite (suwan 1), Hybrid 3x-yx, AMATZER w, TZBRSYN w, AMATZBR y, TZBRSYN y and Ozoro local were evaluated. The current study has objective of evaluating the yield and yield components of seven varieties of maize grown in soil contaminated with crude oil with a view to identifying and selecting the tolerant ones and recommending the same to farmers in the Niger Delta region where oil industrial activities are predominant. The experiment was laid out in a randomized complete block design with five treatments replicated four times. Crude oil application (ring application) was carried out at three weeks after planting (3 WAP), 5 WAP and 7 WAP. Plants were harvested at 112 days after planting and assessed for grain yield, 1000-grain weight, cob girth and shelling percentage. The results showed that crude oil treatment significantly reduced ($P > 0.05$) grain yield, grain weight and cob girth but increased shelling percentage at $P < 0.05$. Significant differences at the 5% probably level in the responses of maize varieties to crude oils were also recorded at 5 and 7 WAP with Hybrid 3x-yx recording highest grain yield and weight whereas Ozoro local produced the highest cob girth and shelling percentage. Based on the results obtained from this investigation, Hybrid 3x-yx appeared to be susceptible to soils affected with crude oil as death was eminent at higher oil doses. The open pollinated varieties (AMATZER w, TZBRSYN w, AMATZBR y, TZBRSYN y) are better in terms of relative tolerance hence should be recommended to farmers and maize growers in the oil producing areas of Nigeria. There is therefore the need to test the open pollinated varieties on farmers' field to determine their adaptability to oil pollution. Furthermore, the need for further studies to determine the level of pollution at which maize growth and yield are adversely affected cannot be overemphasized. The study established varietal differences with respect to maize response to crude oil level and this provides a basis for future breeding work by plant breeders.

Key words: Crude oil, grain yield, yield components, *Zea mays*.

INTRODUCTION

Maize is the third most important cereal crop following wheat and rice in the world production of cereal crops (FAO, 2002). Maize is one of the major staples consumed in Nigeria (Obi, 1991). It has a lot of industrial uses including its use as feed for domestic animals (Watson, 1977; Ronanet, 1992). Maize is grown in most agro ecological areas especially in the Niger Delta region where oil industrial activities are predominant (Agbogidi et al., 2005a, 2006a, 2007). Oil pollution effects on the growth and yield of crop plant species have been variously reported by several workers, to include poor growth, yield reduction and sometimes death (Anoliefo, 1991; Anoliefo and Vwioko, 1994; Agbogidi et al., 2005b; 2006a, b; 2007).

There is however, paucity of information on the effect of crude oil levels on grain yield and yield attributes of maize. A crop yield is a function of a number of factors and processes such as light intercepted by the canopy, metabolic efficiency of the plant, translocation efficiency of photosynthesis from the source (leaves) to economic parts and sink capacity or sink strength (Agbogidi et al., 2006b). The photosynthetic capacity of the plant determines overall productivity and the extent of development

Table 1. Effect of crude oil levels applied at 5 WAP on grain yields (t /ha) and yields components of seven maize varieties in Ozoro, Delta State.

Yield attributes	Maize variety	Crude oil level (ml/plant)					
		0.0	5.2	10.4	20.8	41.6	Means
Grain yield (kg/plant)	Composite (Suwan 1)	0.64b	0.58c	0.52b	0.00c	0.00a	0.35
	Hybrid 3x-yx	1.21a	1.02a	0.90a	0.00c	0.00a	0.63
	AMATZBR w	0.72ab	0.70ab	0.62b	0.59a	0.00a	0.53
	TZBRSYN w	0. 70b	0.60b	0.51b	0.43b	0.00a	0.49
	AMATZBR y	0.68ab	0.63b	0.54b	0.51ab	0.00a	0.47
	TZBRSYN y	0.64b	0.60b	0.51b	0.43b	0.00a	0.44
	Ozoro Local	0.60b	0.54c	0.50b	0.00c	0.00a	0.33
	Means	0.74	0.67	0.60	0.31	0.00	
1000 Grain weight (gm)	Composite (Suwan 1)	160.94c	167.46b	164.57ab	0.00c	0.00a	98.59
	Hybrid 3x-yx	178.46a	174.69a	170.73a	000c	0.00a	104.78
	AMATZBR w	176.74b	164.71c	160.94b	154.91a	0.00a	131.46
	TZBRSYN w	175.96b	171.82ab	162.42ab	151.14b	0.00a	132.27
	AMATZBR y	177.90ab	170.97ab	161.61ab	153.44a	0.00a	132.78
	TZBRSYN y	176.87b	171.08ab	163.42ab	152.18b	0.00a	132.71
	Ozoro Local	156.42b	150.25d	143.34c	0.00c	0.00a	90.00
	Means	171.90	167.28	161.00	87.38	0.00	
Cob girth (cm)	Composite (Suwan 1)	12.00c	11.86b	11.53b	0.00c	0.00a	7.08
	Hybrid 3x-yx	12.68ab	12.26ab	12.22ab	0.00c	0.00a	7.43
	AMATZBR w	12.63AB	12.0AB	12.00B	11.64B	0.00a	9.67
	TZBRSYN w	12.80ab	12.45ab	12.04b	11.66b	0.00a	9.79
	AMATZBR y	13.80a	13.40b	13.21a	13.10a	0.00a	10.70
	TZBRSYN y	12.39b	12.09b	11.92b	11.60b	0.00a	9.60
	Ozoro Local	13.97a	13.63a	13.30a	0.00c	0.00a	8.18
	Means	12.90	12.54	12.34	6.86	0.00	
Shelling percentage	Composite (Suwan 1)	69.64d	69.81c	69.94c	0.00c	0.00a	41.88
	Hybrid 3x-yx	72.45b	72.94b	72.84b	0.00c	0.00a	43.65
	AMATZBR w	71.94c	72.32b	72.65b	73.72b	0.00a	58.13
	TZBRSYN w	72.44b	72.94b	72.98b	73.42b	0.00a	58.36
	AMATZBR y	72.48b	73.84b	74.88b	75.67ab	0.00a	59.37
	TZBRSYN y	74.40ab	75.61ab	76.72a	77.48a	0.00a	60.84
	Ozoro Local	76.39a	76.42a	77.68a	0.00c	0.00a	46.10
	Means	72.82	73.41	73.94	42.90	0.00	

Means with the same letter (s) for each parameter are not significantly different at (P ≥ 0.05) by Duncan's multiple range test (DMRT)

mines overall productivity and the extent of development of each yield character and is dependent on the inter-relationship between the various yield components. The study was conducted to evaluate the effect of various levels of crude oil applied at different stages of growth on maize grain yield and yield components with a view to making appropriate recommendation to maize growers in the oil producing areas of the Niger Delta including Delta State.

MATERIALS AND METHODS

Field experiment using randomized complete block design was carried out during the 2003 and 2004 cropping seasons using seven maize varieties: Composite (suwan 1), Hybrid 3x-yx, AMATZER w, TZBRSYN w, AMATZBR y, TZBRSYN y and Ozoro local and five crude oil levels 0.0 (control), 5.2, 10.4 20.8 and 40.6 ml/ plant at Ozoro, Delta State. NPK (20: 10: 10) fertilizer was incorporated into the soil during the land preparation stage while the crude oil was applied at 3, 5 and 7 WAP on separate sets of maize plants (ring application). The experiment was laid out in three factorial in a randomized complete block design with five treatments replicated four times. Plants were harvested at 112 days after planting and assessed for grain yield, 1000-grain weight, cob grain and shelling percentage. Means were pooled over the years. Data collected were subjected to analysis of variance while the significant means were separated with the Duncan's multiple range tests (DMRT) using SAS (2005).

RESULTS AND DISCUSSION

The results showed that crude oil treatment significantly reduced (P ≤ 0.05) grain yield, grain weight and cob girth. Shelling percentage was however, significantly increased (P ≤ 0.05) as a result of crude oil application to soil (Tables 1 and 2). All the maize seedlings that had crude

Table 2. Grain yield (t/ha) and yields components of seven maize varieties as affected by various levels of crude oil application to soil at 7 WAP.

Yield components	Maize variety	Crude oil level (ml/plant)					
		0.0	5.2	10.4	20.8	41.6	Mean
Grain yield (kg/plant)	Composite (Suwan 1)	0.71d	0.62b	0.60b	0.54ab	0.32b	0.56
	Hybrid 3x-yx	1.42a	1.12a	0.94a	0.90a	0.66a	1.01
	AMATZBR w	0.95b	0.90a	0.84ab	0.82a	0.51ab	0.80
	TZBRSYN w	0.94b	0.84ab	0.64c	0.54ab	0.49ab	0.69
	AMATZBR y	0.97b	0.86ab	0.72b	0.75ab	0.63a	0.77
	TZBRSYN y	0.89c	0.82ab	0.70b	0.62ab	0.68ab	0.74
	Ozoro Local	o.69d	0.60b	0.56c	0.50ab	0.30b	0.53
	Means	0.94	0.82	0.71	0.65	0.51	
1000 Grain weight (gm)	Composite (Suwan 1)	168.56b	166.0b	162.41b	159.32b	155.81b	162.42
	Hybrid 3x-yx	179.40a	177.85a	174.61a	174.32a	172.08a	175.65
	AMATZBR w	177.64a	174.32a	171.46a	164.24a	162.13a	169.96
	TZBRSYN w	176.81a	175.16a	170.82a	165.82a	163.40a	170.36
	AMATZBR y	177.92a	174.81a	170.86a	168.01a	164.32a	171.18
	TZBRSYN y	177.41a	175.62a	172.00a	170.21a	164.62a	171.97
	Ozoro Local	156.43c	154.62c	151.52c	148.41c	145.40c	151.28
	Means	173.45	171.20	167.69	164.30	161.11	
Cob girth (cm)	Composite (Suwan 1)	12.09b	12.02b	11.68c	11.47b	11.20a	11.69
	Hybrid 3x-yx	12.89ab	12.41ab	12.22b	12.01b	11.64b	12.23
	AMATZBR w	12.69ab	12.46ab	12.21b	12.01b	11.44c	12.16
	TZBRSYN w	12.82ab	12.70ab	12.42b	12.13b	12.02b	12.42
	AMATZBR y	13.71a	13.62a	13.36a	13.14a	12.64a	13.29
	TZBRSYN y	12.46ab	12.31b	12.01b	11.94b	11.54b	10.05
	Ozoro Local	13.96a	13.74a	13.52a	13.26a	12.82a	13.46
	Means	12.95	12.75	12.49	12.28	11.90	
Shelling percentage	Composite (Suwan 1)	69.66d	69.84d	70.04d	70.42d	71.64c	70.32
	Hybrid 3x-yx	72.50b	72.96c	73.50c	73.86c	74.38b	73.44
	AMATZBR w	71.61c	72.41c	74.00b	75.08b	75.82ab	73.78
	TZBRSYN w	72.41b	73.50b	74.61b	75.32b	75.45ab	73.93
	AMATZBR y	73.70ab	73.95b	74.08b	74.80b	74.86b	74.28
	TZBRSYN y	74.46ab	75.60ab	75.89ab	76.41ab	76.64ab	75.80
	Ozoro Local	78.41a	76.81a	76.96a	77.81a	78.04a	77.61
	Means	73.25	73.58	74.15	74.81	75.26	

Mean with the same letter (s) for each parameter are not significantly different at (P ≥ 0.05) by Duncan's multiple test (DMRT).

oil treatment at 3 WAP died within 24 h when compared with the control seedlings which grew and developed normally. At 5 WAP, all the maize seedlings from the seven varieties subjected to 41.6 ml crude oil treatment died within 48 h of oil application while only composite (suwan 1), Hybrid 3x-yx and Ozoro local died on exposure to 20.8 ml of the crude oil. The open pollinated varieties (AMATZER w, TZBRSYN w, AMATZBR y, TZBRSYN y) survived at 20.8 ml of the oil treatments. No death was recorded in the four varieties studied (Tables 1 and 2).

The observed significant reduction in the grain yield, grain weight and cob girth with increasing crude oil level could be attributed to the toxic effect of oil. This finding is consistent with the reports of De Jong (1980) and Agbogidi et al. (2006a, 2006c) that crude oil application to soil has a significant effect of reducing the yield characters of plants. The increased shelling percentage of the seven maize varieties as observed in this study may be attributed to the reduction in the other yield parameters as these could have enhanced shelling percentage.

Significant differences ($P \leq 0.05$) in the responses of maize varieties to crude oil level were also recorded at 5 and 7 WAP with the Hybrid 3x-yx generally recording highest grain yield and grain weight whereas Ozoro local recorded the highest cob girth and shelling percentage.

Composite (suwan 1) produced the lowest level of yield responses in terms of cob girth and shelling percentage while Ozoro local had the lowest grain yield and grain weight among the varieties evaluated.

These differences in the yield characters of maize varieties to crude oil levels support the reports of Baker (1970) and Naegele (1974) that the effect of crude oil on plants is dependent on the variety amongst other factors. As reported by Naegele (1974) and Anoliefo (1998), differences in plants reaction to pollution are due to an innate genetic response of the plant system as modified by environmental influences. Based on the results obtained from this investigation, Hybrid 3x-yx appeared to be susceptible to soils affected with crude oil as death was eminent at higher oil doses. The open pollinated varieties (AMATZER w, TZBRSYN w, AMATZBR y, TZBRSYN y) are better in terms of relative tolerance hence should be recommended to farmers in Delta State and maize growers in the oil producing areas of Nigeria. There is therefore the need to test the open pollinated varieties on farmers' field to determine their adaptability to oil pollution. Furthermore, the need for further studies to determine the level of pollution at which maize growth and yield are adversely affected cannot be overemphasized.

The study established varietal differences with respect to maize response to crude oil level and this provides a basis for future breeding work by plant breeders.

REFERENCES

Agbogidi OM, Akparobi SO, Eruotor PG (2006b). Yields of maize (*Zea mays* L) as affected by crude oil contaminated soil. Am. J. Plant Physiol., 1(2): 193-198.

Agbogidi OM, Eruotor PG, Akparobi SO (2005a). Effects of locations and crude oil levels on the growth of seven maize varieties (*Zea mays* L.). In: Uguru MI, Iroegbu CU , Ejere VC (eds.). Proceedings of the 30th Annual Conference of the Genetics Society of Nigeria, held at the University of Nigeria, Nsukka, Enugu State, between 5th and 8th of September. pp. 95-101.

Agbogidi OM, Eruotor PG, Akparobi SO (2006a). Effects of soil contaminated with crude oil on the maturity maize (*Zea mays* L). Afr. J. Environ. Health, 5(2): 63-68.

Agbogidi OM, Eruotor PG, Akparobi SO, Nnaji GU (2007). Evaluation of crude oil contaminated soil on the mineral nutrient elements of maize (*Zea mays*). J. Agron., 6(1): 188-193.

Agbogidi OM, Nweke FU (2005). Impact of gas flaring on the growth and yield of okra (*Abelmoschus sculentus,* L) Moench in Delta State. In: Orheruata AM, Nwokoro SO, Ajayi MT, Adekunle AT, Asumugha GN (eds.). Proceedings of the 39th Annual Conference of the Agricultural Society of Nigeria held at the University of Benin, Benin-City, 9th - 13th October. pp. 231-232.

Agbogidi OM, Nweke FU, Okechukwu EM (2006c). Yield performances of five cultivars of soya bean (*Glycline max* L) as influenced by soil contaminated with crude oil. Nig. J. Trop. Agric., 8: 303-309.

Agbogidi OM, Okonta BC, Dolor DE (2005b). Socio-economic and environmental impact of crude oil exploration and production on agricultural production: a case study of Edjeba and Kokori communities in Delta State of Nigeria. Global J. Environ. Sci., 4(2): 171-176.

Anoliefo GO (1991). Forcados blend crude oil effects on respiratory mechanism, mineral element composition and growth of *Citrullus vulgaris* Schead. Unpublished Ph.D Thesis UNIBEN, Benin-City. p. 293.

Anoliefo GO (1998). Oil spill: effects of spent lubricant on plant life. In "infotect today", October.

Anoliefo GO, Vwioko DE (1994). Effects of spent lubricating oil on the growth of *Capsicum annum* (L) and *Lycopersicon esculentum* (Miller) Environ. Pollut., 88:361-364.

Baker JM (1970). The effects of oil on plants. College of Agriculture Meteorological Station, Ozoro, Delta State. 2003. Environ. Pollut., 1: 27-44.

De Jong E (1980). The effects of a crude oil spill on cereals. Environ. Pollut., 22: 187-196.

FAO (2002). World Agriculture: towards 2015/2030. Summary report, Rome.

Naegele JA (1974). Effect of pollution on plants. In: Sax, I. N. (ed.). Industrial pollution.

Obi IU (1991). Maize: its agronomy, diseases, pests and food values. Optimal Computer Solutions Ltd, Enugu. p. 154.

Ronanet G (1992). Maize Macmillan Edu. Ltd., London. p. 238.

Statistical Software (SAS) (2005). Hargen and Enhanced. SAS, Inst. Inc. USA.

Watson SA (1977). Industrial utilization of corn. In: Sprague, G.F. (ed.). Corn and corn improvement. American Society of Agronomy Inc. Publishers, U.S.A. pp. 721-763.

Breeding for improved organoleptic and nutritionally acceptable green maize varieties by crossing sweet corn (*Zea mays saccharata*): Changes in quantitative and qualitative characteristics in F_1 hybrids and F_2 populations

G. Olaoye*, O. B. Bello, A. K. Ajani and T. K. Ademuwagun

Department of Agronomy, University of Ilorin, P.M.B. 1515, Ilorin, Nigeria.

Reports of improvement in sweet corn (*Zea mays saccharata*) for grain yield and agronomic traits through introgression of genes from the field corn (*Zea mays* L.) are common in literature. However, few studies have reported improvement in field corn varieties for organoleptic and/or nutritional properties through the transfer of recessive alleles that condition sweetness in sweet corn into the field corn. Field and laboratory studies were conducted to determine changes in quantitative and qualitative characteristics of three F_1 hybrids derived by crossing sweet corn (pollen parent) unto field corn and their F_2 populations derived by sibmating approximately 200 ears from each of the F_1 hybrids. Our results showed significant changes in grain yield, grain protein concentration, carbohydrate and calcium contents in the F_1 hybrids and F_2 populations. One hybrid (Pop31DMR x Sweet corn) had 21% grain yield advantage over the maternal parent while another hybrid (Obatanpa x Sweet corn) showed superiority for grain protein (27.3%) with corresponding lower carbohydrate content. Grains obtained from the F_2 populations had lower seed protein content than their respective maternal parents or the F_1 hybrids and with corresponding higher carbohydrate content. Percent (%) loss in grain protein in the F_2 seeds was higher than for other nutrient elements. Modest heterotic response was recorded for grain yield in Pop31DMR x Sweet corn cross and for grain protein concentration in Obatanpa x Sweet corn cross.

Key words: Grain yield, nutritional properties, inbreeding depression, heterotic response.

INTRODUCTION

Maize (*Zea mays* L.) is the third most important cereal crop mainly used as staple food and animal feed in Nigeria (Fakorede et al., 1989; Lovenstein et al., 1995). It is cultivated both as rain-fed and under irrigation on more than five million hectares across agro-ecological zones of Nigeria, with savanna ecology accounting for more than half of total annual production (CRI, 1995). Although maize is mainly considered a carbohydrate source, it is also an important source of protein because of its considerable total protein yield per hectare (Bjarnson and Vasal, 1992). However, from nutritional perspective, protein of maize and that of other cereals is deficient in essential amino acids, particularly lysine and tryptophan that are essential for monogastric animals and humans (Bhatta and Rabson, 1987; Alan et al., 2007). Recent breeding efforts in the west and central Africa (WCA) sub region have focused on improvement in maize-grain quality characteristics of organoleptic and nutritional properties. The justification is premised on the belief that availability of such varieties is likely to increase utilization of maize grains for various forms of dishes, improve the nutritional status of resource-limited rural communities and increase the land area cultivated to maize in the sub

---

*Corresponding author. E-mail: debolaoye@yahoo.com.

region. Genetic improvement of sweet corn (*Zea mays saccharata*) for grain yield has been limited primarily because of its narrow genetic base, the lack of defined heterotic group and the greater effort devoted to improving yield in field corn (Tracy, 1990). Consequently, the focus in field corn x sweet corn crosses is the exploitation of hybrid vigour in the improvement of sweet corn varieties for grain yield, adaptation and genetic diversity (Tracy, 1990; 1994; Cartea et al., 1996). Transfer of mutant genes (su_1, sh_2, bt_2, wx and o^2) from sweet corn into field corn on the other hand, is for the improvement in nutritional contents (protein, lysine, soluble sugar, sucrose, reducing sugar, albumin, globulin and glutelin) of the grains, which has also been accompanied by a corresponding reduction in starch and zein either in the varieties *per se* (Tsyganash et al., 1988; Li and Liu, 1994) or in the resultant F_1 hybrids (Shao and Shao, 1994). Although characters associated with organoleptic and quality characters in maize is well documented (Anonymous, 1997; Ogunbodede, 1999; Olaoye et al., 2001; Akande and Lamidi, 2006; Gupta et al., 2009), there is limited information on the effects of transferring the mutant genes from sweet corn to field corn especially on endosperm attributes of the resulting populations. In an earlier study, Olaoye et al. (2008) reported improvement in flavour and grain texture characteristics of F_1 hybrids derived through the transfer of mutant su^2 genes from sweet corn into three field corn varieties without adverse consequence on grain yield potential of the resulting hybrids. One of the hybrids (Pop31DMR x Sweet corn) had a modest yield advantage of 0.4t/ha^{-1} over the maternal parent while another hybrid (Obatanpa x Sweet corn) also had a 20% increase in grain protein concentration compared to the maternal parent. Analyses of the grains of the F_1 hybrids, their maternal parents and F_2 populations (Olaoye and Ajao, 2008), however showed differences in the suitability of the genotypes for different food values. For example, Obatanpa x Sweet corn cross and its F_2 population had higher amylase contents while one parent (Pop31DMR) had significantly higher starch content indicating its suitability for dry milling processes. Consistent with previous reports on maternal contribution to grain protein and essential amino acid contents in field x sweet corn crosses (Shao and Shao, 1994; Wang et al., 1994), these observations underscore the importance of the genetic background of maternal parents in transfer of recessive mutant traits conditioning grain quality attributes into the field corn. In other words, information on the effects of transferring mutant su^2 genes from sweet corn to field corn on grain yield and quality characteristics will enhance maize breeding efforts in the development of maize varieties for utilization either in the green or processed form. This part of our study reports on changes in grain yield and endosperm quality attributes of F_1 hybrids and F_2 populations resulting from transfer of mutant su^2 genes from the sweet corn into three field corn varieties. The second objective was to dertermine heterotic response in F_1 hybrids for grain yield and nutrient composition.

MATERIALS AND METHODS

Details of the attributes of parents involved in the crosses and the field experimentation have been described elsewhere (Olaoye et al., 2008; Olaoye and Ajao, 2008). Suwan-1-SR and Pop31DMR are high in carotene content and also resistant to the Downy mildew disease. Obatanpa which is a high quality protein maize (QPM) variety developed in Ghana, has the opaque-2-gene which confers softer endosperm and contains almost double the levels of lysine and tryptophan in the normal maize (Pixley and Bjarnson, 1993; Andres et al., 2008). The sweet corn parent was the source of mutant genes (su^2) conditioning sucrose content. Briefly, F_1 hybrids were derived by crossing the three field corn parents unto the standard sweet corn as the pollen parent in the breeding nursery of the Institute of Agricultural Research and Training (IAR&T), Ibadan during the 1996 irrigation season. However, the F_2 populations were derived by sibmating approximately 200 ears from each of the F_1 hybrids at the Teaching and Research (T & R) farm, University of Ilorin during the 1998 growing season. At harvest, the best 150 ears based on cob appearance were selected. Equal number of seeds were thereafter obtained from the selected 150 cobs and mixed together to form a balanced bulk for each of the F_2 populations.

Field study

The 10 entries were planted at two sites. These are: the Faculty of Agriculture (FAG) and the T&R farms of the University of Ilorin in a typical Southern Guinea Savanna ecology (Lat. 8°29'N and 8°30'N; Long. 4°30'E and 4°32'E) during the 1999 cropping season. FAG farm is on a higher elevation while the T&R farm is at lower elevation. FAG farm is also high in %N, organic matter (%) and mineral contents of calcium and potassium while the T&R farm is higher in clay content as well as magnesium and available phosphorus. The experiment was laid out as a randomized complete block design (RCBD) on each site and each plot consisted of four rows 5M long with inter and intra-row spacing of 0.75 and 0.5 M respectively. Plantings were carried out on 28th and 30th of July, 1999 respectively. The plots were over planted but later thinned to two stands per hill to give a plant population of approximately 53,333 plants/ha. Weed was controlled at the FAG farm by hand while a pre-emergence herbicide application supplemented by one hoe weeding was carried out at the T&R farm. Fertilizer application at both sites was in split-dosage at three and seven weeks after planting at the rate of 80 kg N, 60 kg P and 60 kg/ha respectively from compound NPK fertilizer (20-10-10). Data were collected on seedling emergence and days to 50% anthesis and silking, plant and ear heights (cm) and grain yield (kg/plot). Data from seedling emergence were used to compute emergence percentage (E %) while grain yield obtained in kg/plot was converted to tones/hectare (t/ha^{-1}) assuming 85% shelling percentage after adjusting to 12% moisture at harvest.

Laboratory studies

At harvest, five (5) random ears were selected in a plot giving a total of 20 ears per variety, followed by careful removal of the grains by hand. From each variety, equal number of grains were selected from each plot, mixed together to form a balanced bulk and then subjected to proximate analyses in the laboratory. The grains obtained were grounded to form a fine powder and each sample was oven dried to a constant weight at 80°C to obtain grain moisture content (%MC). Two replicate determinations were analyzed for each variety and the mean recorded for each sample. Crude protein (CP) content and ash determination were carried out according to the methods described by AOAC (1980) and Pearson (1973) respectively. Crude lipid (EE) determination was by the Soxhlet method using petroleum ether at boiling point 40 - 60°C as solvent (AOAC, 1975). Crude fibre (CF) determination was by digestion of the defatted sample followed by drying of the residue obtained from washing with boiling distilled water in an oven to a constant weight at 80-85°C. Nitrogen free extract (NFE) was determined by difference as;

%NFE = 100 - [%MC - %CP + %Ash + %CF + %EE]

Total carbohydrate was obtained by the summation of CF which represents the insoluble carbohydrate and the NFE which is the soluble carbohydrate (that is %CF + %NFE).

Two mineral elements viz: calcium and phosphorous were also analyzed for each variety using the standard laboratory procedure and the mean recorded for each sample.

Data analyses

Data collected from field experiments were subjected to analyses of variance (ANOVA) first on individual site basis before a combined ANOVA across sites. Data from proximate analyses and mineral content determinations were pooled across sites. Pertinent means were separated by use of Least Significance Difference (Steel and Torrie, 1980). Estimates of inbreeding depression (F_1-F_2) in actual units and percent inbreeding depression [$100(1-F_2/F_1)$] were computed for each F_1-F_2 pair in each replicate as described by Meghji et al. (1984). F_1 and F_2 are means of first and second filial generation crosses respectively. Mid-parent (MP) and high-parent (HP) heterosis were also computed using the formulae

MP = $100[(F_1-MP)/MP]$ and HP = $100[(F_1-HP)/HP]$

Where MP is the average of the two parents for a trait and HP is the value of the higher parent for the same trait respectively. Significance of inbreeding depression was determined by Least Significance differences (LSD) estimated as

$LSD_{(0.05)} = t_{0.05} [(1/nF_2 + 1/nF_1)Mse]^{1/2}$

While that for both mid- and high parent heterosis was

$LSD_{(0.05)} = t_{0.05} [(1/nF_1 + 1/nP)Mse]^{1/2}$

Where $t_{0.05}$ is the tabular value of t at 5% level, n is the number of observations and Mse is the pooled error mean square. In computing these estimates, emphasis was only on characters for which the genotypes differed significantly.

RESULTS

Effects of planting sites on expression of quantitative and qualitative traits

Grain yield differed significantly between the two planting sites with FAG farm having 33% yield advantage over the T&R farm but with a corresponding higher ear placement than at the T&R farm (Table 1). Conversely, seedling emergence was better at the T&R farm while days to anthesis and mid-silk were similar between the two sites. In consonance with the native nutrient status of the planting sites, FAG farm also had higher grain moisture, grain protein and %Ca than the T&R farm (Table 2). However, Ca content was the only parameter which differed significantly between the planting sites with FAG farm having 11.5% higher Ca than the T&R Farm.

Genotypic performance for quantitative and qualitative traits

The genotypes differed for the quantitative characters including grain yield (Table 1). Seedling emergence was lower in the segregating F_2 populations compared to the parents and hybrids respectively. Days to mid-silk were delayed in the F_2 populations derived from Suwan-1-SR x Sweet corn and Obatanpa x Sweet corn crosses relative to their maternal parents. The sweet corn parent was the latest to attain mid-silk while Pop 31DMR was the earliest by approximately 4 days. Two of the hybrids (Suwan-1-SR x Sweet corn and Obatanpa x Sweet corn) attained mid-silk earlier than either parent and also had lower ear placement. However, Pop 31DMR x Sweet corn cross was similar to the maternal parent for this character. Grain yield in the F_2 populations was reduced by 17.3 and 22% respectively compared to those of the parents and hybrids. The range was from 3.94 - 25.7% of the parent value and 16.99 - 26.97% of the F_1 hybrids. Each of the hybrids except Obatanpa x Sweet corn, yielded higher than their respective maternal parents.

The genotypes also differed significantly for grain moisture, grain protein and carbohydrate contents as well as %Ca but similar for other characteristics (Table 2). Except for Suwan-1-SR, grain moisture in the parents was generally lower than either those of the hybrids or F_2 populations.

The sweet corn parent with significantly higher grain protein content was also superior for %Ca, indicating superiority in P absorption from the soil and assimilating it into the grains. Conversely, the hybrids had higher %P in the grains than their parents. Obatanpa x Sweet corn was the only hybrid with significantly higher grain protein concentration than the maternal parent.

The parents also had lower CF content than either hybrids or F_2 populations but differences were non significant. F_2 populations had lower grain protein concentration compared to their parents or their respective hybrids but with a corresponding higher carbohydrate content and NFE. %P remained relatively unchanged in all genotypes but other parameters especially %Ca showed no consistent trend between the different sets of genotypes. Ca content in the grains of F_2 population derived from Obatanpa x Sweet corn cross was higher than either that of the hybrid or maternal parent but the reverse was the case for the population derived from Suwan-1-SR x Sweet corn cross.

Table 1. Location and entry means (across locations) for grain yield and agronomic traits in parents and F_1 hybrids of Sweet corn x Field corn (Ilorin, Nigeria).

	Emergence percentage	Days to Anthesis	Days to Mid-silk	Ear height (cm)	Grain yield (tha-1)
Location					
Faculty of Agriculture Farm	73.01	60	62	57.45	1.87
Teaching and Research Farm	79.63	60	62	38.76	1.26
LSD α 0.05	2.02**	ns	ns	4.40**	0.20**
SED	1.30	0.35	0.36	1.83	0.09
Parents					
Suwan-1-SR	73.5	63	65	57.6	1.75
Pop31DMR	86.9	57	59	48.4	1.52
Obatanpa	71.4	60	62	45.9	1.57
Sweet corn	77.0	63	66	50.4	1.65
F_1 Hybrids					
Suwan-1-SR x Sweet corn	79.6	60	62	45.0	1.78
Pop31DMR x Sweet corn	84.2	58	60	48.5	1.84
Obatanpa x Sweet corn	83.8	59	60	42.0	1.53
F_2 Populations					
Suwan-1-SR x Sweet corn	72.9	61	63	43.8	1.30
Pop31DMR x Sweet corn	73.6	58	60	52.9	1.46
Obatanpa x Sweet corn	60.4	60	63	46.6	1.27
Grand Mean	76.32	60	62	48.11	1.56
LSD α 0.05	21.9**	5.78**	1.94**	13.11**	0.35*
SED	2.91	0.78	0.81	4.09	0.20
CV (%)	7.68	2.90	2.59	16.56	25.66

*, **; Significant F test at 0.05 and 0.01 levels of probability respectively.
SED = Standard error of mean.

Estimates of inbreeding and heterosis for quantitative and qualitative traits

Mean squares from ANOVA for inbreeding depression (actual units and %) for quantitative characters indicated significant differences among F_2 populations only for seedling emergence (Table 3). Values for these estimates in each population were similar for other characters except grain yield where % inbreeding depress-sion was larger in magnitude than in actual units. Inbreeding in Pop31DMR x Sweet corn cross did not adversely affect any of the traits except ear placement. Conversely, F_2 population from Obatanpa x Sweet corn showed decreased vigour for all characters except for grain yield. Negative estimates were obtained for days to flowering and ear placement in the F_2 populations.

Estimates of MP and HP heterosis among hybrids for quantitative characters studied were significant for days to anthesis (MP) and grain yield (MP and HP) respectively (Table 3). However, MP values were larger in magnitude than for HP. Consistent with estimates of inbreeding, negative estimates were recorded for days to flowering and ear placement indicating that the hybrids matured earlier and had lower ear placement than their field corn for parents. However, of the three hybrids, Pop31DMR x Sweet corn demonstrated superiority for grain yield over the better parent while Suwan-1-SR x Sweet corn was superior only to the mid-parent for grain yield.

Estimates of inbreeding depression (%) for qualitative characters among different F_2 populations were significant for grain moisture and grain protein content but not for carbohydrate content (Table 4). Inbreeding depression for these traits was highest in F_2 population derived from Obatanpa x Sweet corn and ranged from two to three-fold loss compared to other populations. Negative estimates were obtained for F_2 population derived from Suwan-1-SR x Sweet corn cross for grain moisture at harvest. Estimates of % loss in grain protein by growing F_2 seeds when compared to either parent ranged from 6.1% in Obatanpa x Sweet corn to 15.20% in Suwan-1-SR x Sweet corn crosses while in the hybrids, it ranged from 14.6% in Pop31DMR x Sweet corn to 27.14% in Obatanpa x Sweet corn crosses.

MP heterosis for qualitative characters was significant only for grain protein (Table 4).

Heterotic response for grain moisture at harvest was positive in two hybrids and high in Obatanpa x Sweet corn cross. In consonance with the inbreeding estimates, hybrid from Suwan-1-SR x Sweet corn cross showed heterotic advantage for earliness over either parent as indicated by estimates for both MP and HP heterosis for grain moisture while values for the other two hybrids indicated late maturity.

Table 2. Location and entry means (across locations) for nutrient composition and mineral elements in parents and F_1 hybrids of Sweet corn x Field corn (Ilorin, Nigeria).

Location	Grain moisture	Ash	Carbohydrate	Crude protein	Crude fat	Crude fibre	Nitrogen free extract	% P	% Ca
FAG Farm	10.10	1.50	72.83	8.85	6.35	3.58	69.26	0.306	0.243
T and R farm	9.23	1.73	73.54	8.82	6.67	3.74	69.78	0.3277	0.215
LSD α 0.05	ns	ns	Ns	ns	ns	ns	ns	ns	0.028*
SED	0.417	0.289	0.935	0.0517	0.563	0.478	1.003	0.0118	0.053
Parents				Genotypes					
Suwan-1-SR	9.30	1.20	74.05	8.55	6.90	4.05	70.00	0.225	0.380
Pop 31 DMR	8.40	1.95	75.20	8.20	6.25	3.50	71.70	0.225	0.360
Obatanpa	9.15	2.00	74.25	8.25	6.35	3.90	70.35	0.200	0.140
Sweet corn	9.40	1.80	69.10	14.20	5.50	3.74	65.90	0.200	0.720
F_1 Hybrids									
Suwan-1-SR x Sweet corn	8.40	1.45	75.40	8.45	6.30	4.15	71.25	0.250	0.340
Pop 31 DMR x Sweet corn	9.95	2.15	72.35	8.20	7.35	3.15	69.20	0.225	0.200
Obatanpa x Sweet corn	13.90	1.00	68.10	10.50	6.50	3.75	64.40	0.290	0.195
F_2 populations									
Suwan-1-SR x Sweet corn	9.85	1.10	74.10	7.25	6.70	2.80	71.30	0.200	0.090
Pop 31 DMR x Sweet corn	9.00	2.00	74.75	7.00	6.25	4.00	70.75	0.250	0.235
Obatanpa x Sweet corn	9.30	1.50	74.55	7.75	7.00	4.10	70.35	0.225	0.255
Grand Mean	9.66	1.61	73.18	8.83	6.51	3.66	69.52	0.229	0.291
LSD α 0.05	2.11**	ns	4.74*	2.62**	ns	ns	ns	ns	0.27*
SED	0.933	0.647	2.090	1.156	1.263	1.068	2.42	0.026	0.119
CV (%)	9.65	20.20	2.86	13.09	19.36	29.19	3.23	11.54	21.04

*, **; Significant F test at 0.05 and 0.01 levels of probability respectively.
SED = Standard error of mean.

DISCUSSSION

Grain yield and grain protein content in the genotypes were significantly higher at the FAG farm which has higher native N and OM content than at the T&R farm. This is consistent with previous reports which showed that high organic status of the soil (Stanchev and Mitra, 1988; Kling and Okoruwa, 1994; Kamalakumari and Singaram, 1996; Singaram and Kamalakumari, 1999) as well as increase in dosage of inorganic N-fertilizer application (Oikeh et al., 1998; Kramarev et al., 2000) resulted in improved quality of maize grains by enhancing sugar, starch and crude protein contents. Pixley and Bjarnson (1993) also noted that such situation arises when N is limiting in the crop as often the case under low-input farming system. However, our results also showed that grain protein concentration decreased as crude fibre and NFE increased. This was probably due to conversion of protein into insoluble carbohydrate which also corroborates previous reports (Olaoye et al., 2001) which showed decline in grain protein for every day delay in harvesting maize for utilization in the green form. Except for %P, the native nutrient status of planting sites did not significantly alter the expression of other qualitative characters in the test varieties. Although the site with high Ca content produced maize grains with higher Ca content, %P in the genotypes was not consistent with nutrient status of planting site because maize grains from FAG with higher available P were low in P content than those from

Table 3. Inbreeding depression and heterosis for grain yield and related traits in three populations derived from field corn x sweet corn crosses (Ilorin, Nigeria).

Population	Inbreeding depression		(%) Heterosis	
	Actual units	Percent %)	Mid-Parent	High Parent
Emergence (%)				
Suwan-1-SR x Sweet corn	6.68	7.99	6.65	-2.06
Pop 31 DMR x Sweet corn	10.57	11.92	0.38	-2.90
Obatanpa x Sweet corn	23.41	27.86	13.20	7.08
LSD α 005	12.82*	14.28*	15.27	12.94
SED	7.15	8.59	5.23	4.52
Days to Anthesis				
Suwan-1-SR x Sweet corn	-1.40	-2.35	-4.77	-2.03
Pop 31 DMR x Sweet corn	0.20	0.39	-3.40	2.48
Obatanpa x Sweet corn	- 2.0	-1.29	-7.39	-2.87
LSD α 005	2.93	5.15	4.49*	5.72
SED	0.93	1.13	1.66	1.79
Days to Silking				
Suwan-1-SR x Sweet corn	-2.38	-2.27	-4.96	-10.79
Pop 31 DMR x Sweet corn	0.13	0.16	-3.99	- 5.53
Obatanpa x Sweet corn	-2.38	-3.61	-5.50	-11.62
LSD α 005	2.36	6.43	4.49	6.12
SED	1.18	1.56	0.62	3.97
Ear height				
Suwan-1-SR x Sweet corn	1.26	2.84	-2.40	-16.24
Pop 31 DMR x Sweet corn	-4.38	-9.97	-4.28	- 5.14
Obatanpa x Sweet corn	-4.60	-16.90	-8.15	-22.12
LSD α 005	18.75	38.8	3.67	16.74
SED	2.71	8.18	2.39	7.07
Grain yield				
Suwan-1-SR x Sweet corn	0.54	27.73	8.60	-8.48
Pop 31 DMR x Sweet corn	0.38	20.64	21.13	3.92
Obatanpa x Sweet corn	0.26	15.85	-6.96	-21.82
LSD α 005	0.33	18.07	16.03*	4.91*
SED	0.11	4.88	11.49	10.51

*, Significant F test at 0.05 level of probability.
SED = Standard error of mean.

Table 4. Inbreeding depression and heterosis for grain moisture, grain protein and carbohydrate contents in three populations derived from field corn x sweet corn crosses (Ilorin, Nigeria).

Population	Inbreeding depression		(%) Heterosis	
	Actual units	Percent %)	Mid-Parent	High Parent
Grain moisture				
Suwan-1-SR x Sweet corn	-1.45	-18.25	-9.86	-12.74
Pop 31 DMR x Sweet corn	0.95	8.96	11.64	5.85
Obatanpa x Sweet corn	4.6	32.03	49.39	45.28
LSD α 005	3.44*	30.61*	77.31	80.28
SED	2.67	41.36	43.92	44.41
Grain protein				
Suwan-1-SR x Sweet corn	1.20	12.73	-25.86	-40.49

Table 4. Contd.

Pop 31 DMR x Sweet corn	1.20	-15.07	-26.80	-42.26
Obatanpa x Sweet corn	3.10	28.22	- 3.08	-23.59
LSD α 005	5.87	16.63*	17.43*	36.13
SED	0.90	12.22	10.97	8.42
Carbohydrate content				
Suwan-1-SR x Sweet corn	2.60	3.29	5.36	1.82
Pop 31 DMR x Sweet corn	-4.80	-10.38	-6.82	-3.75
Obatanpa x Sweet corn	-12.90	-19.38	-4.99	-8.12
LSD α 005	80.26	19.48	16.59	24.85
SED	45.81	19.69	6.39	8.34

*, Significant F test at 0.05 level of probability.
SED = Standard error of mean

T&R farm.

Estimates of inbreeding depression in the F_2 populations did not follow any consistent pattern either for quantitative or qualitative characters. However, low seedling emergence and grain yield recorded for F_2 populations indicated low vigour due to inbreeding depression which is consistent with previous reports of between 30.1 and 35.5% yield reduction in the F_2 generation compared to the hybrids (Martin and Hallauer, 1976; Lopez-Perez, 1977). Similarly, the two to three-fold loss observed for qualitative attributes in F_2 population derived from Obatanpa x Sweet corn cross when compared with other populations, is probably due to rapid transformation of sugar to starch in F_2 generation especially since both grain protein and sucrose contents are conditioned by recessive alleles and are likely to be expressed in this generation. This population also showed low vigour for all characters except for grain yield probably because neither parent carries alleles for grain yield in the dominant form.

Earliness in days to flowering and reduced plant height, consequently low ear placement, are two attributes which are desired by maize breeders. Negative estimates of inbreeding depression and heterosis recorded for these traits in F_2 populations and hybrids, indicate that the genotypes matured earlier and had lower ear placement than their field corn parents. This also indicates that these traits are inherited through the maternal parent. The negative inbreeding depression obtained for days to flowering in this study, however contradicts earlier reports in literature (Hallauer and Miranda Fhilo, 1988) of positive inbreeding depression estimates for this trait. This may be due to the fact that F_2 populations in the present study were derived through sibmating rather than by selfing.

Estimates of MP heterosis for quantitative attributes which were larger in magnitude than HP suggest a modest increase in heterotic response for such characters in the hybrids. However, Pop 31DMR x Sweet corn cross was the only hybrid which showed heterotic response for grain yield over either the MP or the HP which suggests superiority of the hybrid over the parents for this character. These low estimates are not unexpected since the trait actually being transferred is conditioned by recessive genes.

Differences in genetic background of the field corn parents influenced the results obtained for grain quality attributes in both F_1 hybrids and F_2 populations, depending on the parent used. For example, % loss in grain protein was low in Pop 31DMR x Sweet corn but high in Obatanpa x Sweet corn in the hybrids and also low in Obatanpa x Sweet corn and high Suwan-1-SR x Sweet corn in the F_2 generation. Letchworth and Lambert (1998) noted that endosperm of hybrids from self-pollinated grains were higher in grain protein content than those obtained from open-pollinated crosses while the reverse was the case for oil content and starch concentration. In other words, careful selection of maternal and pollen parent in flint x sweet corn crosses is important when breeding for maize endosperm qualities, since differences in genetic background of flint varieties will likely affect the effectiveness of improving the quality of the resulting hybrids (Shao and Shao, 1994). There was lack of heterosis for protein content in the hybrids obtained from the crosses. Therefore, backcrossing the F_1 hybrids to QPM parents such as Obatanpa followed by selection of progenies with the desired endosperm characteristics in the segregating populations will be necessary in order to enhance protein quality of the green maize varieties. However, Alan et al. (2007) noted that breeding procedure aimed at transferring recessive traits is more complicated because of the requirement for selfing between each backcross generation in order to fix the recessive alleles. The implication is that backcrossing the F_1 hybrids to QPM parents will require alternating with selfing in order to achieve the desired protein contents in future green maize varieties.

REFERENCES

Akande SR, Lamidi GO (2006). Performance of quality protein maize varieties and disease reaction in the derived-savanna agro-ecology of South-West Nigeria. Afr. J. Biotechnol. 5(19): 1744-1748.

Alan FK, Hugo DG, Nilupa SG, Alpha OD, Dennis F (2007). Breeding and disseminating quality protein maize (QPM) for Africa. Afr. J. Biotechnol. 6(4): 312-324.

Andres G, Paul SEM, Otto RL, Monica MD, Javier BD (2008). Phenotypic characterization of Quality Protein Maize Endosperm Modification and Amino Acid Contents in a Segregating Recombinant Inbred Population. Crop Sci. 48: 1714-1722.

Anonymous (1997). IITA Archival Report 1993-1995, Maize improvement Programme. IITA, Ibadan, Nigeria. p. 83.

AOAC (1975). Official methods of analysis of the Association of Official Analytical Chemists. 12th edition (W. Horowized) Washington D.C. Pp. 15-75.

AOAC (1980). Official methods of analysis of the Association of Official Analytical Chemists. 13th edition (W. Horowized) Washington D.C. Pp. 1018.

Bjarnson M, vassal SK (1992). Breeding of quality protein maize (QPM). In Janick (ed.). Plant Breeding Reviews, 9 John Wiley and Sons Inc. New York. pp. 181-216.

Bhatta CR, Rabson R (1987). Relationship of grain yield and nutritional quality. In Olson RA and Frey KJ (ed.). Nutritional quality of cereal grains. Genetic and agronomic improvement. Agron. Monogr. 28, ASA, CSSA, SSSA Madison, WI. pp 11-43.

Cartea ME, Malvar RA, Revilla P, Ordox A (1996). Identification of field corn populations to improve sweet corn for Atlantic European conditions. Crop Sci. 36(6): 1506-1512.

CRI (1995). Crop Research Institute. Research Briefs Bimonthly Bulletin of the Crop Research Institute, Ghana, NO. 1 Council for Scientific and industrial Research. p. 97.

Fakorede MAB, Iken JE, Kim SK, Mareck JH (1989). Empirical result from a study on maize yield potential indifferent ecologies of Nigeria. In Proceedings of maize towards production technologies in semi-arid west and central Africa. Fajemisin JM, Muleba N, Emechebe AM and Daire C (eds.).SAFGRAD and IITA. pp. 79-97.

Gupta HS, Agrawal PK, Mahjan V, Bisht GS, Kumar A, Verma P, Srivastava A, Saha S, Babu R, Pant MC and Mani VP (2009). Quality Protein maize for nutritional security: rapid development of short duration hybrids through molecular marker assisted breeding. Res. Acct. Curr. Sci. 96(2): 12-19.

Hallauer AR, Miranda Filho JB (1988). Quantitative genetics in Maize Breeding. 2nd Edition. Iowa State University Press, Ames Iowa 50010. p. 382.

Kamalakumari K, Singaram P (1996). Quality parameters of maize as influenced by application of fertilizers and manures. Madras-Agricultural-J. 83(1): 32-33.

Kling JG, Okoruwa AE (1994). Influence of variety and environment on maize grain quality for food uses. Paper presented at the American Society of Agronomy meetings, Seattle, WA. pp. 95-127.

Kramarev SM, Skripnik LN, Khorseva YL, Shevchenko VN, Vasil'-eva VV (2000). Increasing the protein content in maize grain by optimizing the nitrogen nutrition of the plants. Kukuruza-i-Sorgo. 1: 13-16.

Letchworth MB, Lambert RJ (1998). Pollen parent effects on oil, protein, and starch concentration in maize kernels. Crop-Sci. 38 (2): 363-367.

Li XY, Liu JL (1994). The effects of maize endosperm mutant genes and gene interactions on kernel components. IV. Relations among kernel components and their values for quality improvement. Acta-Agronomica-Sinica. 20(4): 439-445.

Lopez-Perez E (1977). Comparisons among maize hybrids made from unselected lines developed by selfing and Full sibbing. M. Sc. Thesis, Iowa State University. In Hallauer and Miranda Fihlo (eds.). Quantitative Genetics in Maize Breeding 2nd Edition. Iowa State University Press. Ames, Iowa 50010. pp. 299-336.

Lovenstein H, Lafinga EA, Rabbinge R, vankeulen HV (1995). Principle of Production Ecology. Wageningen agricultural University. The Nertherlands, p. 201.

Martin JM, Hallauer AR (1976). Relation between heterozygosis and yield for four types of maize inbred lines. Egyptian J. Genet. Cytol. 5: 119-135.

Meghji MR, Dudley JW, Lambert RJ, Sprague GF (1984). Inbreeding, inbred and hybrid grain yields and other traits of maize genotypes representing three eras. Crop Sci. 24: 545-549.

Ogunbodede BA (1999). Green maize production in Nigeria in the new millennium – Prospects and Problems. In Olaoye G and Ladipo DO (eds.) Genetics and food security in Nigeria on the Twenty-first century. Genetics Society of Nigeria special publication. pp. 33-37.

Oikeh SO, Kling JG, Okoruwa AE (1998). Nitrogen fertilizer management effects on maize grain quality in the West African moist Savanna. Crop-Science. 38(4): 1056-1061.

Olaoye G, Adegbesan NO, Onaolapo IO (2001). Interactive effects of Location x sampling time on organoleptic and nutritional qualities of six maize genotypes. Nig. J. Exp. Biol., 1(1): 99-108.

Olaoye G, Ogunbodede BA, Olakojo SA (2008). Breeding for improved organoleptic and nutritionally acceptable green maize varieties by crossing unto sweet corn (*Zea mays saccharata*). I. Comparative analyses of grain yield, organoleptic and nutritional properties of F_1 hybrids and their maternal parents. Nig. J. Genet. 22: 131- 148.

Olaoye G, Ajao FA (2008). Breeding for improved organoleptic and nutritionally acceptable green maize varieties by crossing unto sweet corn (*Zea mays saccharata*).II. Grain quality characteristics in F_1 hybrids and their F_2 segregating populations. Nig. J. Genet. 22: 149-166.

Pearson D (1973). Laboratory techniques of food analysis Butterworth, London. p. 7-20.

Pixley KV, Bjarnson MS (1993). Combining ability for yield and protein quality among modified-endosperm opaques-2-tropical maize inbred. Crop Sci. 33: 1229-1234.

Shao J, Shao JP (1994). Quality deficiency of maize F_1 hybrids from flint x sweet crosses for fresh edible maize. Acta-Agriculturae-Shanghai. 10(3): 32-36.

Singaram P, Kamalakumari K (1999). Effect of continuous manuring and fertilization on maize grain quality and nutrient soil enzyme relationship. Madras Agric. J. 86(1-3): 51-54.

Stanchev L and Mitra M (1987. Effect of fertilizers on grain quality of maize hybrid H-708. Pochvoznanie Agrokhimiya I Rastitelna Zashcita 22(1): 59-63.

Steel RGD, Torrie JH (1980). Principles and Procedures of Statistics. A Biometrical Approach. 2nd Edition. Mc Graw Hill Book Inc, New York. p. 580.

Tracy WF (1990). Potential of field corn germplasm for the improvement of sweet corn. Crop Sci. 30: 1041-1045.

Tracy WF (1994). Sweet corn. In A.R. Hallauer (ed.) Specialty types of maize. CRC Press, Boca Raton, FL. Pp. 147-187.

Tsyganash VI, Palii AF, Tsyganash DA, Rotar AI, Manolii VP (1988). Effect of the o^2 and su^2 mutations and their interaction on grain quality in maize.Selektsionno-geneticheskie-metody-povysheniya-urozhainosti,-kachestva-i-ustoichivosti-polevykh-kul'tur. pp. 18-24.

Wang LM, Shi DQ, Liu RD, Bai L (1994). Studies on combining ability of high-lysine maize. Acta-Agronomica-Sinica. 20(4): 446-452.

Variability in sucrose content at grand growth phase in tissues of *Saccharum officinarum* × *Saccharum spontaneum* inter-specific hybrid progeny

Vandana Vinayak*, A. K. Dhawan and V. K. Gupta

CCS Haryana Agricultural University, Regional Research Station, Uchani, Karnal.

The variability for sucrose content in existing commercial sugarcane germplasm is not large and hence crosses were made between two *Saccharum officinarum* clones, 'Gungera' (cross I) and 'Keong' (cross II) with high sucrose content and a clone of *Saccharum spontaneum* 'SES 603' with low sucrose content. A high sucrose commercial var. CoS 8436 was also crossed with a low sucrose var. Co 1148 (cross III). Among the three crosses, only cross I germinated to give healthy twenty nine inter-specific hybrids. Inter-specific hybrids selected from cross I were graded into very high, high, low, and very low sucrose classes on the basis of morphological characters. Also anthrone method of sucrose estimation from internode, Brix value and quality analysis were done on whole sugarcane stem at different growth stages. Rate of photosynthesis, transpiration and stomatal conductance were analyzed using infrared gas analyzer and was found to be high for high sucrose hybrids. Rate of assimilate translocation of sucrose was also estimated with radio-labelled ^{14}C and was found to be highest in the midribs. On this basis, molecular markers for high and low sucrose content ISH-1, ISH-5, ISH-17 and ISH-23 were identified as very high and ISH-10, ISH-11, ISH-12 and ISH-25 as very low sucrose hybrids.

Key words: Inter-specific hybrids, sucrose content, sugarcane, TVD leaf.

INTRODUCTION

All the present day commercial sugarcane cultivars possess parental genes from a few inter-specific hybrids developed at Coimbatore during the early 20th century (Selvi et al., 2005). These cultivars, thus, have a narrow genetic base. Genus *Saccharum* consists of three cultivated species, that is, *Saccharum officinarum*, *Saccharum barberi* and *Saccharum sinensis*, and two wild species of *Saccharum robustum* and *Saccharum spontaneum*. Sugarcane varieties presently under cultivation have been derived through artificial crosses between two or more of these *Saccharum* species as the source of genetic materials. The maturation of sugarcane is characterized by the accumulation of sucrose in developing internodes (Glasziou and Gayler, 1972). It is evident that a cycle of sucrose synthesis and degradation exists in all of the internodes (Whittaker and Botha, 1997). Furthermore, carbon assimilation rate during maturation is influenced by factors like light intensity, plant age, and soil and leaf water contents (Grantz, 1989). Extensive work has been carried out in improving the sucrose accumulation in sugarcane at the CCS HAU Regional Research Station, Karnal. It was observed that sucrose accumula-tion was recorded on the basis of Brix and sucrose content (Dendsay et al., 1992) in sugarcane var. CoJ 64 was three fold more in lower internodes and highest during the peak growth period as compared to sugarcane var. Co1148 made them suitable for selection as parents for crossing so as to obtain hybrids of varying sucrose content. Sehtiya et al. (1991) showed that different zones of internodes vary in the sucrose content and sucrose accumulation was slightly higher in the central core than in the peripheral regions. Parenchyma cells were five to six times larger in the central core than in the periphery of the stalk. Attan et al. (2003) observed that *in vitro*

---

*Corresponding author. E-mail: kapilvinayak@gmail.com.

sucrose uptake from sucrose containing medium by internodal slices of high sugar content var. CoJ 64 was higher than that of low sugar content var. Co 1148, irrespective of the maturity status of the internodes. It was concluded that the high sucrose clones store higher quantities of sucrose in their apparent free space (AFS) during their maturation phase.

Vinayak, et al. (2010) identified molecular markers for high and low sucrose genotypes and for this there was need to create germplasm with range of sucrose content from very low to very high so that inter-specific hybrids selected on the basis of their sucrose content by anthrone method, Brix, quality analysis and morphological findings can be compared to the inter-specific hybrids selected by high and low identifying sucrose markers. Therefore, high sucrose clones of *S. officinarum* were crossed with a low sucrose clone of *S. spontaneum*, to create variabilities. For early selection in breeding work, molecular markers have been employed to determine high sucrose types in the hybrids formed from a cross between high sucrose and low sucrose *Saccharum* genotypes. To study the contribution of *Sus2* to sugar accumulation process, PCR markers have been developed for different *Sus2* alleles (Lingle and Dyer, 2004). Map study of sugarcane shows *Shrunken1* (*Sh1*) gene, a probe from maize (*Zea mays*) is linked to the Brix character in sugarcane (Ming et al., 2001). *Sus2* gene from sugarcane is identical to *Shrunken1* gene from maize. However, northern analysis of diverse *Saccharum* genotypes shows that the *Sus2* gene is differentially expressed among the genotypes (Lingle et al., 2001), probably because of variable indels in promoter region of the gene. Clearly, genetic modification for sucrose accumulation requires a complete knowledge of rate limiting steps that are involved in sucrose synthesis, transport and accumulation. Some advancement towards elucidating of these processes have been made in the past few years and yet clear identification of genes in rate limiting steps is far from complete. The identification of DNA or protein markers related to high sucrose could, therefore, have significant consequences for genetic manipulation. The progeny provides materials with a wide range of sucrose contents. This paper includes observations on morphological characters, photosynthetic rates and assimilates translocation for these hybrid progeny and the parents.

MATERIALS AND METHODS

Crossing experiments were carried out at the Sugarcane Breeding Institute, Coimbatore, where parental clones were maintained. Sugarcane germplasm was planted in the fields on 15th March 2005. Plant materials included in this work were raised at CCS Haryana Agricultural University, Regional Research Station, Karnal. The following two inter-specifics (Cross I and II) and one inter-varietal cross (Cross III) were attempted:

Cross I : *S. officinarum* 'Gungera' × *S. spontaneum* 'SES 603'.
Cross II: *S. officinarum* 'Keong' × *S. spontaneum* 'SES 603'.
Cross III: Sugarcane var. CoS 8436 x Sugarcane var. Co1148.

Seeds packed in polythene laminated aluminum foil were transported from Coimbatore to Karnal.

The seeds were very susceptible to high temperature. Its viability was checked in the laboratory by germination tests in two ways. In test I, seeds were surface sterilized with 0.1% sodium hypochlorite and placed on absorbent cotton wool in petridish, moistened and covered. In test II, a mixture of sand and farm yard manure was sterilized in autoclave bags at 15 psi pressure and 121°C temperature. Sterilized mixture was spread in steel trays; moistened seeds were embedded into the soil and covered with polythene sheets. Samples from both tests were placed in culture room at 25 ± 2°C. Light intensity of 100 µE m^{-2} sec^{-1} was provided using fluorescent tubes and bulbs over a light and dark period of 16 and 8 h, respectively, and observations for germination of seeds were made after 10 days. Seeds from cross II showed no germination and hence were discarded. One hundred seeds from cross I and cross III were planted in the screen house in sterilized potting mixture containing soil, sand and FYM (1:1:1). After one month, the young seedlings were transferred to root trainers and then transferred to the field.

Height of field grown plants was recorded at three stages during grand growth on July 17 (stage 1), August 2 (stage 2) and August 21 (stage 3). Cane girth was recorded at stage 3 with vernier calipers and the number of tillers counted on stage 2. Mother tillers of fifteen randomly chosen plants were used for recording height and girth of genotypes. Photosynthetic rate, stomatal transpiration and stomatal conductance was measured using IRGA (Bioscientific Ltd., England) taking standard leaf of surface area 11.4 cm^2 (length 5.7 cm and breadth 2.0 cm). From the differences in gas concentration and air flow rate, the photosynthetic and transpiration rates were calculated every 20 s.

Rate of ^{14}C fixation into the leaves and its translocation from leaves to internodes in *S. spontaneum* 'SES 603', *S. officinarum* 'Gungera' and their progenies ISH-1, ISH-5, ISH-11, ISH-12, ISH-23 and ISH-25 was measured at about 30 cm middle part after removing 25 cm long tip part of TVD leaf which was inserted in the assimilation chamber, containing a few drops of water to maintain humidity. A disposable vial attached to assimilation chamber was added 1.0 ml of NaH^{14}CO$_3$ (specific gravity 51.9 mCi /mM) and 4.0 ml of 0.1 M phosphate buffer (pH 7.5) with the help of 5 ml syringe. The ^{14}CO$_2$ (400 µCi) was released by injecting 2 ml of 1 N HCl with 5 ml syringe through the rubber cork of high tensile strength fitted on upper surface of chamber just above the attached vial. The test leaf in each experiment was allowed to assimilate ^{14}CO$_2$ in sunlight for 2 h. Translocation rate was determined as radioactive counts after 24 h translocation period in six different samples of plant, namely leaf lamina, midrib, TVD internode, TVD+1 internode, TVD+ 3 internode and TVD+5 internode.

The sample intermodal tissue was boiled in 80% ethanol at 80 ± 2°C and decanted after cooling. The process was repeated thrice. The prepared ethanol soluble extract was then transferred to 20 ml scintillation vials and dried at 50 ± 2°C. 10 ml of Brays scintillation fluid (8 g PPO and 0.5 g POPOP in 1 litre of toluene) was added to the dried vials. The radioactive counts in the sample were determined by using the Liquid Scintillation Analyser (Model No. A 21000 TR, Packard Instrument Company, Neriden, U.S.A.).

Sucrose was estimated from the 6th internode at stage 3 by anthrone method of Van (1968) with O.D measured at 620 nm and sucrose content calculated as mg sucrose g^{-1} tissue. The values were confirmed by measuring Brix with refractometer at great grand growth phase on September 15 (stage 4) and by polarimetry at maturation phase on November 14 (stage 5) from the whole cane juice of a sugarcane stem.

Table 1. Plant height, Average number of tillers and cane girth in sugarcane clones *S. officinarum* "Gungera", *S. spontaneum* "SES 603" and their twenty-nine interspecific hybrids. Observations of the number of tillers and cane girth were recorded on June 15 and plant height on August 21. Values in the table are mean ± S.E.

Genotype	Average number of tillers	Cane girth (cm)	Plant height (cm)
'SES 603'	13.67 ± 1.19	0.62 ± 0.04	149.30 ±15.37
'Gungera'	2.00 ± 0.47	2.61 ± 0.12	108.30 ±10.90
ISH-1	4.67 ± 1.19	1.72 ± 0.07	199.40 ±13.58
ISH-2	8.67 ± 1.91	1.63 ± 0.08	132.10 ±10.48
ISH-3	4.67 ± 0.98	1.62 ± 0.06	139.60 ± 7.93
ISH-4	4.33 ± 0.54	1.82 ± 0.09	155.50 ±13.60
ISH-5	6.33 ± 1.09	1.71 ± 0.04	138.70 ± 5.78
ISH-6	5.00 ± 0.47	0.40 ± 1.19	133.40 ±13.48
ISH-7	9.67 ± 2.13	1.58 ± 0.03	145.80 ±14.24
ISH-8	2.67 ± 0.72	1.28 ± 0.02	134.70 ±11.62
ISH-9	4.67 ± 0.72	1.82 ± 0.06	154.00 ± 7.31
ISH-10	10.33 ±0.27	2.15 ± 0.04	146.40 ±10.96
ISH-11	3.33 ± 0.27	1.27 ± 0.04	126.00 ± 8.82
ISH-12	4.67 ± 0.27	1.52 ± 0.06	150.00 ±11.99
ISH-13	5.00 ± 0.47	1.55 ± 0.05	199.30 ± 5.70
ISH-14	6.00 ± 2.16	1.61 ± 0.05	144.30 ±12.43
ISH-15	5.33 ± 0.54	1.73 ± 0.06	160.50 ± 7.58
ISH-16	6.33 ± 0.98	1.62 ± 0.06	133.20 ±11.27
ISH-17	3.00 ± 0.94	1.50 ± 0.06	103.10 ±15.69
ISH-18	9.33 ± 2.37	1.56 ± 0.07	129.20 ±11.82
ISH-19	3.00 ± 0.82	2.12 ± 0.14	151.30 ± 9.67
ISH-20	0.67 ± 0.27	1.40 ± 0.00	-
ISH-21	5.33 ± 0.27	1.65 ± 0.05	144.40 ± 6.86
ISH-22	4.33 ± 0.72	1.52 ± 0.04	155.50 ± 9.75
ISH-23	3.67 ± 0.72	1.68 ± 0.05	143.90 ± 9.92
ISH-24	5.00 ± 0.82	1.61 ± 0.06	142.70 ±16.54
ISH-25	4.00 ± 1.25	1.53 ± 0.07	158.50 ± 9.27
ISH-26	6.33 ± 0.98	1.47 ± 0.08	175.50 ±20.04
ISH-27	5.33 ± 0.98	1.61 ± 0.06	168.40 ±14.31
ISH-28	2.33 ± 0.54	1.35 ± 0.03	134.90 ± 9.14
ISH-29	4.33 ± 0.54	1.93 ± 0.07	136.10 ± 6.38

RESULTS

Sets of *S. spontaneum* 'SES 603', *S. officinarum* 'Gungera' and *S. officinarum* 'Keong' showed 88.88, 80.95 and 72.22% germination, respectively, in the screen house and 83.33, 82.35 and 61.53% germination, respectively, in the field. Seed germination rates were zero in Cross II, while it was 50 and 60% in Cross I and Cross III, respectively, in the net house and 58 and 56.66% germination in the field (data not shown), indicating no differences in the germination rates between the two crosses. The plantlets of Cross I were named as interspecific hybrids (ISH) and those from Cross III were named as commercial hybrids (CH). A total of twenty-nine ISH and thirty-four CH survived in the field. However, CH showed relatively poor and stunted growth, only twenty-nine ISH from cross I were selected for further work.

Of the two parental species, *S. spontaneum* 'SES 603' showed much higher tillering (13.67) compared to *S. officinarum* 'Gungera' (2.00) (Table 1). Among the inter-specific hybrids, profuse tillering (average number of tillers) was observed in ISH-2 (8.67), ISH-7 (9.67), ISH-10 (10.33) and ISH-18 (9.33), while very poor tillering was seen in ISH-8 (2.67), ISH-11(3.33), ISH-17 (3.00), ISH-19 (3.0), ISH-20 (0.67) and ISH-28 (2.33). Cane girth of second internode from bottom was only 0.62 cm in *S. spontaneum* clone 'SES 603' compared to 2.61 cm in *S. officinarum* clone 'Gungera' (Table 1). Even though none of the progeny showed cane girth

Table 2. Photosynthetic rate (µ mole m–2s–1), stomatal transpiration (mole m–2s–1) and stomatal conductance (mole m–2s–1) in sugarcane clones S. spontaneum 'SES 603', S. officinarum 'Gungera' and their interspecific hybrids observed using infra red gas analyzer. Values in the table are mean ± S.E.

Genotype	Photosynthetic Rate (µ mole m–2s–1)	Stomatal transpiration (mole m–2s–1)	Stomatal conductance (mole m–2s–1)
'SES 603'	67.78 ± 4.67	11.68 ± 2.08	0.94 ± 0.18
'Gungera'	22.62 ± 0.71	3.84 ± 0.28	0.35 ± 0.06
ISH-1	22.20 ± 1.31	5.58 ± 0.31	0.23 ± 0.03
ISH-2	21.63 ± 1.72	5.75 ± 0.35	0.23 ± 0.03
ISH-4	25.66 ± 2.03	6.76 ± 0.40	0.36 ± 0.07
ISH-5	22.03 ± 3.22	6.17 ± 0.52	0.26 ± 0.06
ISH-6	25.78 ± 1.55	7.99 ± 0.39	0.23 ± 0.03
ISH-7	29.88 ± 1.16	9.79 ± 0.40	0.24 ± 0.02
ISH-8	29.16 ± 0.30	10.16 ± 0.80	0.23 ± 0.04
ISH-9	24.55 ± 1.03	6.22 ± 0.25	0.26 ± 0.03
ISH-10	23.56 ± 2.90	8.57 ± 0.73	0.18 ± 0.04
ISH-11	22.69 ± 2.51	7.88 ±0.70	0.18 ± 0.03
ISH-12	22.45 ± 0.85	6.32 ± 0.30	0.26 ± 0.04
ISH-13	23.90 ± 0.92	6.84 ±0.22	0.20 ± 0.01
ISH-14	25.28 ± 0.67	6.56 ± 0.18	0.28 ± 0.03
ISH-15	25.78 ± 1.51	6.71 ± 0.19	0.28 ± 0.02
ISH-16	20.78 ± 2.06	7.30 ± 0.39	0.14 ± 0.02
ISH-17	25.22 ± 0.67	6.69 ± 0.18	0.25 ± 0.02
ISH-19	23.43 ± 1.36	6.01 ± 0.14	0.28 ± 0.02
ISH-20	21.69 ± 1.47	6.07 ± 0.27	0.22 ± 0.03
ISH-21	33.26 ± 3.75	9.81 ± 0.88	0.30 ± 0.05
ISH-22	28.83 ± 3.52	9.09 ± 0.78	0.25 ± 0.04
ISH-23	22.68 ± 3.02	7.24 ± 0.83	0.21 ± 0.05
ISH-24	24.84 ± 1.32	6.94 ± 0.16	0.22 ± 0.03
ISH-25	21.81 ± 2.14	6.54 ± 0.45	0.18 ± 0.03
ISH-26	25.56 ± 1.50	7.56 ± 0.32	0.21 ±0.02
ISH-27	24.02 ± 2.35	8.09 ± 0.48	0.18 ± 0.02
ISH-28	22.16 ± 0.73	7.49 ± 0.23	0.21 ± 0.01

comparable to Gungera, hybrids ISH-10, ISH-19 and ISH-29 showed cane girth of more than 1.90 cm, while ISH-1, ISH-4, ISH-5, ISH-9 and ISH-15 showed cane girth between 1.70 to 1.90 cm morphologically indicating high sucrose hybrids like clone 'Gungera' as cane girth is directly proportional to sucrose accumulation. The remaining progeny had girth shorter than 1.70 cm morphologically falling into low sucrose hybrids like clones 'SES 603'. Clone S. spontaneum 'SES 603' attained a cane height of 149.30 cm at stage 3, whereas clone S. officinarum 'Gungera' attained cane height of 108.3 cm at this stage. Tall hybrids as indicated by cane height at this stage include ISH-1 (199.40 cm); ISH-4 (155.50 cm); ISH-9 (154.00 cm); ISH-12 (150.00 cm); ISH-13 (199.30 cm); ISH-15 (160.50 cm); ISH-19 (151.30 cm); ISH-22 (155.50 cm); ISH-25 (158.50 cm); ISH-26 (175.50 cm) and ISH-27 (168.40 cm). The remaining hybrids showed cane height between 103 and 148 cm.

S. spontaneum 'SES 603' showed highest photosynthetic rate (µmole m^{-2}s^{-1}) of 67.78 which is three times greater than S. officinarum 'Gungera' which showed low photosynthetic rate of 22.62 (Table 2). The inter-specific hybrids; ISH-4, ISH-6, ISH-7, ISH-8, ISH-14, ISH-15, ISH-17, ISH-21, ISH-22 and ISH-26 showed photosynthetic rate between 25.22 to 33.26 µmole m^{-2}s^{-1} while the rest of the hybrids had less than that. The stomatal transpiration rate was also higher for S. spontaneum 'SES 603' when compared to S. officinarum 'Gungera'. The inter-specific hybrids

Table 3. Percent of 14C counts in different plant parts in clones *S. spontaneum* 'SES 603', *S. officinarum* 'Gungera' and their six interspecific hybrids (ISH-1, ISH-5, ISH-11, ISH-12, ISH-23 and ISH-25). Leaves were allowed to photosynthesize in the assimilation chamber in radiolabelled $^{14}CO_2$ air for 2 h in the morning. Samples were collected after 24 h of translocation period. Percent 14C counts in lamina, midrib and internodes were calculated.

S/no	Percent	SES 603	Gungera	ISH-1	ISH-5	ISH-11	ISH-12	ISH-23	ISH-25
					Percent ^{14}C counts				
1.	Leaf lamina	46.02	31.33	34.39	43.60	27.25	45.15	31.30	38.68
2.	Midrib	36.16	50.20	63.43	47.63	62.19	44.12	55.31	58.99
3.	TVD internode	3.23	5.84	0.70	2.67	2.37	2.44	1.37	0.50
4.	TVD+1 internode	7.93	8.76	1.00	3.13	6.27	6.23	10.5	1.04
5.	TVD+3 internode	2.45	3.10	0.30	2.55	0.72	1.13	0.77	0.44
6.	TVD+5 internode	4.18	0.66	0.15	0.38	1.16	0.89	0.57	0.32

showing stomatal transpiration rate above 7.0 mole $m^{-2}s^{-1}$ were ISH-6, ISH-7, ISH-8, ISH-10, ISH-11, ISH-16, ISH-21, ISH-22, ISH-23, ISH-26, ISH-27 and ISH-28. The stomatal conductance rate (mole $m^{-2}s^{-1}$) was again greater for clone 'SES 603' (0.94) when compared to 'Gungera' (0.35). Hybrids showing stomatal conductance between 0.25 and 0.36 mole $m^{-2}s^{-1}$ were ISH-4, ISH-5, ISH-9, ISH-12 ISH-14, ISH-15, ISH-17, ISH-19, ISH-21 and ISH-22 and the remaining ISH lines showed stomatal conductance below them.

Rate of translocation of sucrose using radiolabelled ^{14}C showed that radioactive counts were found to be at the highest in lamina and midrib (Table 3). *S. spontaneum* 'SES 603' showed comparatively high total incorporation of ^{14}C compared to high sucrose parent, *S. officinarum* 'Gungera'. The percent incorporation of the midrib was greater in clone 'Gungera' than clone 'SES 603'. Thus, the percentage of radioactivity in midrib of clone 'SES 603' was 36.16% which is far lesser than that in clone 'Gungera' (50.2%). From Table 3, it is also clear that the percentage of photosynthates entering translocation stream in clone 'SES 603' and clone 'Gungera' was 53.97 and 68.6%, respectively. The pattern of translocation of ^{14}C was same in all the genotypes studied, that is, ISH-1, ISH-6, ISH-11, ISH-12, ISH-23 and ISH-25.

They was a very high increase in midrib and a decrease towards lower internodes, that is, from TVD internode to TVD+1, TVD+3 and TVD+5.

Sucrose content estimated by Van Handel method at stage 3 in 6th internode from bottom for all the sugarcane genotypes is shown in Table 4. *S. spontaneum* 'SES 603' and *S. officinarum* 'Gungera' had 45.5 mg sucrose g tissue^{-1} and 56.1 mg sucrose g tissue^{-1}, respectively. The ISH hybrids having sucrose content above 70 mg g tissue^{-1} were graded into very high sucrose types, viz., ISH-1, ISH-5, ISH-9, ISH-13, ISH-17 and ISH-23. High sucrose types having sucrose content between 50 to 70 mg g tissue^{-1} include hybrids like ISH-3, ISH-4, ISH-6, ISH-26 and ISH-28. On the other hand, ISH hybrids having sucrose content between 30 to 50 mg g tissue^{-1} were graded into low sucrose types, they were hybrids like ISH-8, ISH-14, ISH-16, ISH-18, ISH-20, ISH-21 and ISH-22, whereas ISH hybrids which showed very low sucrose content less than 30 mg g tissue^{-1} were ISH-2, ISH-7, ISH-10, ISH-11, ISH-12, ISH-15, ISH-19, ISH-24, ISH-25, ISH-27 and ISH-29. Brix values in juice of sugarcane were estimated at alternate internodes at stage 4 (Table 5). Also quality analysis of sugarcane clones and its hybrids was done at stage 5 which showed variable values of Brix, Sucrose percentage and Commercial Cane Sugar (CCS) (Table 6). Thus, the two clones of 'SES 603' and 'Gungera' had large difference in their growth and morphological characters. Indeed, this was the basis of selecting these two as parents in the present study. The present studies showed that the two parents and twenty nine inter-specific hybrids obtained satisfy the initial goal proposed and present a complete range of variation for tillering, cane girth and other characters.

DISCUSSION

This study attempts to develop materials in specifically designed crosses to obtain a complete range of genotypes. Vinayak, et al. (2010) identified four high sucrose PCR markers: AI +AIR (Acid invertases); MSSCIRI + MSSCIRIR; A +AR; and B + BR and one low sucrose PCR marker SMC226CG + SMC226CGR which graded ISH-1, ISH-5, ISH-17 and ISH-23 into high sucrose inter-specific hybrids and ISH-10, ISH-11, ISH-12 and ISH-25 into low sucrose inter-specific hybrids with variable sucrose content for the study of sucrose accumulation process.

Seeds from cross II showed no germination and hence could not be continued. Cross I and cross III showed good germination, but plants from cross III showed poor and stunted growth in the field and hence progeny of only cross I formed experimental material for further work. Sugarcane has sexually developed true

Table 4. Sucrose content (mg sucrose g tissue^{-1}) in internode number 6 from bottom in sugarcane clones *S. officinarum* "Gungera", *S. spontaneum* "SES 603" and their twenty-nine inter-specific hybrids. Values in the table are mean ± S.E. Observations were recorded on September 15 at stage 3.

Genotypes	Sucrose (mg sucrose g tissue^{-1})
'SES 603'	45.5 ± 6.79
'Gungera'	56.1 ± 0.55
ISH-1	78.9 ± 2.16
ISH-2	28.6 ± 0.60
ISH-3	57.7 ± 0.55
ISH-4	53.9 ± 0.32
ISH-5	91.8 ± 2.56
ISH-6	58.1 ± 2.27
ISH-7	18.2 ± 0.99
ISH-8	44.3 ± 1.15
ISH-9	80.6 ± 1.13
ISH-10	26.8 ± 0.57
ISH-11	4.4 ± 1.60
ISH-12	12.9 ± 0.76
ISH-13	78.1 ± 2.71
ISH-14	45.0 ± 1.82
ISH-15	26.2 ± 0.37
ISH-16	49.3 ± 0.64
ISH-17	74.4 ± 3.23
ISH-18	45.5 ± 3.25
ISH-19	15.3 ± 2.85
ISH-20	37.6 ± 2.22
ISH-21	44.8 ± 0.94
ISH-22	46.3 ± 2.25
ISH-23	71.4 ± 1.72
ISH-24	16.8 ± 1.93
ISH-25	17.9 ± 0.78
ISH-26	55.4 ± 0.54
ISH-27	4.7 ± 1.81
ISH-28	68.3 ± 2.64
ISH-29	17.2 ± 1.00

seed like any other grain crop and follows Mendel's law of inheritance in crossing. *S. spontaneum*, a wild type cane, has a very poor sink and extremely low sucrose accumulation capacity; whereas the cultivated form of sugarcane, viz. *S. officinarum,* has efficient sucrose accumulating sinks. The cross between clones of these two divergent species, *S. officinarum* 'Gungera' with *S. spontaneum* 'SES 603', in the present work resulted in progeny that provided ideal study material with a wide range of sucrose content.

'SES 603', clone from wild *S. spontaneum*, had profuse tillering, greater plant height and hence thin canes. 'Gungera', on the other hand, a clone of noble species of *S. officinarum,* had lesser tillering, less plant height and thicker canes. The progeny of the two selected as ISH-1 to ISH-29 hybrids had variable cane girth, number of tillers and height. MacColl (1976) has described the relationship of growth patterns with sucrose content and had shown that tillering at times when early formed tillers are elongated, reduces the sugar content.

Thus, since determination of sucrose accumulation capacity of parents and hybrids used in this study was very critical to interpretation of results, different methods (Brix, Anthrone reagent, Polarimetry) were employed several times to arrive at a conclusion. Even though there were occasional variations in results, an overall conclusion of the repeated measurement of Brix, sucrose and juice quality assigned parent clone *S. officinarum* 'Gungera' and four hybrids, viz., ISH-1, ISH-5,

Table 5. Brix values in juice in sugarcane clones *S. officinarum* "Gungera" and *S. spontaneum* "SES603" and their twenty-nine interspecific hybrids. Internodes were numbered from 1 to 9 with 1 being the uppermost newly formed (≥ 1 cm long) internode. Observations were recorded on September 15. Values in the table are mean ± S.E.

Genotype	Internodes number				
	1	3	5	7	9
'SES 603'	6.9 ± 0.42	9.0 ± 0.05	10.9 ± 0.05	10.8 ± 0.16	12.3 ± 0.19
'Gungera'	6.2 ± 0.30	6.0 ± 0.00	8.9 ± 0.05	11.0 ± 0.47	13.0 ± 0.09
ISH-1	7.1 ± 0.19	6.2 ± 0.76	10.2 ± 0.14	15.2 ± 0.56	18.2 ± 0.21
ISH-2	6.2 ± 0.30	5.0 ± 0.09	9.2 ± 0.24	12.8 ± 0.43	15.2 ± 0.09
ISH-3	5.2 ± 0.11	8.2 ± 0.68	11.6 ± 0.51	16.8 ± 0.46	18.6 ± 0.80
ISH-4	6.8 ± 0.33	7.6 ± 0.18	11.7 ± 0.64	13.2 ± 0.09	12.3 ± 0.27
ISH-5	4.8 ± 0.46	6.5 ± 0.09	13.5 ± 0.43	15.6 ± 0.35	16.4 ± 0.05
ISH-6	9.2 ± 0.97	8.4 ± 0.19	13.4 ± 0.35	10.2 ± 0.11	13.1 ± 0.44
ISH-7	7.0 ± 0.05	6.9 ± 0.33	10.1 ± 0.05	12.0 ± 0.37	–
ISH-8	5.9 ± 0.05	4.4 ± 0.23	6.4 ± 0.24	9.9 ± 0.39	11.2 ± 0.14
ISH-9	6.3 ± 0.12	6.8 ± 0.48	10.3 ± 0.12	12.3 ± 0.14	14.2 ± 0.09
ISH-10	5.1 ± 0.07	8.9 ± 0.78	9.1 ± 0.04	11.1 ± 0.08	–
ISH-11	5.0 ± 0.47	5.4 ± 0.24	9.8 ± 0.15	–	–
ISH-12	5.0 ± 0.02	5.7 ± 0.15	8.3 ± 0.07	10.5 ± 0.17	–
ISH-13	5.1 ± 0.07	7.7 ± 0.46	13.5 ± 0.23	16.0 ± 0.26	14.0 ± 0.05
ISH-14	7.0 ± 0.94	5.7 ± 0.21	7.3 ± 0.54	10.4 ± 0.24	11.2 ± 0.09
ISH-15	4.6 ± 0.27	5.0 ± 0.42	8.3 ± 0.10	12.8 ± 0.16	–
ISH-16	5.6 ± 0.24	5.0 ± 0.09	7.1 ± 0.10	10.5 ± 0.23	12.5 ± 0.30
ISH-17	6.7 ± 0.54	8.1 ± 0.11	11.6 ± 0.24	14.2 ± 0.33	12.3 ± 0.72
ISH-18	7.6 ± 0.54	8.0 ± 1.24	11.4 ± 0.23	14.2 ± 0.33	12.3 ± 0.72
ISH-19	6.1 ± 0.10	4.1 ± 0.13	6.8 ± 0.24	12.0 ± 0.47	14.8 ± 0.33
ISH-20	4.8 ± 0.49	6.8 ± 0.05	11.3 ± 0.12	15.4 ± 0.28	15.3 ± 0.14
ISH-21	4.2 ± 0.30	5.9 ± 0.42	8.5 ± 0.23	12.4 ± 0.23	14.4 ± 0.19
ISH-22	8.0 ± 0.79	4.0 ± 0.05	8.6 ± 0.98	12.1 ± 0.46	13.2 ± 0.09
ISH-23	9.8 ± 0.16	9.5 ± 0.48	13.3 ± 0.27	16.4 ± 0.23	17.0 ± 0.00
ISH-24	5.0 ± 0.47	6.0 ± 0.47	7.9 ± 0.05	1.6 ± 0.24	10.5 ± 1.03
ISH-25	4.3 ± 0.27	5.0 ± 0.00	6.9 ± 0.16	7.0 ± 0.05	8.0 ± 0.00
ISH-26	5.8 ± 0.46	5.0 ± 0.48	7.6 ± 0.24	12.5 ± 0.40	14.4 ± 0.19
ISH-27	7.1 ± 0.10	7.3 ± 0.54	8.3 ± 0.27	11.0 ± 0.47	–
ISH-28	5.1 ±0.37	5.0 ±0.14	8.6 ±0.47	13.2 ±0.21	16.8 ±6.13
ISH-29	5.3 ±6.72	4.2 ±0.30	5.2 ±0.16	9.0 ±0.25	12.5 ±0.23

ISH-17, ISH-23 as high sucrose types and parent clone *S. spontaneum* 'SES 603' and four hybrids ISH-10, ISH-11, ISH-12 and ISH-25 as low sucrose types for experimental work.

From the radioactive counts and the percentage radioactivity determined in six different sample parts of each selected variety, it appears that the photosynthates entered and accumulated in the midrib and were then translocated to the other regions; primarily to lower internodes. The TVD internode above the TVD leaf showed lesser radioactive counts and percentage radioactivity, since the translocation of photosynthetic radiolabelled assimilates is mainly downwards. TVD + 1 internode had higher radioactivity than TVD internode. The lower internodes like TVD+3 and TVD+5 have very low percentage of radioactivity and radioactive counts since the TVD leaf may accumulate its photosynthates only to nearby upper internodes. Therefore, percentage of radioactivity estimated in TVD+5 internodes is less than 1%. Interestingly, the wild type parent 'SES 603' has higher number of total radioactive counts incorporated (47578) compared to the high sucrose parent 'Gungera' (14099). This is perhaps because in clone 'SES 603' the leaves are very narrow and the total surface area of leaf is very small, so the amount of radiolabelled sucrose accumulated per cell in the leaves is more than the amount of radiolabelled sucrose accumulated in 'Gungera'. Thus, for sucrose accumulation, the capacity of the sink to accumulate seems more important than

Table 6. Quality analysis showing Brix, sucrose percent and CCS percent in sugarcane clones *S. officinarum* "Gungera", *S. spontaneum* "SES 603" and their twenty-nine inter-specific hybrids. Observations were recorded on November 14.

Genotypes	Brix	Suc %	CCS %
'SES 603'	12.32	8.94	3.79
'Gungera'	13.02	9.40	3.98
ISH-1	15.24	9.83	3.83
ISH-2	15.34	8.84	3.12
ISH-3	14.64	8.88	3.29
ISH-4	15.34	9.33	3.46
ISH-5	14.64	8.38	2.94
ISH-6	15.14	8.35	2.82
ISH-7	16.14	10.27	3.96
ISH-8	15.94	9.83	3.69
ISH-9	15.34	9.83	3.81
ISH-10	14.64	8.35	2.92
ISH-11	14.44	8.38	2.98
ISH-12	12.24	6.46	2.07
ISH-13	17.44	10.23	3.67
ISH-14	16.24	9.79	3.61
ISH-15	16.74	10.27	3.84
ISH-16	13.44	7.42	2.51
ISH-17	15.94	10.32	4.04
ISH-18	14.24	8.38	3.02
ISH-19	14.64	8.88	3.29
ISH-20	–	–	–
ISH-21	12.44	6.46	2.03
ISH-22	14.24	7.89	2.68
ISH-23	14.34	8.38	3.00
ISH-24	13.74	8.91	3.49
ISH-25	13.94	8.41	3.10
ISH-26	13.84	7.42	2.43
ISH-27	15.24	9.33	3.48
ISH-28	13.34	7.42	2.53
ISH-29	12.24	6.46	2.07

the rate of photosynthesis. Thus, the two clones 'SES 603' and 'Gungera' had large difference in their growth and morphological characters. Indeed, this was the basis of selecting these two as parents in the present study. Twenty-nine inter-specific hybrids obtained from crossing of the two parents showed a complete range of variation for tillering, cane girth and other characters. Thus, the two parents and twenty-nine ISH hybrids provided the complete variation in sucrose content and selection of high and low sucrose inter-specific hybrids.

REFERENCES

Attan PS, Dendsay JPS, Dhawan AK (2003). *In vitro* sucrose uptake by internodal slices of sugarcane genotypes varying in maturity and sucrose accumulation potential. Proc. S.T.A.I. 65: 97-109.

Dendsay JPS, Dhawan AK, Sehtiya HL (1992). Biochemical approaches for predicting high sucrose content in sugarcane. Coop. Sugar, 23: 597-604.

Glasziou KT, Gayler KR (1972). Storage of sugars in stalks of sugarcane. Bot. Rev. 38: 471-490.

Grantz DA (1989). Effect of cool temperature on photosynthesis and stomatal conductance in field grown sugarcane in Hawaii. Field Crops Res. 22: 143-155.

Lingle SE, Dyer JM (2001). Cloning and expression of sucrose synthase-1 cDNA from sugarcane. J. Plant Physiol. 158: 129-131.

Lingle SE, Dyer JM (2004). Polymorphism in the promoter region of the sucrose synthase-2 gene of Saccharum genotypes. J. Am. Soc. Sugarcane Technol. 24: 241-249.

MacColl D (1976). Growth and sugar accumulation of sugarcane. I. Percentage of sugar in relation to pattern of growth. Exp. Agric. 12: 369-377.

Ming R, Liu SC, Moore PH, Irvine JW, Peterson AH (2001). QTL Ming, R., Liu, S.C., Moore, P.H., Irvine, J.W. and Peterson, A.H. (2001). QTL analysis in a complex autopolyploid : genetic control of sugar content in sugarcane. Genome Res. 11: 2075-2084.

Sehtiya HL, Dendsay JPS, Dhawan AK (1991). Internodal invertases and stalk maturity in sugarcane. J. Agric. Sci. Cambridge 116: 239-243.

Selvi A, Nair NV, Noyer JL, Singh NK, Balasundaram N, Bansal KC, Koundal KR, Mohapatra T (2005). Genomic constitution and Genetic Relationship among the tropical and subtropical Indian Sugarcane Cultivars revealed by AFLP. Crop Sci. 45: 1750-1758

sucrose Saccharum genotypes. Int. J. Physiol. Mol. Biol. Plants Springer 2010; 16(1): 107-111.

Van HE (1968). Direct microdetermination of sucrose. Anal. Biochem., 22: 280-283.

Vandana V, Dhawan AK (2010). PCR Primers for identification of high Whittaker A, Botha FC (1997). Carbon partitioning during sucrose accumulation in sugarcane internodal tissue. Plant Physiol. 115: 1651-1659.

Evaluation of head yield and participatory selection of horticultural characters in cabbage (*Brassica oleraceae* var. *capitata*)

O. T. Adeniji[1]*, I. Swai[1], M. O. Oluoch[1], R. Tanyongana[1] and A. Aloyce[2]

[1]AVRDC – The World Vegetable Center, Regional Center for Africa, PMB 25, Duluti, Arusha, Tanzania.
[2]Horticultural Research and Training Institute, Tengeru, Arusha, Tanzania.

Cabbage is an important leafy vegetable in sub-Saharan Africa, yet breeding and seed production activities take place in Europe and Asia. Participatory evaluation of varieties alongside head yield was carried out to select preferred varieties. Field experiments were conducted during the 2008 long rainy season and the 2008/09 short rainy season at AVRDC–The World Vegetable Center, Regional Center for Africa, Arusha, Tanzania. A randomized complete block design with three replications was adopted for the study. Results indicated that for each season marketable head yield differed significantly ($P < 0.05$) among the varieties. Performance for head yield varied in response to the biotic and abiotic environment. During the long rainy season Gloria F_1 and Victoria F_1 were best for head yield, while Quick Start and Rotan performed best for head yield during the short rainy season. For taste, head shape and firmness, Summer Summit F_1 was best. Tropical Delight was preferred for head size, firmness, and low incidence of loose heads. Gloria F_1 was identified as an early maturing variety with good head solidity. Good-tasting cabbage varieties are Summer Summit, Summer Glory and Bonus. Research on stability of performance for head yield over seasons, years and locations is important for future selection.

Key words: Cabbage, marketable head yield, farmer participatory research, sensory evaluation, Tanzania, vegetable production.

INTRODUCTION

Cabbage (*Brassica oleraceae* var. *capitata*) is one of the most important leafy vegetables worldwide (Talekar, 2000). It originated in Northern Europe, the Baltic Sea coast (Monteiro and Luan, 1998) and the Mediterranean region (Vural et al., 2000), where it has been grown for more than 3000 years and is adapted to cool moist conditions (Tindall, 1993; Thompson, 2002). Cabbage is cultivated for its head, which consists of water (92.8%), protein (1.4 mg), calcium (55.0 mg) and iron (0.8 mg); the leaves are eaten raw in salads or cooked. The optimum mean temperature for growth and quality head development is 15 - 18°C, with a minimum temperature of 4°C and a maximum of 24°C. Cabbage grows well on a range of soils with adequate moisture and fertility. It tolerates a soil pH range of 5.5 - 6.8 and it is a heavy feeder. To maintain growth, cabbage requires a consistent supply of moisture, and should as a general rule receive a minimum of 2.5 cm of water per week. With proper management, cabbage can produce 25 - 30 t/ha. Cabbage heads are ready for harvest 80 - 120 days after germination, depending on genotype and climate.

The importance of head cabbage in tropical and subtropical regions has increased considerably in recent decades. Recent estimates indicate Africa has 100,000 ha planted with head cabbage (van der Vossen et al., 2004). Based on sales of commercial seed, at least 40,000 ha of white-headed cabbage is grown in Kenya, Uganda and Tanzania; 10,000 ha in Malawi, Zambia and Zimbabwe; 4000 ha in Ethiopia; and 3000 ha in Cameroon. Almost all white-headed cabbage is produced for local urban markets. Mozambique imports considerable quantities of headed cabbage from South Africa and until recently also did so from Zimbabwe. Mwasha (2000) reported a total production of head cabbage in

*Corresponding author. E-mail: wale.adeniji@worldveg.org.

Tanzania to be about 208,919 t in 1996. However, in the 2004/05 cropping season, head cabbage production was 250,000 t and Chinese cabbage production was 4,000 t (Zoss, 2006). In Tanzania, cabbage is grown mostly in the cooler highland areas of Arusha, Tanga, Iringa, Mbeya and Morogoro. In a survey of Arumeru district Massomo (2002) noted that cabbage cultivation started in the mid-1970s, with a sharp increase in the number of cabbage growers from 1997 to 2001. Cultivation occurs throughout the year in places where irrigation facilities are available (Massomo et al., 2004a). Farm are less than 0.5 ha and most farmers grow one or two cabbage crops per year, mostly in monoculture, with the exception of a few villages with irrigation facilities where up to four cabbage crops are produced annually (Massomo et al., 2004a). Farmers in Arusha apply farmyard manure and inorganic fertilizers to remedy declining soil fertility (Massomo et al., 2004b). More than 30 cabbage varieties (open pollinated and hybrids) are cultivated in Tanzania; popular varieties are Gloria F_1, Glory of Enkhuizen and Romenco (Massomo et al., 2003).

Cabbage breeding and seed production takes place outside Africa and no landrace exists. Seed is imported from Europe and Asia and is sold by seed companies and other retail outlets. Most of these varieties are described to be high yielding and with varying levels of resistance to black rot (*Xanthomonas campestris*). Important horticultural characters in cabbage are head size, shape and firmness, taste, resistance to bolting, late flowering and maturity. Evaluating cabbage varieties for adaptation and yield will help farmers, breeders and seed companies select and develop varieties best suited to the local environment and market.

Farmer participatory research is increasingly recognized as a demand-driven process, where the end users (farmers) participate in the design, evaluation and implementation of new technology rather than just the final testing. Traditional top-down, prescriptive approaches to agricultural research and extension have been criticized (Ceccarelli et al., 1996). Participatory research is a simple, more direct way of evaluating multiple traits to assess the value of a variety to farmers. The approach is client-oriented, not controversial and based on the pretext that farmers are researchers in their own right and have indigenous knowledge of local conditions (Ceccarelli et al., 1996). The involvement of farmers in the selection process facilitates acceptance and adoption. This approach has been used extensively in sorghum (Nkongolo et al., 2008) and barley (Ceccarelli et al., 2001). Farmers' participatory selection has been applied to rice (Oyemanmi et al., 2008).

Farmers' choice of a variety depends on a combination of characters (high and stable head yield, resistance to insect pests and diseases, good taste, storability and firmness, head shape and size). Information on the local adaptation and performance of cabbage varieties for yield is limited, as is participatory research on horticultural characters and sensory evaluation. This paper describes the performance of cabbage varieties for yield and other characters; it explains how farmers participate in research to identify and select cabbage varieties with preferred horticultural characters for cultivation in Arusha district, Tanzania.

MATERIALS AND METHODS

Materials, location, experimental design and statistical analysis

Thirty-two hybrid and open pollinated cabbage varieties marketed by seed companies in Tanzania were used in the study (Table 1). A widely cultivated variety (Gloria F_1) was included as check. Field trials were conducted during the 2008 long rainy season and the 2008/09 short rainy season at AVRDC – The World Vegetable Center, Regional Center for Africa, Arusha, Tanzania (lat 4.8°S long 3.7°E; alt. 1290 m). In Arusha, the long rains start in March and continue through May/June; the short rains begin in September/November and continue through December. The dry season occurs from January to March. The mean minimum and maximum temperatures during the long rainy season are 13.2 and 26.4°C respectively. The mean minimum and maximum temperature for the short rainy season are 18.5 and 31.9°C. Annual rainfall of 1085 mm was recorded on-station.

Field experiments were conducted in May and repeated in October 2008. A randomized complete block design with three replications was used; each plot consisted of two ridges 8 m long and 0.30 m high, with 0.60 m between ridges. Cabbage seedlings were raised in multipot seedling trays for four weeks and then transplanted to the sides of the ridges with 0.50 m space between plants. The experiment was furrow-irrigated every two days for the first two weeks after transplanting, then once a week thereafter. Weeding was carried out manually and frequently to maintain weed-free plots. Fertilizer NPK (20-10-10) was applied at the rate of 200 kg/ha during transplanting. Urea was applied at the rate of 90 kg/ha in a split application 30 days after transplanting and 30 days thereafter. Both evaluations were maintained as no-spray (insecticides and fungicides). Head yield was computed from a net plot of 4 × 1.2 m and head characters (head length and width) were determined at harvest. Head length and width were measured in centimeters; five heads per plot were selected at random and each head was cut into two halves for measurement. The number of bolt heads and loose heads were counted and expressed as a percentage of the total heads harvested per plot. Head yield was separated into marketable and non marketable yield (t/ha).

Participatory variety selection

Participatory selection was conducted in the 2008 long rainy season to select cabbage varieties with horticultural characters preferred by farmers. Two groups participated in the evaluation. The first group comprised 35 cabbage farmers randomly selected from Moshi, Arumeru and Arusha, the major cabbage growing areas in Northern Tanzania's Arumeru district. The second group comprised of 29 researchers drawn from sub-Saharan Africa. Characters preferred by farmers (head size, shape and firmness; resistance to insect pests and diseases; resistance to bolting and loose heads; medium and late maturity; taste) were used to evaluate the cabbage varieties. Head solidity (firmness) was determined by applying pressure on the head using the thumb. Head shape and size were determined through visual observation. Farmers made their selections by dropping between one and five seeds (1= extremely poor and 5 = excellent) in a container in front of the cabbage row. The research team evaluated cabbage

Table 1. Identity, type and source of cabbage used in the trials.

Sn	Cabbage varieties	Type[a]	Breeder/supplier of seeds
1	Gloria F_1	H	East Africa Seeds Co., Tanzania
2	02-12609 (hybrid)	H	East Africa Seeds Co., Tanzania
3	Cheers Hybrid	H	East Africa Seeds Co., Tanzania
4	Globe Master (F_1)	H	East Africa Seeds Co., Tanzania
5	Glory of Enkhuizen	OP	East Africa Seeds Co., Tanzania
6	Riana F_1	H	East Africa Seeds Co., Tanzania
7	Green Coronet F_1	H	East Africa Seeds Co., Tanzania
8	Copenhagen Market	OP	East Africa Seeds Co., Tanzania
9	Super Master	H	East Africa Seeds Co., Tanzania
10	Chou F_1 Tropical Cross	H	Technisem, France
11	Chou F_1 Santa	H	Technisem, France
12	Chou F_1 KK Cross	H	Technisem, France
13	Africana F_1	H	Takil and Co.,,Kyoto, Japan
14	Bonus	H	Takil and Co., Kyoto, Japan
15	Quick Start	H	Takil and Co., Kyoto, Japan
16	Victoria F_1	H	Mukpar Tanzania Ltd.
17	Dragon F_1	H	Nongwoo Bio Co., Korea
18	Drum Head	OP	Hygrothech, South Africa
19	BEJO 2658 F_1	H	Kibo Seeds, Tanzania
20	Pruktor F_1	H	Kibo Seeds, Tanzania
21	Thomas F_1	H	Kibo Seeds, Tanzania
22	Glory of Enkhuizen	OP	Kibo Seeds, Tanzania
23	Rotan F_1	H	Kibo Seeds, Tanzania
24	Riana	H	Kibo Seeds, Tanzania
25	Spring Light	H	Known-You Seed, Taiwan
26	Besta	H	Known-You Seed, Taiwan
27	Good Season	H	Known-You Seed, Taiwan
28	Summer Tide	H	Known-You Seed, Taiwan
29	Summer Summit (Hybrid)	H	Known-You Seed, Taiwan
30	Tropical Delight (Hybrid)	H	Known-You Seed, Taiwan
31	Summer Glory (Hybrid)	H	Known-You Seed, Taiwan
32	Summer Autumn (Hybrid)	H	Known-You Seed, Taiwan

Type OP = Open pollinated; H = F_1 hybrid variety.

varieties for horticultural characters and overall performance using a five-point Likert scale, with 1 (least preferred) and 5 (most preferred) (Olowu and Oladeji, 2004). The mean score of performance for each variety was calculated. A variety with a mean score of equal and above the cut-off mean of 2.5 was declared as being preferred, while a mean score less than 2.5 was classified as least preferred. The mean score of the check was compared with the other varieties.

For each growing season, data was collected on head yield (t ha^{-1}), head length and width, leaf length and width. For each character, analysis of variance for each season and combined analysis of variance over seasons was conducted in a conventional manner, with genotype being a fixed factor and season (a combination of season and year) being a random factor (Gomez and Gomez, 1984). Data for each season and pooled over seasons, was subjected to statistical analysis as PROC GLM procedure of SAS Institute (1998). Entry means were separated using Duncan's Multiple Range Test (DMRT). Percentages and ranking were used to determine varieties preferred by cabbage farmers.

RESULTS AND DISCUSSION

Evaluation of head yield and yield components in 2008 trial

Significant differences ($p < 0.05$) were found among the entries during the long and short rainy seasons for marketable head yield (t/ha), nonmarketable head yield (t/ha), head length and width, leaf length and width (Table 2). During the long rainy season (2008), marketable head yield ranged between 23 and 49 t/ha (Table 3). Dragon F_1, Cheers HYB, Gloria F_1, and Victoria F_1 significantly out-yielded other varieties (Table 3). Head length was high in Gloria F_1 during the long rainy season. Summer Summit, Bejo, Thomas F_1, Victoria F_1, Glory of Enkhuizen, Summer Glory and Riana recorded the

Table 2. Mean squares for head yield and yield characters in cabbage during 2008 and 2008/09 and pooled for 2008 and 2008/09.

Source of variation	Df	Head length	Head width	Marketable weight – head yield (t/ha)	Nonmarketable weight – head yield (t/ha)	Leaf length (cm)	Leaf width (cm)
2008 season							
Varieties	31	18.50**	10.78*	108.94*	15.74*	81.07**	84.17***
Replication	2	0.64	10.44*	3.20	8.28**	4.34	5.41
Error	62	2.56	1.37	9.63	1.98	4.55	2.86
R2		0.78	0.81	0.85	0.80	0.90	0.97
CV		9.19	5.20	9.2	34.67	5.77	4.34
Mean		17.44	22.51	33.60	4.05	36.89	38.52
2008/09 season							
Varieties	31	7.26**	9.95**	64.82**	14.41**	43.17***	36.73***
Rep	2	4.21	2.07	2.23	4.11	11.74	5.8
Error	62	1.77	1.96	7.48	3.72	9.13	7.62
R2		0.67	0.72	0.81	0.66	0.70	0.71
CV(%		8.55	6.67	9.65	37.88	9.80	8.63
Mean		15.58	20.95	28.33	5.00	30.80	31.97
Pooled data over seasons							
Varieties	31	22.02***	15.95**	158.91**	24.91**	117.62***	105.50***
Rep within seasons	4	3.90	9.86	5.17	3.43	11.69	9.80
Season	1	159.80***	110.31*	1265.0***	50.03*	2812.35**	1168.09***
Seasons x varieties	31	3.91**	4.68*	13.09*	5.26**	10.04*	17.05***
Error	124	2.17	1.80	8.38	2.92	6.01	6.00
R2		0.77	0.77	0.86	0.72	0.90	0.88
CV		8.94	9.19	9.34	37.57	12.00	7.12
Mean		16.51	21.73	30.97	4.56	34.64	36.24

* = significant at 5%, ** = significant at 1%.

highest marketable head yields during the short rainy season (Table 3). High head yield recorded for some varieties on-station is consistent with reports of Talekar (2000) and Ijoyah and Rakotomavo (2007) for head yield in cabbage.

The analysis of variance over seasons (trials) revealed highly significant entry means squares ($p < 0.01$) for seasons, varieties and varieties × season interaction effects. These results suggest large genetic difference among the entries and that environmental differences between seasons influence the expression of head yield and yield characters. The relative differences for head yield and head characters are inconsistent within seasons and over seasons. The entry means over seasons for marketable head yield ranged between 21 and 42 t/ha and was best in Quick Start, Bejo, Thomas F_1 and Globe master. Entries, seasons and entries × seasons interaction summarized 63, 16 and 6% respectively of total variation observed in the treatment sums. The ranking of the entries for head yield, length and width was inconsistent over seasons. The top ten entries (head weight, length and width) during the long rainy season ranked differently than for the short rainy season and for both seasons. The percentage increase and decrease in head yield ranged between 1 and 16% and 0.3 and 49% respectively (Table 3). Gloria F_1 and Dragon F_1 recorded 49 and 42% for yield reduction, indicating their unsuitability for cultivation during the short rainy season. However, the short rainy season is characterized by inconsistent rainfall and high temperatures (31 °C). The adaptation of Quick Start and Bejo to the short rainy season was evident in the head yields. Our study showed that the long rainy season was best for cabbage production in Arusha and the cultivation of any of the top five varieties evaluated during the long rainy season could provide considerable head yield. For production during the short rainy season, Quick Start and Bejo will provide appreciable head yield. The large difference for seasons and seasons × entries interaction means squares indicate considerable influence of the environment. The foregoing indicates that head yield and characters are dependent on season. Hence the search for varieties with stable performance for head yield and characters over seasons and years is important.

Table 3. Means separation for head yield and head characteristics for 2008, 2008/09 and pooled for both year and percent increase and decrease of in head yield.

Variety	Hdl 2008	Hdl 2009	Hdl pooled	Hw 2008	Hw 2009	Hw Pooled	Mkwt (t/ha) 2008	Mkwt (t/ha) 2008/09	Mkwt (t/ha) Pooled	% Yield reduction	% Yield increase	Nmkwt (t/ha) 2008	Nmkwt (t/ha) 2008/09	Nmktwt pooled
Gloria F₁	23ᵃ	14ᵃ⁻ᵉ	15ᶠᵍⁱ	21ᶠ⁻ⁱ	17ᵍʰ	18ᵍ	44.0ᵃᵇ	23.33ᶠᵍ	26.00ⁱʲ	47	-	0.75ʰⁱ	4.67ᵇ⁻ʰ	4.73ᶠ⁻ᵏ
02-12609 (hybrid)	22ᵃ	14ᶜ⁻ᵉ	16ᶠᵍ	24ᵇ⁻ᵉ	19ᶜ⁻ᵍ	23ᵃ⁻ᵈ	32.67ᵉ⁻ⁱ	22.67ᶠᵍ	22.83ʲᵏ	31	-	0.67ʰⁱ	2.67ᶠ⁻ʰ	4.00ᵍʰ
Pruktor F₁	21ᵃᵇ	13ᵍʰ	13ⁱ	23ᶜ⁻ʰ	19ᵃ⁻ᵍ	20ᵉᶠ	36.33ᵈ⁻ⁱ	22.33ᶠᵍ	23.83ʲᵏ	39	-	4.03ᶜ⁻ᵍ	7.66ᵃᵇᶜ	6.50ᵃ⁻ᶠ
Glory of Enkhuizen-KB	20ᵃᵇᶜ	15ᵃ⁻ᶠ	16ᵉ⁻ᵍ	23ᶜ⁻ʰ	19ᶜ⁻ᵍ	20ᵉᶠ	34.00ᵈ⁻ⁱ	32.33ᵃ⁻ᵈ	34.00ᵈ⁻ᶠ	5	-	2.77ᵉ⁻ⁱ	11.00ᵃ	7.33ᵈ⁻ᵃ
Summer Glory	20ᵃ⁻ᵉ	16ᵃ⁻ᵈ	16ᵉ⁻ᵍ	22ᵈ⁻ʰ	16ʰ	17ᵍ	34.0ᵈ⁻ⁱ	30.33ᵇ⁻ᵉ	32.50ᵈ⁻ᵍ	11	-	5.33ᵇ⁻ᵉ	2.47ᶠ⁻ʰ	1.55ᵐᵐ
Thomas F₁	20ᵃ⁻ᵉ	17ᵃᵇᶜ	18ᵃᵇᶜ	25ᵃᵇᶜ	21ᵃ⁻ᵉ	21ᵈ⁻ᶠ	35.67ᵈ⁻ᵍ	35.00ᵃᵇ	38.67ᵇᶜ	2	-	3.21ᵉ⁻ʰ	2.16ᵍʰ	1.43ᵐᵐ
Cheers Hybrid	20ᵃ⁻ᶠ	16ᵃ⁻ᵈ	19ᵃᵇ	21ᶠ⁻ⁱ	22ᵃᵇᶜ	23ᵃ⁻ᵈ	44.33ᵃᵇ	28.67ᶜᵈᵉ	30.67ᶠ⁻ʰ	35	-	0.69ʰⁱ	2.10ᵍʰ	1.34ᵐᵐ
Bejo F₁	19ᵃ⁻ᵍ	17ᵃᵇᶜ	19ᵃ	21ᵍʰⁱ	21ᵃ⁻ᵉ	21ᵈ⁻ᶠ	32.67ᵉ⁻ⁱ	35.00ᵃᵇ	39.67ᵉ⁻ⁱ	-	7	2.53ᶠ⁻ⁱ	2.4ᶠ⁻ʰ	1.54ᵐⁿ
DragonF₁	18ᵇ⁻ʰ	14ᶠᵍʰ	13ʰⁱ	18ᵏ	23ᵃᵇ	22ᵇ⁻ᵈ	49.00ᵃ	28.50ᶜᵈᵉ	30.86ᶠ⁻ʰ	42	-	0.12ˡ	5.75ᵇ⁻ᵍ	5.86ᵇ⁻ᵍ
Globe Master (F₁)	18ᵇ⁻ʰ	15ᶜ⁻ᶠ	16ᶜ⁻ᵍ	21ᶠ⁻ⁱ	23ᵃᵇ	24ᵃᵇ	35.67ᵈ⁻ᵍ	31.50ᶜ⁻ᵉ	36.00ᵃ⁻ᶜ	12	-	3.67ᵈ⁻ᵍ	5.35ᵇ⁻ᵍ	4.42ᶠ⁻ᵏ
Summer Summit F₁	18ᵇ⁻ʰ	16ᵃ⁻ᵈ	16ᶜ⁻ᵍ	23ᶜ⁻ʰ	22ᵃᵇᶜ	22ᵇ⁻ᵈ	34.67ᵈ⁻ⁱ	36.33ᵃ	38.20ᵇᶜ	-	4.56	6.35ᵇ⁻ᵈ	6.35ᵇ⁻ᶠ	4.38ᶠ⁻ᵏ
Sky AceF₁	17ᵈ⁻ʰ	16ᵃ⁻ᵈ	16ᵉ⁻ᵍ	22ᶠ⁻ⁱ	19ᶜ⁻ᵍ	20ᵉᶠ	23.33ᵏ⁻ᵐ	23.25ᶠᵍ	23.29ᵐ	0.3	-	7.00ᵃᵇ	7.00ᵇ⁻ᵉ	7.00ᵃ⁻ᵉ
Quick Start	17ᵉ⁻ⁱ	18ᵃ	18ᵃᵇᶜ	22ᶠ⁻ⁱ	18ᶠᵍʰ	18ᵍ	26.33ᵉ⁻ⁱ	31.50ᵃ⁻ᵉ	42.67ᶠ⁻ʰ	-	16.4	9.33ᵃ	1.00ʰ	0.56ⁿ
Summer Autumn	17ᵉ⁻ⁱ	16ᵃ⁻ᵈ	18ᵃᵇᵉ	22ᵈ⁻ʰ	21ᵃ⁻ᵉ	21ᵈ⁻ᶠ	32.33ᵉ⁻ⁱ	28.67ᶜ⁻ᵉ	30.67ᶠ⁻ʰ	11	-	6.67ᵇᶜ	4.00ᶜ⁻ʰ	3.27ʰ⁻ⁱ
Summer Tide	17ᵉ⁻ⁱ	17ᵃᵇᶜ	19ᵃ	23ᶜ⁻ʰ	22ᵃᵇᶜ	22ᵇ⁻ᵈ	35.00ᵈ⁻ʰ	32.67ᵃ⁻ᵈ	34.50ᵈ⁻ᶠ	7	-	4.00ᶜ⁻ᵍ	5.00ᵇ⁻ᵍ	4.52ᶠ⁻ᵏ
Chou F₁ KK Cross	17ᵉ⁻ⁱ	17ᵃᵇᶜ	18ᵃᵇᶜ	23ᶜ⁻ʰ	24ᵃ	24ᵃ	35.00ᵈ⁻ʰ	28.67ᶜ⁻ᵉ	32.17ᵈ⁻ᵍ	18	-	5.33ᵇ⁻ᵉ	5.00ᵇ⁻ᵍ	4.10ᶠ⁻ᵏ
Rotan	17ᵉ⁻ⁱ	17ᵃᵇᶜ	19ᵃ	24ᵇ⁻ᵉ	23ᵃ⁻ᵈ	23ᵃ⁻ᵈ	26.33ᵏ⁻ᵐ	30.33ᵇ⁻ᵉ	32.17ᵈ⁻ᵍ	-	13.0	2.12ᵍ⁻ⁱ	4.17ᵇ⁻ʰ	3.46ʰ⁻ⁱ
Glory of Enkhuizen	17ᶠ⁻ⁱ	15ᶜ⁻ᶠ	16ᵉ⁻ᵍ	27ᵃ	23ᵃᵇ	23ᵃᵇ	23.0ᵐ	21.67ᶠᵍ	24.00ʲᵏ	6	-	5.33ᵇ⁻ᵉ	3.67ᵈ⁻ʰ	2.89ʲᵏˡ
Victoria F₁	16ᶠ⁻ⁱ	12ʰ	12ⁱ	23ᶜ⁻ʰ	21ᵃ⁻ᵉ	21ᵈ⁻ᶠ	42.67ᵇᶜ	32.86ᵃ⁻ᵈ	35.33ᶜ⁻ᵉ	23	-	3.07ᵉ⁻ʰ	3.33ᵈ⁻ʰ	2.77ᵏ⁻ⁿ
Riana F₁	16ᵍ⁻ⁱ	12ʰ	12ⁱ	24ᵇ⁻ᵉ	20ᶜ⁻ᵈ	23ᵃ⁻ᵈ	39.00ᵇ⁻ᵈ	30.00ᵇ⁻ᵉ	32.17ᵈ⁻ᵍ	23	-	3.80ᵈ⁻ᵍ	3.83ᶜ⁻ʰ	2.56ᵏ⁻ⁿ
Green Coronet F₁	16ᵍ⁻ⁱ	15ᶜ⁻ᶠ	15ᶠᵍʰ	18ᵏ	22ᵃᵇᶜ	22ᵇ⁻ᵈ	34.67ᵈ⁻ⁱ	27.33ᵈ⁻ᶠ	28.83ᵍ⁻ⁱ	20	-	0.63ʰⁱ	4.67ᵇ⁻ʰ	4.17ᶠ⁻ᵏ
Copenhagen Mkt	16ʰⁱʲ	16ᵃ⁻ᵈ	16ᶜ⁻ᵍ	19ʲᵏ	23ᵃᵇ	23ᵃ⁻ᵈ	28.67ᵇ⁻ᶠ	31.67ᵃ⁻ᵉ	33.33ᵈ⁻ᶠ	-	9.4	5.00ᵇ⁻ᶠ	7.33ᶜ⁻ʰ	5.67ᶜ⁻ʰ
Africana F₁	16ʰⁱʲ	16ᵃ⁻ᵈ	17ᵇ⁻ᵉ	20ⁱʲᵏ	21ᵃ⁻ᵉ	22ᵇ⁻ᵈ	32.33ᵉ⁻ⁱ	29.67ᶜ⁻ᵉ	32.17ᵈ⁻ᵍ	8	-	5.00ᵇ⁻ᶠ	8.00ᵃᵇ	7.18ᵃ⁻ᵈ
Bonus	16ʰⁱʲ	17ᵃ⁻ᵈ	16ᵉ⁻ᵍ	21ᶠ⁻ⁱ	20ᶜ⁻ᵈ	21ᵈ⁻ᶠ	27.33ⁱ⁻ᵐ	30.33ᵇ⁻ᵉ	33.00ᵈ⁻ᶠ	-	9.89	7.67ᵃᵇ	7.33ᶜ⁻ʰ	7.18ᵃ⁻ᵈ
Chou F₁ Tropical Cross	15ʰ⁻ᵏ	17ᵃᵇ	19ᵃᵇ	21ᵍʰⁱ	20ᶜ⁻ᵈ	21ᵈ⁻ᶠ	29.00ʰ⁻ᵏ	29.33ᶜ⁻ᵈᵉ	31.64ᵉ⁻ʰ	-	1.12	5.00ᵇ⁻ᶠ	5.67ᵇ⁻ʰ	5.50ᶠ⁻ʰ
Chou F₁ Santa	16ᵍ⁻ⁱ	15ᵃ⁻ᶠ	16ᶜ⁻ᵍ	23ᶜ⁻ʰ	23ᵃᵇ	23ᵃ⁻ᵈ	29.67ᵍ⁻ᵏ	29.00ᶜ⁻ᵉ	30.67ᶠ⁻ʰ	-	2.26	6.00ᵇ⁻ᵈ	4.67ᵇ⁻ʰ	5.67ᶜ⁻ʰ
Good Season	15ʰ⁻ᵏ	13ᶠᵍʰ	15ᶠᵍʰ	22ᵈ⁻ʰ	22ᵃᵇᶜ	22ᵇ⁻ᵈ	30.33ᵈ⁻ᵍ	34.00ᵃᵇ	34.50ᵈ⁻ᶠ	2	-	3.67ᵈ⁻ᵍ	4.33ᵇ⁻ʰ	4.83ᶜ⁻ʰ
Tropical Delight	14ⁱ⁻ˡ	14ᶠ⁻ʰ	15ᶠᵍʰ	22ᶠ⁻ⁱ	21ᵇ⁻ᶜ	20ᵉᶠ	35.67ᵈ⁻ᵍ	28.67ᶜ⁻ᵉ	30.50ᶠ⁻ʰ	20	-	5.31ᵇ⁻ᵉ	6.67ᵇ⁻ᵈᵉ	5.83ᵇ⁻ʰ
Super Master	13ʲ⁻ˡ	14ᶠ⁻ʰ	16ᶠᵍʰ	23ᶜ⁻ʰ	21ᵇ⁻ᶜ	21ᵈ⁻ᶠ	34.00ᵈ⁻ⁱ	21.67ᶠᵍ	24.00ʲᵏ	3136	-	6.00ᵇ⁻ᵈ	7.33ᵇᶜᵈ	8.33ᵃ
Drum Head	13ˡ	15ᵃ⁻ᶠ	15ᶠᵍʰ	22ᶠᵍʰ	18ᶠᵍʰ	23ᵃ⁻ᵈ	25.33ᵏ⁻ᵐ	16.00ʰ	21.67ᵏ	37	-	5.33ᵇ⁻ᵉ	8.00ᵃᵇ	7.83ᵃᵇ
Besta	13ᵏ⁻ˡ	15ᵃ⁻ᶠ	15ᶠᵍʰ	26ᵃᵇ	20ᶜ⁻ᵍ	20ᵉᶠ	34.33ᵈ⁻ⁱ	22.67ᶠᵍ	25.83ʲⁱ	34	-	1.29ᵍʰⁱ	8.33ᵇ⁻ᵍ	5.17ᶜ⁻ⁱ
Spring Light	12ˡ	16ᵃ⁻ᵈ	16ᵉ⁻ᵍ	22ᶠ⁻ⁱ	21ᵇ⁻ᶜ	22ᵇ⁻ᵈ	38.00ᶜ⁻ᵉ	26.33ᵈ⁻ᵍ	28.0ⁱʰ	31	-	2.21ᵍʰⁱ	4.16ᵇ⁻ʰ	5.08ᶜ⁻ʰ
Mean	17.44ᵃ	15.58ᵇ		22.59ᵃ	20.95ᵇ		33.60ᵃ	28.33ᵇ				4.05ᵇ	5.09ᵃ	

Hdl= Head length; Hdw = head width; Mkwt = marketable weight; Nmkwt = nonmarketable weight. Means within the same column followed by the same letter(s) are not significantly different at 5% probability level by DMRT.

Table 4. Percentage distribution of farmers' preference and rank (superscript) for characters in cabbage varieties during the 2008 long rainy season.

Cabbage varieties	Head Size (%)	Taste (%)	Earliness (%)	Firmness (%)	Head shape (%)	Head splitting (%)	Loose Head (%)	Days to harvest
Copenhagen Market	24	36	34	64	64	35	25	98
Glory of Enkhuizen	33	33	33	53	40	53	26	100
Drum Head	33	65	34	37	6	59	25	100
Globe Master (F_1)	55	62	66	55	63	65	43	111
Green Coronet F_1	30	53	52	59	56	71	45	113
Cheers Hybrid	80^1	62	68^4	78	72^3	73	43	101
02-12609 (hybrid)	35	50	33	29	28	55	43	113
Gloria F_1	54	64^4	74^2	88^4	77^2	80^3	51	85
Super Master	55	42	55	67	48	82^1	44	98
Riana F_1	57	52	68	78	64	75	45	118
Victoria F_1	48	53	53	76	56	54	50	100
Sky Ace F_1	19	61	35	46	50	82^1	50	102
Dragon F_1	79^2	32	70	75	52	77	53	100
Bejo F_1	47	51	48	62	58	56	52	128
Pruktor F_1	44	62	35	45	56	70	52	104
Thomas F_1	45	51	47	78	53	47	45	108
Glory of Enkhuizen-KB	35	48	44	48	43	36	44	126
Rotan	26	46	31	47	23	52	40	122
Spring Light	40	47	52	47	33	55	50	111
Besta	41	35	51	79	53	55	55	112
Good Season	52	59	60	62	63	56	53	111
Summer Tide	52	64	70^3	75	63	78	54	99
Summer Summit F_1	64	66^3	65	95^2	79^1	68	60^2	100
Tropical Delight	73^4	54	28	88^3	51	21	58^3	118
Summer Glory	22	73^1	78^1	67	68	49	42	82
Summer Autumn	64	50	55	75	66	36	64^1	100
Chou F_1 KK Cross	76^3	36	67	150^1	55	36	52	112
Africana F_1	45	51	41	67	50	43	23	102
Quick Start	39	56	40	54	48	43	46	98
Bonus	46	68^2	48	63	57	19	55	99
Chou F_1 Tropical Cross	41	65^4	39	68	57	19	45	123
Chou F_1 Santa	43	55	51	58	66	57	44	118

Participatory variety selection

Demographic variables of the respondents

The mean age of the respondents was 25 years, with a gender distribution of 67, 33% male and female respectively. Tsoho (2004) noted that young adult farmers have a higher aspiration to accept new technologies than old conservative farmers. Most of the cabbage farmers operate at a small scale with fields less than one hectare and they have between 5 - 15 years' experience in cabbage cultivation.

Using head shape as selection criteria, farmers' participatory research indicated that round-headed varieties Cheers Hybrid, Gloria F_1, and Summer Summit are preferred for head shape. Cheers Hybrid, Dragon F_1, Tropical Delight and Chou F_1/KK Cross recorded high preference percentage (Table 4) for head size. Earliness in cabbage could be a desirable character when the need arises for varieties with a short growth cycle to meet early market demands. Conversely, early maturing varieties do not store well in the field because they are liable to bolt; yield loss could be higher if market demand falls. Summer Glory, Gloria F_1, Summer Tide and Cheers Hybrid recorded high preference percentage for earliness and were categorized as early maturing varieties (Table

Table 5. Likert scale test of overall performance for horticultural characteristics among cabbage varieties by the research team during the 2008 long rainy season.

Cabbage variety	1	2	3	4	5	Mean score
Copenhagen Market	0.36	1.14	0.11	0.14	0.00	1.75
Glory of Enkhuizen	0.00	0.36	1.60	0.71	0.18	2.85
Drum Head	0.00	0.43	1.50	0.86	0.18	1.33
Globe Master F_1	0.04	0.21	1.18	1.00	0.71	3.14
Green Coronet F_1	0.00	0.21	1.18	1.00	0.36	2.75
Cheers Hybrid	0.04	0.00	0.96	1.71	0.71	3.42
02-1260 9 (hybrid)	0.14	0.14	0.53	1.71	0.36	2.88
Gloria F_1	0.18	0.36	1.18	0.43	0.00	2.15
Super Master	0.04	0.21	1.28	0.57	0.71	2.81
Riana F_1	0.04	0.21	0.78	1.57	1.07	3.67
Victoria F_1	0.04	0.50	0.86	1.57	1.07	4.04
Sky Ace F_1	0.14	0.86	0.53	0.43	0.18	2.14
Dragon F_1	0.04	0.00	0.43	2.00	0.89	3.36
Bejo F_1	0.00	0.21	1.18	1.43	0.71	3.53
Pruktor F_1	0.07	0.29	1.28	0.71	0.18	2.53
Thomas F_1	0.00	0.86	1.18	0.14	0.00	2.18
Glory of Enkhuizen-KB	0.07	0.79	0.86	0.43	0.00	2.15
Rotan	0.07	0.64	0.78	0.71	0.00	2.20
Spring Light	0.07	0.79	0.96	0.43	0.00	2.25
Besta	0.00	0.29	1.18	1.00	0.00	2.47
Good Season	0.00	0.57	1.18	0.43	0.36	2.54
Summer Tide	0.07	0.74	0.85	0.29	0.18	2.14
Summer Summit F_1	0.04	0.29	1.07	1.14	0.18	2.72
Tropical Delight	0.07	0.14	0.64	1.00	1.25	3.10
Summer Glory	0.54	0.36	0.21	0.14	0.00	1.05
Summer Autumn	0.18	0.86	0.43	0.43	0.18	2.05
Chou F_1 KK Cross	0.11	0.36	1.18	0.71	0.36	2.72
Africana F_1	0.21	0.29	1.18	0.43	0.16	1.68
Quick Start	0.43	0.29	0.53	0.43	0.00	2.86
Bonus	0.04	0.29	1.28	0.71	0.54	2.00
Chou F_1 Tropical Cross	0.29	0.57	0.10	0.14	1.00	3.10
Chou F_1 Santa	0.00	0.74	0.74	0.43	1.00	2.93

Score: 1 = Poor, 2 = Fair, 3=Good, 4 = Very good, 5 = Excellent. Mean score (2.50) = 5.00 (maximum score)/2.

5). Loose heads, an undesirable horticultural characteristic, is associated with a high incidence of cabbage head caterpillar and occurs if cabbage is attacked at an early growth stage. Preference (expressed as percentage) for Summer Autumn F_1 Summer Tide, Tropical Delight and Bonus was high due to low incidence of loose heads. These varieties performed better for tolerance to bolting when compared with the check. Sky Ace, Super Master and Gloria F_1 were preferred for low incidence of bolting (Table 5). Sensory evaluation (taste) showed that Summer Summit, Summer Glory, Bonus and Chou F_1 Tropical Cross were preferred for good taste.

Summer Summit F_1 was preferred by the respondents for multiple characters (taste, firmness, head shape and reduced percentage of loose heads). Tropical Delight was better for head size, insect pests' tolerance, firmness and low incidence of loose heads. Gloria F_1 was identified as an early maturing variety with good head solidity and shape, but with high incidence of bolting, which may be associated with earliness and poor field storability.

The responses of the research team from sub-Saharan Africa on the overall performance of the cabbage varieties during the 2008 long rainy season were analyzed on a Likert scale (Table 5). Victoria F_1, Riana F_1, Bejo F_1, and Tropical F_1 had high mean scores above 2.50, and were selected as best for overall field performance. Gloria F_1 recorded a mean score less than 2.15; this was below the average and it implies that some characters the research team found in other varieties are lacking in Gloria F_1. Both sampling units used for the participatory research differed in the preference for Gloria F_1 (Table 5).

Conclusion

In this study we evaluated cabbage varieties for heat yield and yield characters over seasons.

Participatory selection was used to select preferred varieties and characters. It was evident that sufficient genetic variation exists among the cabbage varieties evaluated. Relative difference for head yield, head length and width were highly inconsistent within season and over seasons as shown by significant entries and entry × seasons interact-tion, with no similarity among best performing entries within and over seasons. The late rainy season was best for cabbage production in Arumeru district.

Farmer assessment of cabbage varieties has the potential to improve the relevance of on-station researcher-designed trials to identify preferred characters. Farmers from Arumeru district identified head yield and size, head firmness, shape and taste as important characteristics for a good cabbage. Promising varieties that could perform as the check for head yield are Cheers HYB, Victoria F_1, Pruktor, Riana, Green Coronet, Tropical Delight, Summer Tide, Glory of Enkhuizen, Spring Delight and Besta. Preference and selection for varieties might differ within and between locations in the country, hence the need to conduct similar studies at other locations to note farmers' preferences and selections.

REFERENCES

Ceccarelli S, Grando S, Booth RH (1996). International breeding programmes and resource-poor farmers: Crop improvement in difficult environments. In: Eyzaguirre P, Iwanaga M, eds. Participatory plant breeding: Proceeding of a workshop on participatory plant breeding, 26-29 July 1995, Wageningen, The Netherlands. IPGRI, Italy. pp. 99-116.

Ceccarelli S, Grando S, Bailey E, Amri A, El-Felah M, Nassif F, Rezgui S, Yahyaoul A (2001). Farmer participation in barley breeding in Syria, Morocco, and Tunisia. Euphytica, 122: 521-536.

Gomez KA, Gomez AA (1984). Statistical Procedures for Agricultural Research. New York: John Wiley and Sons.

Ijoyah MO, Rakotomavo H (2007). Yield performance of five cabbage (*Brassica oleraceae* var. *capitata*) varieties compared with local variety under open field conditions in Seychelles. J. Sustain. Dev. Agric., 34: 76-80.

Massomo SMS (2002). Black rot of Cabbage in Tanzania; characterization of *Xanthomonas camprestris* pv.camprestris and disease management strategies. Ph.D Thesis. The Royal Veterinary and Agricultural University, Copenhagen, Denmark. p. 215.

Massomo SMS, RB Mabagala RB, Swai IS, Mortensen CN (2004a). Evaluation of varietal resistance in cabbage against the black rot pathogen *Xanthomonas campestris* pv. Campestris in Tanzania. Crop Prot., 23: 315-325.

Massomo SMS, Nielsen H, Mansfield-Giese K, Mabagala RB, Hockenhull J, Mortensen CN (2003). Identification and characterization of black rot (*Xanthomonas campestris* pv. Campestris from Tanznia by pathogenicity tests, Biolog, rep-PCR and fatty acid methyl ester analysis. Eur. J. Plant Pathol., 109:775-789.

Massomo SMS, Mabagala RB, Swai IS, Mortensen CN (2004b). Biological control of black rot (*Xanthomonas campestris* pv. Campestris) in Tanzania with bacillus strains. J. Phytopathol., 152: 98-105.

Monteiro A, Lunn T (1998). Trends and perspectives of vegetable brassica breeding. World Conference on Horticultural Research. 17-20 June 1998. Rome, Italy.

Mwasha AM (2000). Status of vegetable production in Tanzania In: Chada ML, Nono-Womdim R, Swai I, eds. Proceedings of the Second National Vegetable Research and Development Planning Workshop held at HORTI-Tengeru, Arusha, Tanzania, 25-26 June 1998. AVRDC. pp. 22-27.

Nkongolo KK, Chinthu KKL, Mulusi M, Vokhiwa Z (2008). Participatory variety selection and characterization of sorghum (*Sorghum bicolour* (L.) Moench) elite accessions from Malawi gene pool using farmers and breeders knowledge. Afr. J. Agric. Res., 3(4): 273-283.

Olowu T, Oladeji JO (2004). Statistics for agricultural extension research. In: Olowu T, ed. Research Methods in Agricultural Extension. AESON (ARMTI), Ilorin. p. 20.

Talekar NS (2000). Chinese cabbage. Proceedings of the 1st International Symposium on Chinese Cabbages. AVRDC, Shanhua, Tainan, Taiwan. pp. 67-69.

Thompson JK (2002). Yield evaluation of cabbage varieties. J. Agric. Technol., 5:15-19.

Tindall HD (1993). Vegetables in the Tropics Macmillan International College. 3rd Edition, London, UK. pp. 354-356.

Tsoho BA (2004). Economics of cassava based cropping system under small-scale irrigation in Sokoto state, Nigeria. Unpublished M.sc Thesis, Department of Agricultural Economics and Farm Management, University of Ilorin, Nigeria. pp. 30-31.

Van der Vossen HAM, Seif AA (2004). *Brassica oleracea* L. (headed cabbage) In: Grubben GJH, Denton OA, eds. PROTA 2: Vegetables/Légumes. [CD-Rom]. PROTA, Wageningen, Netherlands.

Vural H, Esiyok D, Duman I (2000). The culture of vegetables (Vegetable growing). Izmir, Turkey. p. 440.

Zoss M (2006). Vegetable value chain in Northern Tanzania: Mode of governance, collective actions and facilitators intervention Swiss Federal Institute of Technology Zurich. [Internet] [cited: 18 June 2009] Available from: http://www.northsouth.ethz.ch/capacity.

Effect of different sowing dates on the yield and yield components of direct seeded fine rice (Oryza sativa L.)

Nadeem Akbar[1], Asif Iqbal[1]*, Haroon Zaman Khan[1], Muhammad Kashif Hanif[1] and Muhammad Usman Bashir[1]

[1]Department of Agronomy, University of Agriculture, Faisalabad, Pakistan.
[2]House No. 631, Block 12, Rail Bazar, Jhang Sadar, Pakistan.

The research was conducted to study the response of yield and yield components of direct seeded fine rice (Oryza sativa L.) to different sowing dates during 2008 at Agronomic Research Area, University of Agriculture Faisalabad. Experiment was laid out in a randomized complete block design with three replications having net plot size of 4 x 3 m. Experiment comprised of six different sowing dates that is 31st May, 10th June, 20th June, 30th June, 10th July and 20th July. Experiment was sown in 20 cm apart rows with single row hand drill using seed rate of 75 kgha^{-1}. Nitrogen and phosphorous were applied at the rate of 120 and 60 kgha^{-1} respectively. Urea and diammonium phosphate (DAP) were used as sources of nitrogen and phosphorus respectively. Full dose of the phosphorous and half dose of the nitrogen were applied at sowing while the remaining half dose of nitrogen was applied in two equal splits, at 20-30 days and 45-55 days after sowing (DAS). Yield components like tillers per meter square, number of kernel per panicle and 1000-kernel showed significant response to different sowing dates. The crop sown on 20th June (D_3) produced the maximum number of productive tillers per meter square (328.7), kernel per panicle (94.47). 1000-kernel weight (18.82 g) and paddy yield (4468 kgha^{-1}). The result showed that the direct seeding of super basmati on 20th June gave the best results in term of entire yield and yield components.

Key words: Direct seeding, fine rice, sowing dates, yield and yield components.

INTRODUCTION

Rice (Oryza sativa L.) is a major food of the world and more than half of the population subsists on it. It is the main livelihood of rural population living in subtropical and tropical Asia and hundreds of millions people living in Africa and Latin America. It contains a number of energy rich compounds such as carbohydrates, fat, protein and reasonable amount of iron, calcium, thiamine, riboflavin and niacin (Juliano, 1993). In Pakistan next to wheat, rice is the second important food crop. Pakistan is 12th largest rice producing country. The area and production under rice is 2,515,000 ha and 5,563,000 tones respectively and the average yield is 2211 Kgha^{-1} (GOP, 2008). The common method of rice cultivation in Punjab is transplanting the nursery which is very laborious and time consuming job. The high cost of farm labor invariably delays transplanting and often leads to the use of aged seedling. To overcome these problems, the method of direct seeding is evolved. Direct seeding is an attractive and sustainable alternative to traditional transplanting of rice.

Direct seeding offers such advantages as faster and easier planting, reduced labor, earlier crop maturity by 7–10 days, more efficient water use and higher tolerance of water deficit, less methane emission and often higher profit in areas with an assured water supply. The sowing date for direct seeding of rice plays vital role in improving its growth and increasing the yield. The sowing time of the rice crop is important for three major reasons. Firstly, it ensures that vegetative growth occurs during a period of satisfactory temperatures and high levels of solar radiation. Secondly, the optimum sowing time for each cultivar ensures the cold sensitive stage occurs when the minimum night temperatures are historically the warmest (late January to early February). Thirdly, sowing on time guarantees that

---

*Corresponding author. E-mail: asifiqbaldasahi@yahoo.com.

Table 1. Effect of different sowing dates on yield and yield components of direct seeded fine rice.

Sowing dates	Productive tillers per meter square	Kernel per panicle	1000-kernel weight	Grain yield (kgha^{-1})
31st May	301.0 c	84.6 c	16.5 c	3562 c
10th June	317.0 b	90.2 b	17.4 b	3964 b
20th June	328.7 a	94.4 a	18.8 a	4468 a
30th June	287.0 d	79.2 d	15.2 d	2852 d
10th July	270.0 e	73.2 e	14.4 e	1908 e
20th July	256.3 f	67.2 f	13.4 f	1347 f
LSD value	3.404	1.85	0.28	94.65
CV (%)	0.64%	1.25%	0.99%	1.72%

grain filling occurs when milder autumn temperatures are more likely, hence good grain quality is achieved (Farrell et al., 2003). The delayed sowing result in the poor emergence and reduced heading panicle per meter square and spikelets per panicle and ultimately yield is affected (Hayat et al., 2003). The present study was therefore conducted to determine the effect of different sowing dates on the yield and yield components of direct seeded fine rice.

MATERIALS AND METHODS

Experiment was laid out in a randomized complete block design with three replications. Net plot size was of 4 x 3 m. Experiment was comprised of six different sowing dates that is 31st May, 10th June, 20th June, 30th June, 10th July and 20th July. Crop was sown with a single row hand drill using a seed rate of 75 kgha^{-1} in 20 cm apart rows. Nitrogen and phosphorus were applied at the rate of 120-60 kgha^{-1} to all the treatments. Urea and diammonium phosphate (DAP) were used as sources for nitrogen and phosphorus. Half of nitrogen along with full dose of phosphorus was applied in two equal splits, at 30 days and 55 DAS. Data on following observations was recorded during the course of study by using standard procedures. The parameters include the number of productive tillers per meter square, number of kernel per panicle, 1000-kernel weight (g), and paddy yield (kgha^{-1}) and economic analysis.

RESULTS AND DISCUSSION

Number of productive tillers per meter square

The data regarding the number of productive tillers per meter square is given in Table 1. It is evident from the data that number of productive tillers per meter square was affected significantly by different sowing dates. The crop sown on 20th June (D_3) produced the maximum number of productive tillers per meter square (328.7) followed by 10th June (D_2) sowing with 317.0 number of tillers per meter square. The lowest number of productive tillers per meter square (256.3) was observed when rice was sown on 20th July (D_6). This showed that total number of productive tillers gradually decreases as the sowing was done before 20th June or delayed after 20th June. This increase in number of productive tillers per meter square at 20th June sowing was attributed due to the favorable environmental conditions which enabled the plant to improve its growth and development as compared to other sowing dates.

These results are similar to that of Pandey and Agarwal, 1991 and they indicated that different sowing dates had significant effect on number of fertile tillers per meter square. These results are also in line with Rakesh and Sharma (2004) who are of the opinion that delay in planting resulted in significant decrease in number of productive tillers per meter square and ultimately the paddy yield.

Number of kernels per panicle

The data regarding the number of kernels per panicle is presented in Table 1. It is evident from the data that number of kernels per panicle was affected significantly by different sowing dates. The crop sown on 20th June (D_3) produced the maximum number of kernels per panicle (94.4) followed by 10th June (D_2) sowing giving 90.2 numbers of kernels per panicle. The lowest number of kernels per panicle (67.2) was observed when rice was sown on 20th July (D_6). This showed that number of kernels per panicle gradually decreases as the sowing was done before 20th June or delayed after 20th June. These results are similar to that of Akram et al. (2007) and Kameswara and Jackson (1997) who reported that number of kernels per panicle were significantly affected as sowing date is delayed. However these results are contrary to that of Habibullah et al. (2007), who reported that sowing date had no significant effect on number of grains per panicle.

1000-Kernel weight (g)

The data regarding the 1000-kernel weight (g) are given in Table 1. It is evident from Table 1 that 1000-kernel weight

Table 2. Economic analysis regarding the effect of different sowing dates on yield and yield components of direct seeded fine rice.

Sowing dates	Gross income (Rs)	Total expenditure (Rs)	Net benefit (Rs)
31st May	96179	45070	51109
10th June	107025	44205	62820
20th June	120623	43341	77282
30th June	77004	42476	34528
10th July	51520	41612	9908
20th July	36369	40747	-4378

(g) was affected significantly by different sowing dates. The crop sown on 20th June (D_3) produced the maximum 1000-kernel weight (g) (18.8) followed by 10th June (D_2) sowing giving 17.4 1000-kernel weight (g). The lowest 1000-kernel weight (g) 13.4 was observed when rice was sown on 20th July (D_6). This showed that 1000-kernel weight gradually decreases as the sowing was done before 20th June or delayed after 20th June. This showed that the environmental conditions like temperature, humidity was most favorable for grain development during 20th June as compared to other sowing date. These results are in line with that of Tari et al. (2007) and they reported that effect of sowing date on 1000-kernel weight (g) was significant at 0.01 probability level.

Grain yield (kgha^{-1})

The data regarding the grain yield (kgha^{-1}) are given in Table 1. It is evident from Table 1 that grain yield (kgha^{-1}) was affected significantly by different sowing dates. The crop sown on 20th June (D_3) produced the maximum grain yield (4468 kgha^{-1}) followed by 10th June (D_2) sowing giving 3964 kgha^{-1} grain yield. The lowest grain yield (1347 kgha^{-1}) was observed when rice was sown on 20th July (D_6). This showed that the grain yield gradually decreases as the sowing was done before 20th June or delayed after 20th June. The higher yield in case of sowing on 20th June was attributed to increased cumulative mean value of temperature and sunshine hour due to early sowing, more number of productive tillers, more number of kernels per panicle, and increase 1000-kernal weight.

These results are in line with that of Iqbal et al. (2008) who reported that the highest yield (4-5 tha^{-1}) was obtained when the rice crop was sown earlier in the season. Similarly, according to Baloch et al. (2006) among planting dates, June 20th planted crop gave highest paddy yield.

Economic analysis

Data pertaining to economic analysis is given in Table 2. A perusal of Table 2 indicated that maximum net income of Rs 77282 was recorded when rice was sown on 20th June. High net income was due to high paddy yield and straw yield as well as due to fewer pest attacks thereby reducing the additional cost of inputs in terms of pesticide application. This was followed by the rice sown at 10th June and 31st May giving net income of Rs 62820 and 51109 respectively.

The minimum net income of Rs 9907 was observed when rice was sown on 10th July. However a loss of Rs 4378 was observed when the rice was sown on 10th July. This showed that the net benefit gradually decreased as sowing was done after 20th June and later on. Seeding rice before the predicted optimum periods would lengthen the time between seeding and emergence; increase production costs from the use of recommended seed treatments, higher seeding rates; a longer period for pest control and possibly result in poor stand establishment (Slaton et al., 2007).

Conclusion

The conclusion drawn from this investigation is that, the sowing dates significantly affects the entire yield and yield components of direct seeded fine rice. The result showed that the direct seeding of super basmati on 20th June gave the best results in term of entire yield and yield components.

REFRENCES

Akram HM, Ali A, Nadeem MA, Iqbal MS. (2007). Yield and Yield components of rice varieties as affected by transplanting dates. J. Agric. Res., 45(2): 105-111.

Baloch MS, Ullah AI, Gul H (2006). Growth and yield of rice as affected by transplanting dates and seedlings per hill under high temperature of Dera Ismail Khan, Pakistan. J. Zhejiang Univ. Sci., 7(7):572-579.

Farrell TC, Fox K, Williams RL, Fukai S, Lewin LG (2003). Avoiding low temperature damage in Australia's rice industry with photoperiod sensitive cultivars. Proceedings of the Australian Agronomy Conference, Australian Society of Agronomy.

Govt. of Pakistan (2008). Economic survey of Pakistan 2007-08. Pakistan Food Agriculture and Livestock Division. (Economic wing), pp. 17-23.

Hayat K, Awan IU, Hassan G (2003). Impact of seeding dates and varieties on weed infestation yield and yield components of rice under direct wet seeded culture. Pak. J. Weed Sci. Res., 9(1-2):59-65.

Habibullah N, Shah H, Ahmad N, Iqbal F (2007). Response of rice varieties to various transplanting dates. Pak. J. Plant Sci., 13 (1): 1-4.

Iqbal S, Ahmad A, Hussain A, Ali MA, Khaliq T, Wajid SA (2008). Influence of transplanting date and nitrogen management on productivity of paddy cultivars under variable environments. Int. J. Agric. Biol., 10(3): 288-292.

Juliano BO (1993). Rice in human nutrition (FAO Food and Nutrition Series No.26) Int. Rice Res. Inst. Manila, Philippines, pp. 40-41.

Kameswara RN, Jackson MT (1997). Effect of sowing date and harvest time on longevity of rice seeds. Seed Sci. Res., 7(1): 13-20.

Pandey R, Agarwal MM (1991). Influence of fertility levels, varieties and transplanting time on rice (Oryza sativa). Indian J. Agron., 36: 459-463.

Rakesh K, Sharma HL (2004). Effect of dates of transplanting and varieties on dry matter accumulation, yields attributes and yields of rice (*Oryza sativa* L.). Himachal J. Agric. Res., 30 (1): 1-7.

Tari DB, Pirdashti H, Nasiri M, Gazanchian A, Hoseini SS (2007). Determination of Morphological Characteristics Affected by Different Agronomical Treatments in Rice (IR6874-3-2 Promising Line). Asian J. Plant Sci., 6 (1): 61-65.

Combining the yield ability and secondary traits of selected cassava genotypes in the semi-arid areas of Eastern Kenya

Joseph Kamau[1], Rob Melis[2], Mark Laing[2], John Derera[2], Paul Shanahan[2] and Eliud Ngugi[3]

[1]Kenya Agricultural Research Institute- Katumani, Nairobi, Kenya.
[2]University of KwazuluNatal – Petermaritzburg, South Africa.
[3]University of Nairobi, Kabete Campus, Nairobi, Kenya.

Despite the importance of cassava for food security in semi-arid areas of Kenya, there is a lack of information regarding gene action determining yield of local varieties. Therefore the objective of this study was to estimate the combining ability for yield and associated secondary traits by crossing popular local varieties with some varieties from IITA using a NC II mating design. The F1 progenies were evaluated in a seedling trial laid out as a 7 × 7 simple lattice with two replicates. Results indicated significant variation among progenies for shoot weight, root number, root weight, root yield, biomass, harvest index, percentage dry matter, dry matter yield, cyanide content, and resistance to cassava mosaic disease and green mites. Average fresh root weight at 6 mo ranged from 1.1 kg to 1.4 kg plant^{-1}. To a great extent SCA effects (57 to 75%) explained variation for shoot weight, root weight, harvest index, dry matter content, root cyanide content and resistance to cassava mosaic, while GCA effects (55%) were more important for root number. Thus, our results suggested that non-additive gene action was more important than additive gene action in influencing yield and most of its associated traits in this cassava population. Overall, the results suggested that the success of cassava breeding in the semi-arid areas would depend on the ability of breeders to assemble heterotic groups of germplasm that combine well in order to achieve early vigour, disease and pest resistance, root quality and high yield potential.

Key words: Cassava, yield, secondary traits.

INTRODUCTION

Cassava is one of the leading staples in the world with a global total production of 233 million metric tones from 18.6 million ha (FAOSTAT, 2008). It was long ago established as the fourth most important staple food in the tropics (De Vries et al., 1967). For, example in topical Africa, total cassava production was about 118 million metric tones in 2008 (FAOSTAT, 2008). It is commonly cultivated in areas considered marginal for most other crops and is adaptable to low soil fertility and erratic rainfall ranging from less than 600 mm in semi-arid tropics to more than 1000 mm in the humid tropics. It survives prolonged drought of 4 to 7 mo during the growing cycle in north eastern Brazil (El-Sharkawy, 2003). It requires minimum inputs, which makes it ideal for drought prone areas in tropical and sub-tropical Africa, Asia and America (El-Sharkawy, 2003).

In Kenya, cassava is grown in both semi-arid and high rainfall areas for food security and as a cash crop. Surplus cassava is sold to earn income for the family. However, the varieties grown by farmers in this region are landraces that are late bulking and have low root yield potential. In order to improve the yield potential of these landraces, an understanding of the gene effects controlling root yield and secondary traits affecting yield is important. Such knowledge would assist in devising the best breeding strategy to improve early bulking and yield potential (Kariuki et al., 2002).

Improving the local landraces requires a hybridisation programme to generate hybrid progenies for selection and recombination (Fehr, 1984). Population improvement and recurrent selection in cross-pollinated crops progressively increases the frequencies of genes for specific

*Corresponding author. E-mail: jkamauw@yahoo.com.

specific desirable traits (Hahn et al., 1980). However, the success of population breeding depends largely on the choice of parents. Parental genotypes are usually selected on the basis of their performance or the performance of their F_1 progenies (Banziger and Paterson, 1992). In maize, selection of parental genotypes to produce F_1 hybrids is usually based on the performance of their progenies (Fehr, 1984; Lee, 1995). However, experienced breeders with fully characterised core germplasm, also use direct evaluation of parents, when breeding, for simply knowing the inherited traits in maize (Lee, 1995). Cassava breeders have traditionally used performance per se of parental genotypes (CIAT, 2004). In the current study, parental genotypes were selected based on their performance per se in semi-arid Eastern Kenya. The local varieties, though late bulking, have good root qualities and are popular with the farmers in the area (Kiarie et al., 1991). The IITA varieties, used to cross with the local produced popular varieties that were early bulking, but lacked certain attributes acceptable to farmers (Kamau et al., 1998). It was assumed that crossing the two groups (local and IITA), would result in new genotypes, which combine early bulking with acceptable root qualities.

Plant breeders and geneticists frequently use diallel-mating design to obtain genetic information (Sprague and Tatum, 1942; Griffing, 1956; Eberhart and Gardner, 1966). Analyses of broad based populations were generally conducted according to Eberhart and Gardner (1966) analyses I, II and III. Apart from diallel design, breeders also use factorial mating designs such as the North Carolina (NC) mating designs I, II, and III (Comstock and Robinson 1948; 1952) to generate genetic information on parents based on progeny performance. Genetic information generated by these mating designs is used to estimate general combing ability of the parental genotypes and specific combining ability of the progenies (Sprague and Tatum, 1942; Haulauer and Miranda, 1995).

In this study, the NC II mating design was used to generate the progenies from crosses between two groups of parents (local versus IITA varieties). Several researchers have used this design in, for example, sugar cane (Hogarth et al., 1981), variety crosses in maize (Eberhart and Gardner, 1966), maize (Pixley and Bjamason, 1993; Derera et al., 2000) and even feed conversion in broiler rabbits (Dedkova et al., 2002). In cassava, the design has been used to study resistance to cassava mosaic disease (Lokko et al., 2004). Combining abilities in cassava were creatively estimated, because of the difficulties of obtaining reliable family (cross combination) mean values for traits. In most cases, data collected on plants were selected from the seedling and later selection stages. Thus, the combining ability information on cassava lines is estimated from a small group of superior progenies, which germinated or a few advanced into the clonal trials (Ceballos et al., 2004). In addition, the problem with this approach is that the combining ability estimates will not be based on a random, unselected progeny population and will therefore be biased. With selection, non-additive effects tend to increase.

MATERIALS AND METHODS

Selection of parents

The selection of parents, to build the population of the future cassava breeding work for the mid-altitude eastern semi-arid areas of Kenya, began when open pollinated derived seeds were introduced from IITA, Ibadan, Nigeria from 1994 to 2000. The seeds were mainly bulk collections from the trials. Selected genotypes were evaluated on station trials at KARI-Katumani main centre, Kampi Ya Mawe, and Ithookwe sub-centres over several seasons. The superior genotypes were advanced by subjecting them to on-farm testing by farmers. Farmers used their experience to observe the growing habit of the various genotypes and performed a palatability test at the end of each trial. Palatability tests of raw and boiled roots were based on appearance of fresh and boiled roots, taste (bitter or sweet) and fibre (presence or absence) (Kamau et al., 1998; Githunguri and Migwa, 2003). The four local entries were popular local varieties with high root yield, good root quality and were tolerant to cassava mosaic disease. Their selection for this research was based on their performance per se and not on the performance of their progenies.

Crossing block

A crossing block was established at KARI-Kiboko farm in 2004 with four popular, but late, bulking varieties and six early bulking varieties from the IITA germplasm. The varieties were crossed following the NC II mating design. The local varieties were used as the females and the IITA as the males. The method of pollination was a modification of that employed by IITA (IITA, 1982).

Seedling nursery

Preliminary experiments were done at KARI-Katumani to establish optimum conditions for uniform germination of the cassava seeds. The hybrid seeds were germinated at 36°C in the laboratory. The germinated seeds were planted in 5 cm × 8 cm black polythene bags and grouped according to family. The bags were filled with forest soil that had been cooked for 4 d to kill most of the microorganisms. Soil analysis was conducted to determine the mineral composition of the forest soil (Table 1).

The seedbeds were covered with a clear polythene sheet that created a humidity chamber. The temperature inside a seedbed without the seedlings rose up to 50°C when the outside air temperature was 30°C. Therefore, to keep the seedbeds temperature at 2°C above the air temperature, the sides of the seedbeds were lifted between 9.00 am and 4.00 pm every day.

After 21 d the seedlings were transported to KARI-Kiboko farm where they were, once again, arranged into family groups. They were left in the open for 4 d to harden up and were watered twice daily.

Seedling field trial

KARI-Kiboko farm is located along the Mombasa-Nairobi road located at 2° 10'S; 37° 40' E and 975 m altitude. The KARI-Kiboko farm, at which the F_1 seedling trial was conducted, receives bimodal rainfall although there are yearly variations with peaks usually between March - May and from October – December. The monthly

Table 1. Mineral composition of the forest soil.

Sample Description	pH	P	K	Ca	Mg	Na	Zn	Fe	Mn	Cu	N
Unit					(ppm)						(%)
Forest soil	4.1	10	248	585	140	27	1.54	168	26	0.98	0.31

Analysed at the Del Monte Kenya limited, Thika.

Table 2. KARI-Kiboko farm monthly rainfall data (mm) between November 2003 and June 2006

Months	Period of experimentation			
	2003	2004	2005	2006
January		143.0	6.5	12.4
February		49.0	0.0	6.0
March		22.5	40.5	85.7
April		70.8	186.5	205.8
May		0.0	13.8	43.5
June		0.0	0.0	0.0
July		0.0	0.0	0.0
Aug		0.0	2.5	0.0
September		5.0	0.5	
October		15.0	20.5	
November		49.5	57.5	
December	31.5	113.6	9.2	
Total	31.5	468.4	337.5	353.4

rainfall for the period of experimentation, December 2003 to August 2006, is provided in Table 2. The soil at Kiboko farm is ferric luvisols (Hornetz et al., 2000).

The seedlings were planted at Kiboko farm in a 7 × 7 simple lattice design with two replications on 2nd December 2004, where only the families were replicated. Sixteen full-sibs from a family were planted in each plot per replication at the commercial spacing of 1 m × 1 m. The plots and blocks were separated by 1.5 and 2.0 m wide alleys, respectively, to avoid competition from neighbouring families. Stakes were used to plant the parental genotypes in the trial. No mineral fertilizer was applied at planting and during growing period. Sprinkler irrigation was used to supplement the rains when necessary. The experiment was weeded once every month and no fertiliser was added.

The trial was harvested by hand when the plants were 6 mo old. The individual plants were assessed for their number of storage roots per plant and root yield per plant. Shoot weight was determined by weighing the stems and leaves of each plant. Plot data on number of tuberous roots and yield was averaged over the plants harvested in each plot.

Specific gravity of root samples was measured on an individual plant basis. Dry matter content was determined indirectly based on the correlation between root specific gravity and dry matter (Kawano et al, 1987). Measurement of specific gravity was obtained by weighing roots in air and then in water. The weight in water was measured by submerging the roots in a net into a 200 L container with water. Dry matter content (DM %) percentage was determined using the formula:

DM % = 158.3 × weight in air / (weight in air − weight in water) − 142.

Dry matter yield (DMY) per hectare was estimated by multiplying the fresh root yield per hectare by the dry matter content (Kawano et al., 1987):

DMY= (DM % /100) × fresh root yield

Harvest index (HI %) was computed as the ratio of root weight to the total harvested biomass per genotype on fresh weight basis:

HI %= (root weight/ biomass) × 100

Cyanide content in the roots of each genotype was estimated using the semi-quantitative determination (O'Brien et al., 1994). Cyanide content was determined by colour change from pale green to dark brown of the picrate on the paper strip (125 mm Whatman® filter paper).

Reaction to cassava mosaic disease and green mites were assessed on individual F_1 genotypes at 3, 4 and 5 mo after planting. A scale of 1 − no apparent symptoms, 2 − mild symptoms and 3 − severe symptoms was used to rate the genotypes for resistance to cassava mosaic disease (CMD) and green mites.

Data analysis

The parental varieties were considered as a fixed reference population; consequently results only pertain to this set of heterozygous genotypes. Even though the selected parents represent the superior groups for the breeding programme at KARI-Katumani, the inferences drawn from this study are not to be generalised. The REML (residual maximum likelihood) procedure in the Genstat Version 9 statistical software package was used to analyse the data. General combining ability (GCA), effects and

Table 3. Mean square values for yield, secondary traits, disease and pests

Source	df	Mean square value										
		TSW	RTN	RTW	RTY	Biomass	HI	DM	DMY	RCNP	CMD	CGM
Crosses	23	1.80*	2.30**	2.53**	2.53**	1.82**	2.36**	1.50*	2.32**	1.84**	2.73**	1.46
GCA (IITA)	5	2.24	5.92**	2.98*	2.98*	2.24*	1.80	2.28*	3.35**	0.19	1.04	2.43*
GCA (LOCAL)	3	1.89*	0.29	0.77	0.77	2.16	0.58	1.17	0.36	0.07	5.17**	0.38
SCA (Local × IITA)	15	0.04*	1.66*	1.74*	1.74	1.61	2.19**	1.34	1.83*	0.02*	2.83**	1.25
ERROR	23	0.30	0.66	0.02	2.21	0.36	6.31	6.25	0.44	0.05	0.01	0.01

*, ** Significant at P≤ 0.05 and P ≤ 0.01 probability, respectively. Shoot weight (TSW kg plant^{-1}), root number (RTN count), root weight (RTW kg plant^{-1}), root yield (RTY t ha^{-1}), total biomass (kg plant^{-1}), harvest index (HI %), dry matter (DM %), dry matter yield (DMY t ha^{-1}), root cyanide content (RCNP score), cassava mosaic disease (CMD score) and cassava green mite attack (CGM score)

Table 4. Proportion (%) of GCA and SCA effects relative to the sum of squares for the crosses.

Trait	GCA (%)		SCA (%)
	IITA	Local	Local × IITA
Shoot weight (kg plant^{-1})	26.38	13.33	60.30
Root number (count)	53.49	1.55	44.95
Root weight (kg plant^{-1})	34.43	5.32	60.25
Total biomass (kg plant^{-1})	26.76	15.47	57.77
Harvest index (%)	20.69	3.97	75.34
Dry matter content (%)	32.49	9.99	57.52
Dry matter yield (t ha^{-1})	37.03	2.41	60.57
Root cyanide content (score)	17.78	16.61	65.61
Cassava mosaic disease (score)	8.22	24.57	67.21
Cassava green mites (score)	37.94	3.53	58.52

GCA – general combing ability, SCA – specific combing ability

specific combining ability (SCA) effects were estimated using the following model:

$$Y_{ijk} = \mu + Fg_i + Mg_j + FMs_{ij} + R_k + E_{ijk,}$$

Where,

Y_{ijk} is the observed value for a cross between the ith and jth parents in the kth replication;
μ is the general population mean;
Fg_i is the GCA value of the ith maternal parent;
Mg_j is the GCA value of the jth paternal parent;
FMs_{ij} is the SCA value for the cross between the ith and jth parent;
R_k is the replication effect;
E_{ijk} is the experimental error.

In this model, the terms Fg_i and Mg_j estimated GCA effects due to the local varieties and IITA varieties, respectively, while the interaction term, FMs_{ij}, estimated SCA effects. The GCA and SCA variances provide an indication of the levels of additive and non-additive variance in a population respectively (Falconer and Mackay, 1996). Pearson's phenotypic correlation coefficients were also calculated between root yield and the following: shoot weight, root number, root weight per plant, root yield, biomass yield, harvest index, dry matter content and dry matter yield.

RESULTS

REML analysis of variance for agronomic traits

Among the crosses, significant differences (p<0.05) were identified for shoot weight (TSW) and percentage dry matter content (Table 3). Other traits that were significantly different (p<0.01) were root weight (RTW kg plant^{-1}), root yield (RTY t ha^{-1}), total biomass, harvest index, root dry matter yield and resistance to cassava mosaic disease. However, resistance to cassava green mites was not significantly different. The IITA varieties did not differ significantly for shoot weight, harvest index, root cyanogenic potential and reaction to cassava mosaic disease. Also, the local varieties differed significantly in shoot weight (p<0.05) and reaction to cassava mosaic disease (p<0.01). General combining ability (GCA) effects were estimated for those traits that were significant (Table 4). The SCA effects were significant (p<0.05) for shoot weight, root number, dry matter yield and root cyanide content, while harvest index and reaction to cassava mosaic disease were highly significant

Table 5. GCA effects of genotypes for shoot weight and root number.

Genotypes	Source	Shoot weight			Root number		
		Mean (kg plant^{-1})	GCA (kg plant^{-1})	GCA (SE)	Mean	GCA	GCA (SE)
820001	Local	3.35	-0.37*	0.16	8.63	-0.05	0.23
820058	Local	3.58	-0.14	0.16	8.35	-0.16	0.23
990010	Local	4.18	0.46*	0.16	8.73	0.06	0.23
990014	Local	3.78	0.06	0.16	8.67	0.15	0.23
960249	IITA	3.32	-0.41*	0.19	8.03	-0.48	0.29
990056	IITA	4.26	0.53*	0.19	9.82	1.03*	0.29
990067	IITA	4.11	0.39*	0.19	9.25	0.54*	0.29
990072	IITA	3.51	-0.21	0.19	7.51	-0.99**	0.29
990127	IITA	3.82	0.10	0.19	8.75	0.19	0.29
990183	IITA	3.31	-0.41*	0.19	8.21	-0.30	0.29

*, **, *** Significant at 0.05, 0.01 and 0.001; GCA – general combining ability, SE - standard error.

Table 6. The genotypes GCA effect for root weight.

Genotypes	Source	Root weight		
		Mean (kg plant^{-1})	GCA (kg plant^{-1})	GCA (SE)
820001	Local	1.06	-0.04	0.04
820058	Local	1.04	-0.05	0.04
990010	Local	1.17	0.08	0.04
990014	Local	1.11	0.01	0.04
960249	IITA	1.20	0.10	0.05
990056	IITA	1.03	-0.07	0.05
990067	IITA	1.24	0.14*	0.05
990072	IITA	1.10	0.00	0.05
990127	IITA	1.05	-0.05	0.05
990183	IITA	0.97	-0.13*	0.05

*, **, *** Significant at 0.05, 0.01 and 0.001; GCA – specific combining ability, SE – standard error.

the sum of squares for crosses between the local and IITA varieties were very variable. The local varieties contributed less GCA effects for most of the traits, except for the reaction to cassava mosaic disease, for which they contributed 24.57%. The SCA effects were more important for most of the traits except for root number, which had 53.49% GCA from the IITA lines (Table 4).

General combining ability

Among the parental genotypes, 990056 from IITA had the highest GCA effects for shoot weight and root number. Although the GCA values were low, some progenies from the crosses of the 990010 (Local) and 990056 and 990067 (both IITA) had positive and significant (P<0.005) GCAs for shoot weight, while 990056 and 990067 had significant GCAs for root number (Table 5). The GCA effects for the mean root number were significant for the IITA varieties 990056 and 990067 (Table 5).

The local parental genotypes had non-significant GCA effects for root yield (Table 6). However, among the IITA varieties, 990067 had the highest, significant (P<0.005) GCA effects for root weight per plant (Table 6).

Total biomass GCA effects for 990056 and 990067 (IITA) were significant (p=0.05). Genotype 960249 had the highest and significant (P<0.01) GCA effects for harvest index (Table 7).

The GCA effects for dry matter content were quite low except for 990127, which was significant (P<0.05) and positive. In dry matter yield, only the highest yielding genotype, 990067, had positive and significant (P<0.01) GCA effects (Table 8).

The local varieties, 820058 and 990010, had significant GCA effects of (P<0.05) and (P<0.01), respectively, for low and high root cyanide content respectively (Table 9). The two local cultivars, 990014 and 620001, had significant (negative and positive, respectively) GCA effects for reaction to cassava mosaic diseases.

Table 7. Parental varieties GCA effects and standard errors for biomass and harvest index.

Genotypes	Source	Biomass (kg plant^{-1})			Harvest index (%)		
		Mean	GCA	GCA (SE)	Mean	GCA	GCA (SE)
820001	Local	4.41	-0.41*	0.17	25.17	1.24	0.73
820058	Local	4.62	-0.20	0.17	23.24	-0.7	0.73
990010	Local	5.36	0.54**	0.17	24.00	0.07	0.73
990014	Local	4.89	0.06	0.17	23.33	-0.60	0.73
960249	IITA	4.52	-0.30	0.21	26.66	2.72**	0.89
990056	IITA	5.28	0.46*	0.21	22.95	-0.98	0.89
990067	IITA	5.35	0.53*	0.21	23.85	-0.09	0.89
990072	IITA	4.61	-0.21	0.21	24.30	0.36	0.89
990127	IITA	4.87	0.05	0.21	22.62	-1.32	0.89
990183	IITA	4.29	-0.53*	0.21	23.23	-0.71	0.89

*, **, *** Significant at 0.05, 0.01 and 0.001; GCA – general combining ability, SE – standard error.

Table 8. The genotype GCA effects for dry matter content and dry matter yield.

Genotypes	Source	Dry matter content (%)			Dry matter yield (t ha^{-1})		
		Mean	GCA	GCA (SE)	Mean	GCA	GCA(SE)
820001	Local	38.23	0.07	0.72	4.30	-0.13	0.20
820058	Local	38.95	0.79	0.72	4.30	-0.11	0.20
990010	Local	38.04	-0.12	0.72	4.70	0.28	0.20
990014	Local	37.43	-0.73	0.72	4.40	-0.10	0.20
960249	IITA	37.77	-0.39	0.88	5.00	0.06	0.20
990056	IITA	38.26	0.10	0.88	4.20	-0.20	0.20
990067	IITA	38.75	0.59	0.88	5.10	0.60**	0.20
990072	IITA	35.08	-3.08**	0.88	4.20	-0.30	0.20
990127	IITA	40.01	1.85*	0.88	4.20	-0.20	0.20
990183	IITA	39.09	0.93	0.88	4.00	-0.50*	0.20

*, **, *** Significant at 0.05, 0.01 and 0.001; GCA – general combining ability, SE – standard error.

Table 9. Genotype GCA for root cyanide ratio and reaction to cassava mosaic disease.

Genotypes	Source	Root cyanide ratio			Cassava mosaic disease		
		Mean	GCA	GCA (SE)	Mean	GCA	GC (SE)
820001	Local	4.26	-0.07	0.07	1.22	0.11**	0.03
820058	Local	4.16	-0.17*	0.07	1.09	-0.02	0.03
990010	Local	4.55	0.22**	0.07	1.08	-0.03	0.03
990014	Local	4.36	0.03	0.07	1.05	-0.06*	0.03
960249	IITA	4.21	-0.12	0.08	1.07	-0.04	0.03
990056	IITA	4.49	0.15	0.08	1.13	0.03	0.03
990067	IITA	4.37	0.04	0.08	1.09	-0.02	0.03
990072	IITA	4.31	-0.02	0.08	1.13	0.02	0.03
990127	IITA	4.20	-0.13	0.08	1.16	0.05	0.03
990183	IITA	4.41	0.08	0.08	1.06	-0.04	0.03

*, **, *** Significant at 0.05, 0.01 and 0.001; GCA – general combining ability, SE – standard error.

Table 10. Mean and SCA effects of crosses for shoot weight (kg plant^{-1}), root number and root weight (kg plant^{-1}).

Cross	Shoot weight		Root number		Root weight	
	Mean	SCA	Mean	SCA	Mean	SCA
820001 × 960249	3.43	0.49	8.47	0.50	1.14	-0.02
820001 × 990056	3.33	-0.55	9.69	0.13	0.94	-0.05
820001 × 990067	3.46	-0.28	7.96	-1.83**	0.97	-0.23*
820001 × 990072	3.62	0.49	7.82	0.36	1.12	0.05
820001 × 990127	2.95	-0.50	10.38	1.53*	1.11	0.10
820001 × 990183	3.30	0.37	7.47	-0.69	1.09	0.16
820058 × 960249	3.88	0.71	7.00	-0.87	1.11	-0.03
820058 × 900056	3.58	-0.53	10.47	1.09	1.24	0.26**
820058 × 900067	3.96	0.00	9.41	0.52	1.17	-0.02
820058 × 990072	3.24	-0.12	6.87	-0.49	1.02	-0.03
820058 × 990127	3.75	0.07	7.74	-0.81	0.80	-0.19
820058 × 990183	3.04	-0.12	8.60	0.55	0.92	0.01
990010 × 960249	2.58	-1.19*	8.82	0.73	1.35	0.07
990010 × 990056	5.70	0.99*	8.73	-1.87**	0.99	-0.12
990010 × 990067	5.38	0.81*	9.92	0.81	1.46	0.15
990010 × 990072	4.11	0.14	8.47	0.90	1.27	0.10
990010 × 990127	3.68	-0.60	8.11	-0.65	0.96	-0.16
990010 × 990183	3.63	-0.14	8.33	0.07	1.01	-0.04
990014 × 960249	3.37	-0.01	7.82	-0.36	1.20	-0.02
990014 × 990056	4.41	0.10	10.39	0.65	0.95	-0.09
990014 × 990067	3.65	-0.52	9.69	0.49	1.36	0.11
990014 × 990072	3.05	-0.51	6.89	-0.78	0.99	-0.12
990014 × 990127	4.92	1.04*	8.78	-0.08	1.32	0.26*
990014 × 990183	3.26	-0.10	8.43	0.07	0.86	-0.13
Statistics						
Mean	3.72		8.59		1.10	
SED	0.68		1.01		0.18	
SCA SE		0.39		0.57		0.11
Correlation		0.75		0.66		0.77

*, ** Significant at 5 and 1%, respectively; SCA - specific combining ability.

Specific combining ability effects

The crosses had an average TSW of 3.72 kg plant^{-1}, with a range between 2.58 and 5.70 kg plant^{-1} (Table 10). The SCA effects of crosses were significant (P<0.05) and positive for crosses 990014 × 990127, 990010 × 990056 and 990010 × 990067. Other significantly different SCA effects were negative, for example, cross 990010 × 960249 (Table 10). There was significant interaction (p<0.05) between the local and IITA varieties in RTN (Table 3). The specific combining abilities (SCA) effects of RTN were significant (P<0.01) but negative for 990010 × 990056 and 820001 × 990067, while cross, 820001 × 990127 had a positive and significant SCA (P<0.05) (Table 10). Root weight per plant ranged from 0.80 to 1.46 kg/plant (Table 10). A few of the crosses, 990014 × 990127 and 820058 × 900056, had positive and significant (P<0.05 and 0.01, respectively) SCA effects, while for cross 820001 × 990067, this was negative (Table 10).

Biomass of the crosses ranged from 3.93 to 6.85 kg plant^{-1} (Table 11). The SCA effects for biomass were significant and positive (P<0.01) for the crosses 990014 × 990127; 990010 × 990056; and 990010 × 990067. The SCA effects of 990010 × 960249 were negative and significant (Table 11). The harvest index of all the crosses was low, ranging from 18.08 to 32.85% with an overall average of 23.93% (Table 11). The SCA effects for harvest index were significant and positive for 820001 × 990127, 820058 × 990056 and 990010 × 960249 but negative for 820058 × 990127 and 820058 × 960249.

Dry matter content of all the crosses ranged from 32 to 42% with an overall average of 38% (Table 12). The SCA effects for dry matter among the crosses were all significant

Table 11. Mean and SCA effects of the crosses for the agronomic traits, total biomass (kg plant^{-1}) and percentage harvest index.

Cross	Biomass		Harvest index	
	Mean	SCA	Mean	SCA
820001 × 960249	4.58	0.47	25.38	-2.51
820001 × 990056	4.27	-0.60	23.17	-1.02
820001 × 990067	4.43	-0.52	23.54	-1.54
820001 × 990072	4.74	0.54	24.68	-0.85
820001 × 990127	4.06	-0.41	29.54	5.69**
820001 × 990183	4.39	0.51	24.70	0.24
820058 × 960249	5.03	0.70	22.43	-3.53*
820058 × 900056	4.81	-0.28	27.65	5.40**
820058 × 900067	5.13	-0.02	23.90	0.75
820058 × 990072	4.26	-0.15	24.76	1.16
820058 × 990127	4.55	-0.13	18.08	-3.84*
820058 × 990183	3.96	-0.13	22.58	0.05
990010 × 960249	3.93	-1.13*	32.85	6.13**
990010 × 990056	6.69	0.87*	21.08	-1.94
990010 × 990067	6.85	0.95*	21.98	-1.94
990010 × 990072	5.38	0.23	23.76	-0.60
990010 × 990127	4.64	-0.77	20.27	-2.42
990010 × 990183	4.67	-0.15	24.06	0.77
990014 × 960249	4.54	-0.04	25.96	-0.09
990014 × 990056	5.36	0.01	19.91	-2.44
990014 × 990067	5.01	-0.41	25.97	2.73*
990014 × 990072	4.04	-0.63	23.98	0.29
990014 × 990127	6.24	1.30**	22.58	0.57
990014 × 990183	4.12	-0.23	21.57	-1.05
Statistics				
Mean	4.82		23.93	
SED	0.74		3.11	
SCA SE		0.42		1.78
Correlation		0.74		0.86

*, ** Significant at 5 and 1% respectively; SCA – specific combining ability.

significant (P<0.01) except for 820001 × 990072 and 820058 × 9900127 (Table 12). At 6 mo, the crosses produced from 3.20 to 6.20 t ha^{-1} had root dry matter yield (Table 12). The SCA effects for dry matter yield of most crosses were not significant except for crosses 990014 × 990067 and 990014 × 990127 (P<0.05). The root cyanide content of 6 mo old cassava plants had a range of 4 to 5. The SCA effects for root cyanide content were significant (P<0.05) and positive for 820001 × 990127, and 820058 × 900056; and negative for 820058 × 990127 (Table 12).

Phenotypic correlations

The phenotypic correlations among the family averages for shoot weight, root yield, root weight and number, dry matter and biomass evaluated in this study are presented in Table 13. Most of the traits were positively and significantly correlated, except for dry matter content with harvest index, cyanide content, harvest index with biomass and shoot weight, which were negatively correlated. Biomass was highly correlated with shoot weight (0.969). However, root weight was highly correlated with dry matter yield and harvest index.

DISCUSSION AND CONCLUSION

The aim of the study was to generate a segregated population from crosses between the late bulking local and the early IITA varieties to study gene action for root yield and related traits. Crosses were segregated for shoot weight, root number, root weight, root yield, early

Table 12. Mean and SCA effects of the crosses for the agronomic traits, dry matter content (%), dry matter yield (t ha^{-1}) and root cyanide content SCA - specific combining ability.

Cross	Dry matter content		Dry matter yield		Root cyanide content	
	Mean	SCA	Mean	SCA	Mean	SCA
820001 × 960249	39.44	1.61**	4.80	-0.10	4.19	0.05
820001 × 990056	40.74	2.41**	4.00	-0.10	4.33	-0.08
820001 × 990067	39.79	0.97**	4.00	-0.90	4.10	-0.19
820001 × 990072	35.25	0.10	4.50	0.40	3.94	-0.30
820001 × 990127	37.12	-2.96**	4.30	0.20	4.53	0.40*
820001 × 990183	37.02	-2.13**	4.50	0.60	4.48	0.13
820058 × 960249	36.80	-1.76**	4.90	0.00	4.15	0.11
820058 × 900056	37.74	-1.31**	4.60	0.50	4.71	0.40*
820058 × 900067	40.99	1.45**	5.10	0.10	3.98	-0.22
820058 × 990072	35.09	-0.78**	3.90	-0.20	4.38	0.24
820058 × 990127	40.85	0.05	3.50	-0.70	3.51	-0.51*
820058 × 990183	42.22	2.35**	4.10	0.20	4.22	-0.02
990010 × 960249	39.53	1.88**	5.90	0.60	4.47	0.05
990010 × 990056	35.35	-2.79**	4.50	0.00	4.55	-0.15
990010 × 990067	33.99	-4.64**	5.00	-0.30	4.83	0.25
990010 × 990072	38.28	3.32**	5.10	0.60	4.37	-0.16
990010 × 990127	40.53	0.64*	3.80	-0.70	4.64	0.23
990010 × 990183	40.56	1.60**	4.10	-0.20	4.43	-0.21
990014 × 960249	35.30	-1.73**	4.60	-0.40	4.03	-0.20
990014 × 990056	39.22	1.69**	3.80	-0.30	4.35	-0.16
990014 × 990067	40.23	2.22**	6.20	1.10*	4.56	0.16
990014 × 990072	31.72	-2.63**	3.30	-0.80	4.55	0.21
990014 × 990127	41.54	2.27**	5.30	1.20*	4.12	-0.11
990014 × 990183	36.54	-1.81**	3.20	-0.70	4.54	0.10
Statistics						
Mean	38.16		4.50		4.33	
SED	1.61		0.85		0.29	
SCA SE		0.11		0.50		0.17
Correlation		0.79		0.78		0.79

*, ** Significant at 5 and 1%, respectively.

early shoot vigour, the breeder should select and identify lines that combine well. Families 990010 and 990014 had the highest shoot weight and their crosses, 990010 × 990056; 990014 × 990072; 990014 × 990056; and 990014 × 990127 had the highest positive SCA effects, suggesting that these two parents could belong to two separate heterotic groups that can be used for future breeding for early shoot vigour.

GCA effects (53%) were mainly responsible for determining the root numbers; but a breeder should also consider SCA effects, which accounted for 45% of the variation. The small difference between GCA and SCA effects suggested that it is possible to breed for increased root yield by selecting parents with high GCA for root number. Alternatively, the breeder can use germplasm that combines well for increased root numbers. Our results were in agreement with previous studies. Whyte (1985) reported that both additive and non-additive gene action influenced root number. Root number was found to be positively correlated (r = 0.33) with root yield, indicating that selection for large number of roots would increase root yield. Kawano et al. (1987) obtained similar results.

Predominantly, SCA (at 60% of crosses variance) controlled root yield, indicating the importance of non-additive gene action in influencing yield. The GCA, due to IITA varieties, accounted for 34% of crosses variance, indicating that these genotypes made a significant ($p<0.05$) contribution to early root bulking in the crosses. The proportionally higher SCA effects indicated that the individual genotypes of the two groups of parents, IITA and local, combined specifically well for root yield. Perez

Table 13. Phenotypic correlations between yield and secondary traits.

	DMY	%DM	%HI	Biomass	RTW	RTN	TSW
RCNP (score)	0.102**	-0.015**	0.034ns	0.069*	0.104**	-0.044ns	0.043ns
DMY (t ha^{-1})		0.44***	0.501***	0.378***	0.873***	0.361ns	0.186***
DM (%)			-0.026***	0.042ns	0.04ns	0.102**	0.039ns
HI (%)				-0.186***	0.602***	0.089*	-0.353***
Biomass					0.429***	0.29***	0.969***
RTY (t ha^{-1})					1.00***	0.382***	0.206***
RTW (kg)						0.382***	0.206***
RTN (count)							0.22***

RCNP – root cyanide content, DMY - dry matter yield, DM -dry matter content, HI- harvest index, RWT – root weight (kg plant^{-1}), RTN –root number per plant, TSW – shoot weight (kg plant^{-1}). *, ** - Significantly different from zero at the 0.05 and 0.01 probability levels, respectively (two-tailed test).

et al. (2005) also reported predominance of SCA, while GCA was not significant for yield in a diallel analysis. Jaramillo et al. (2005) reported 59% and 41% for SCA and GCA, respectively, for root yield, which is highly consistent with these findings.

For dry matter yield, 37% of the crosses sum of squares were accounted for by GCA effects mainly due to the IITA varieties, and were significant (P<0.05) (Table 19), while SCA was responsible for 61% of the crosses sum of squares, again suggesting the predominance of non-additive gene action. A similar trend was observed with cyanide content with SCA effects accounting for 66% of crosses' sums of squares. Jaramillo et al. (2005) did not measure cyanide content, but reported that SCA accounted for 37%, while GCA explained 63% of the crosses variation in a diallel analysis. However, Perez et al. (2005) reported that GCA was not significant for dry matter content, which supports the predominance of SCA, and thus has no additive effects in determining dry matter content in cassava.

The local varieties had more GCA effects for reaction to cassava mosaic compared to the IITA varieties. In particular, the local genotype 990014 had negative GCA effects which reflect the involvement of additive genes in the resistance it expresses. Resistance was, however, mostly explained by SCA, 67% of crosses sum of squares, suggested predominance of non-additive gene action for disease resistance. There is a need for continuous improvement of genotypes for disease resistance, because the disease is prevalent in all cassava-growing areas in Africa.

Significantly, high correlations between SCA and mean values for all traits of the F_1 progeny indicated that performance of crosses *per se* could be used to predict their SCA values (Tables 10, 11, 12 and 13). Jaramillo et al. (2005) reported similar results.

Harvest index was positively associated with root yield, indicating that selecting high harvest index will not compromise yield. There will be declining returns on selected harvest index in order to increase root yield until a fall off occurs. Redesigning the crop morphology and physiology then becomes necessary. Harvest index is an important trait as it measures the efficiency of a genotype in partitioning dry matter to the storage roots. Positive correlation between root yield and harvest index confirmed previous results (Kawano et al, 1978). There were positive associations among shoot weight, root yield, root weight, root number, dry matter content and biomass, suggesting that breeding for any of these traits will not reduce the desired level of the other.

REFERENCES

Banziger PS, Paterson CJ (1992). Genetic variation: Its origin and use for breeding self-pollinated species. In: Stalker TM, MurphyJP (eds). Plant Breeding in the 1990s: March 1991 Symposium Proceedings, Wallingford, United Kingdom: C.A.B. International pp. 69-92.

Ceballos H, Iglessias CA, Perez JC, Dixon AGO (2004). Cassava breeding: opportunities and challenges. Plant Molecul. Bio. 56: 503-516.

CIAT (Centro Internacional de Agricultura Tropical) (2004). Annual Report. Output 3, 24 Paper presented at the 5th symposium of ISTRC, Manila, Philippines, Sept 1979.

Comstock CC, Robinson HF (1948). The component of genetic variance of populations of biparental progenies and their use in estimating the average degree of dominance. Biometrics. 4: 254-266.

Comstock CC, Robinson HF (1952). Estimation of average dominance of genes. In: J.W. Gowen (ed.). *Heterosis*. Iowa State University Press: Ames, pp. 494-516.

Dedkova L, Mach K, Mohsen A (2002). Analysis of growth and feed conversion in broiler rabbits by factorial crossing. Czech J. Anim. Sci. 47: 133-140.

Derera J, Denash GP, Pixley KV (2000). Resistance of maize to the maize weevil: II. Non-preference. J. Crop Sci. 9: 441-450.

De Vries CA, Ferwerda JD, Flack M (1967). Choice of food crops in relation to actual and potential production in the tropics. Netherlands J. Agric. Sci. 15: 241-248.

Eberhart SA, Gardner CO (1966). A general model for genetic effects. Biometrics. 22: 864-881.

Falconer DS, Mackay TFC (1996). Introduction to Quantitative Genetics. 4th ed. Longman Group.

Githunguri CM, Migwa YN (2003). Farmer participatory perspectives on cassava clones developed in KARI-Katumani in three divisions in Machakos district. First KARI Adaptive Research Conference, p. 36.

Griffing B (1956). Concept of general and specific combining ability in relation to diallel cross systems. Australian J. Biol. Sci. 9: 463-493.

Hahn SK, Howland AK, Terry ER (1980). Correlated resistance of cassava to mosaic and bacterial blight diseases. Euphytica. 29: 305-311.

Halauer AR, Miranda JB (1995). Quantitative Genetics in Maize

Breeding. 2nd ed. Iowa State University Press, Ames, Iowa, pp. 267-298.

Hogarth DM, Wu KK, Heinz DJ (1981). Estimating genetic variance in sugarcane using a factorial cross design. Crop Sci. 21: 21-25.

Hornetz B, Shisanya CA, Gitonga NM (2000). Studies on the ecophysiology of locally suitable cultivars of food crops and soil fertility, monitoring in the semi-arid areas of Southeast Kenya. Final Report on a Collaborative Research Project between Kenyatta University, Nairobi, Kenya and University of Trier, Germany, p. 33.

IITA (International Institute of Tropical Agriculture) (1982). Tuber and Root Crops Production Manual. Manual Series No.9. May 1982.

Jaramillo G, Marante N, Perez JC, Calle F, Ceballos H, Arias B, Bellotti AC (2005). Diallel analysis in cassava adapted to the mid-altitude valleys environment. Crop Sci. 45: 1-12.

Kamau JW, Kinama JM, Nguluu SN, Muhammad L, Whyte JBA, Ragwa SM, Migwa EN, Simiyu PM (1998). Farmers' evaluation of cassava varieties in semi-arid areas of Kenya. In: Akoroda MO, Ngeve JM (eds). Root Crops in the 21st Century: Proceedings of the 7th Triennial Symposium of the ISTRC-AB, pp. 378-383.

Kariuki CW, Kamau JW, Mbwika J, Munga T, Makhoha AO, Tunje T, Nzioki S, Gatheru, Njaimwe, Wambua, Odendo M, Lutta M, Karuri EG (2002). A report on cassava sub-sector analysis for Kenya. In: Mbwika JM, Ntawuruhunga P, Kariuki C, Makhoka A (eds). Proceedings of the Regional Workshop on Improving the Cassava Sub-Sector, Nairobi Kenya. April 2002, p. 35.

Kawano K (1987). Inherent and environmental factors related to cassava varietal selection. In: C. Hershey (ed.). Cassava Breeding: a Multidisciplinary Review. CIAT, Cali, Colombia.

Kawano K, Daza P, Amaya A, Rios M, Goncalves WMF (1978). Evaluation of cassava germplasm for productivity. Crop Sci. 17: 377-382.

Kawano K, Fukunda WMG, Cenpukdee U (1987). Genetic and environmental effects on dry matter content of cassava roots. Crop Sci. 27: 69-74.

Kiarie AW, Omari F, Kusewa F, Shakoor A (1991). Variety improvement of cassava for dry areas of Kenya with emphasis on utilization. In: Recent advances in KARI's research programmes: Proceeding of the 2nd KARI annual scientific conference held at Panafric Hotel, Nairobi, Kenya on 5-7 September 1990, pp. 20-24.

Lee M (1995). DNA markers and plant breeding programmes. Advances in Agronomy, 55: 265-344.

Lokko Y, Dixon AGO, Offei SK (2004). Combining ability analysis of field resistance in cassava to the African cassava mosaic disease. Int. Crop Sci. http://www.cropscience.org.au/icsc2004/ poster.

O'Brien GM, Wheatley CC, Iglesias C, Poulter H (1994). Evaluation, modification and comparison of two rapid assays for cyanogens in cassava. J. Food Sci. Agric. 65: 391-399.

Perez JC, Ceballos H, Jeramillo G, Marante N, Perez JC, Calle F, Arias B, Bellotti AC (2005). Epistasis in cassava adapted to midaltitude valley environment. Crop Sci. 45: 1-12.

Pixley KV, Bjarnason MS (1993). Combining ability for yield and protein quality among modified endosperm opaque-2 tropical maize inbreds. Crop Sci. 33: 1229-1234.

Sprague GF, Tatum LA (1942). General vs. specific combining ability in single crosses of corn. J. Am. Agron. 34: 23-32.

Whyte JBA (1985). Breeding cassava for adaptation to environmental stress. In: C. H. Hershey (ed.). Cassava breeding: a multidisciplinary review. Proceedings of a workshop in Philippines, 4 – 7 March 1985. CIAT, Cali, Colombia, pp. 147-176.

Evaluation of intra and interspecific rice varieties adapted to valley fringe conditions in Burkina Faso

M. Sié[1]*, S. A. Ogunbayo[1], D. Dakouo[2], I. Sanou[2], Y. Dembélé[2], B. N'dri[1], K. N. Dramé[1], K. A. Sanni[1], B. Toulou[1] and R. K. Glele[3]

[1]Africa Rice Center (WARDA), 01 B.P. 2031, Cotonou, Benin.
[2]Institut de l'Environnement et de Recherches Agricoles (INERA), Programme Riz et Riziculture, Centre Régional de Recherches Environnementales et Agricoles de l'Ouest, BP 910 Bobo-Dioulasso, Burkina Faso.
[3]Universite d'Abomey Calavi, 01 BP 526, Cotonou, Benin.

The immense potential of the lowlands in Burkina Faso for durable intensification of rice cropping have not been realised due to biotic and abiotic stress constraints. To this end, the rice research program in Burkina Faso evaluated 16 intra-and interspecific lowland progenies in 2002 and 2003. The aim of the study is to introduce new lowland NERICAs through a participatory approach and to identify ideotypes that are adapted to lowland conditions. Variability was found among the 16 rice varieties with respect to the 9 variables that were evaluated. A principal components plot and clustering analysis technique were used to group 16 intra-and interspecific lowland progenies. The interspecific varieties formed the most interesting group and showed a better capacity for adaptation to the diversity of lowlands. They had good yields, sometimes higher than those of intraspecific varieties and check. Thus, the results obtained were quite satisfactory as the varieties possess good agronomic traits that are well adapted to intensified lowland rice farming. The recent naming of some of these interspecific varieties as NERICA-L (New Rice for Africa Lowland) by Africa Rice Center confirmed their status. Thus from this study, a new set of interspecific lines that are adapted to lowland conditions and which the national research programs in Burkina Faso can use in various tests for satisfying farmers' needs are discussed.

Key words: Hybridisation, inter-specific, insect, blast, NERICA, *Oryza glaberrima*, *Oryza sativa*, sterility, yield.

INTRODUCTION

Rice consumption is increasing fast in Burkina Faso because the rapidly rising urban population is shifting from traditional cereals to rice. Thus, rice has been coming up as a major staple food in Burkina Faso and demand has grown at an annual rate of 3% between 1973 and 1992 compared with an annual population growth rate of 2.9%, which can be explained by changing consumer preferences (WARDA, 1996; Randolph, 1997). Currently, in-country production covers about 60% of the demand, and 40% is met from imports (Segda et al., 2005). Hence, there is an urgent need to increase and improve the production of rice in Burkina Faso and in Africa as a whole, in order to meet up with the high demand (Ogunbayo et al., 2005; 2007). Burkina Faso has three major rice ecologies - upland (10% of land area with 5% of the country's rice production), irrigated (23% of area and 53% of production) and rainfed lowland (67% of area and 42% of production) (Sié, 1999). Irrigated systems were introduced in the 1960s, and the development was accentuated from the 1970s onward. Average yields of irrigated rice in Burkina Faso were estimated at 4.0 to 4.5 t ha^{-1} and in general two crops per year are grown (Illy, 1997; Wopereis et al., 1999). Thus, rainfed lowland is the major rice ecology in the country, combining the characteristics of upland and irrigated systems. The declining and unpredictable rainfall pattern has led to the disappearance of traditional *Oryza sativa* cultivars. However, some farmers still grow *Oryza glaberrima*, which has good agronomic traits, that is, acceptable grain quality, plant vigor, and resistance to major biotic and abiotic stresses (Pham, 1992; Besançon,

---
*Corresponding author. E-mail: m.sie@cgiar.org

1993; Adeyemi and Vodouhe, 1996; Sié, 1999). In 1992, the Africa Rice Center (WARDA) and its partners started the Interspecific Hybridization Project (IHP) in an attempt to combine the useful traits of both cultivated rice species (*O. sativa* and *O. glaberrima*). Crossing the two species is complicated by their incompatibility, which leads to hybrid mortality. This problem was overcome through backcrossing with the *O. sativa* parent coupled with anther culture, resulting in the first interspecific rice progenies from cultivated varieties (Jones et al., 1997a, b, c). In addition to the upland NERICA varieties, WARDA and national programs developed NERICA varieties well adapted to irrigated and rainfed lowlands, one of the most complex rice ecologies in the world. Key to this success was the unique research and development partnership model forged between WARDA and the national programs through the Rice Research and Development Network for West and Central Africa (ROCARIZ), which facilitated the shuttle-breeding approach to accelerate the selection process and achieve wide adaptability of the lowland NERICAs (WARDA, 2006). To meet the demand of rice farmers and consumers, the rice research program in Burkina Faso started evaluating intra-and interspecific lowland progenies obtained from WARDA, Senegal, in 2000. This study aimed to identify, through a participatory approach, high yielding varieties with resistance to biotic and abiotic stresses.

MATERIALS AND METHODS

The plant materials comprised of 16 genotypes which include nine interspecific lines (*O. glaberrima* × *Oryza sativa indica*), six intraspecific (*O. sativa* × *O. sativa*) lines and one check. The check variety (TOX 3055-10-1-1-1) had been released and extensively used in Burkina Faso. The experiment was carried out in valley fringe at the Banfora research station in the southern region of Bobo-Dioulasso in Burkina Faso during 2002 - 2003 wet season. Seeds were sown directly, three seeds per hill and thinned to one seedling at a spacing of 0.25 m within and between rows. The randomized complete block design with three replications was used with 16 rows of 5 m and plot area was 20 m^2.

A pre-drilling base application of 200 kg.ha^{-1} of NPK (15-15-15) was made, followed by a total of 100 kg.ha^{-1} of urea in two applications of 35 kg/ha at 14 days after seeding and 65 kg.ha^{-1} at the panicle initiation. Two manual weeding were carried out and no chemical treatment was applied. Plants in the middle rows in each plot were harvested. The IRRI Standard evaluation system (IRRI, 1996) for rice was used to score quantitative traits, disease and insect pest damage. Agronomic traits evaluated were plant height at maturity; tillering; days to flowering; number of panicles p/m^2; sterility and yield. Reaction to the specific diseases and insects that were observed were: Stem borer; gall midge and leaf blast. The data collected were subjected to statistical analysis using SAS (SAS, 1999) and GGE biplot version 5.2 (Yan, 2003). A GGE biplot was constructed using the first two principal components (PC1 and PC2) derived from subjecting the environment-centered data to singular-value decomposition and it resulted in analysis from several angles: (i) The polygon view of a GGE biplot allowed visualization of the which-won-where pattern (that is, which variety had the highest yield in which environment); (ii) the average environment coordination view allowed simultaneous visualization of the mean performance and stability of the treatments, the discriminating ability vs. representativeness of the environments; and (iii) the environment vector view allowed visualization of the interrelationship among environments (Yan et al., 2000; Yan, 2001, 2002; Yan and Kang, 2003). In addition, attempting to characterize the environments and to relate the mean yield of the environments to the ecology, a biplot based on an environment × factor two-way table was constructed, which was similar to that based on a genotype × trait two-way table described by Yan and Rajcan (2002). All biplots presented in this paper were generated using the software GGEbiplot package that runs in a Windows environment. Principal components grouping of the traits was used to examine the percentage contribution of each trait to total genetic variation and to spot characters that reflected the greatest proportion of variations among the 9 variables. This is because the PCA has been reported to be able to choose independent (orthogonal) axes that are minimally correlated and then represent linear combinations of the original characters (Akoroda, 1983).

The relative discriminating power of the axes and their associated characters were measured by the Eigen values and factor scores, respectively. The similarity coefficient was used to construct a dendrogram by the unweighted pair group method with arithmetic average (UPGMA) according to (Sneath et al., 1973; Swofford et al., 1990).

RESULTS

2002 wet season

Table 1 presents the means of nine characters measured in sixteen rice varieties. Highly significant differences were observed in plant height, flowering date, leaf blast and sterility while non-significant difference was observed in tiller number, stemborer, AfRGM and yield was significant. However, ten varieties including four intraspecifics and six interspecifics had higher number of tillers compared to average (137 tillers). Plant height varied from 73 to 118 cm. Out of the materials tested, nine (9) varieties including two intraspecific (V3 and V4) and 7 interspecifics (V8, V10, V11, V12, V13, V14 and V15) had values below the average (91cm) thus, they were semi-dwarf varieties. Based on IRRI Standard Evaluation System (1996) all varieties tested with the exception of 9 varieties mentioned above have medium height. The average value for panicle number was 46 and seven lines including one intra-specific and 6 interspecifics recorded high panicle number which were above the average number. Out of the materials tested four varieties had flowering days above an average (87) while the rest were equal or below the average. For sterility, eight varieties had value below average. The average yield recorded was 1642 kg/ha.

The correlation matrix showed that plant height was positively and significantly associated with yield. However, plant height had negative but significant association with sterility. While flowering date were also positively and highly significant correlated to leaf blast. Sterility was positively and significantly correlated to leaf blast (Table 2).

Table 3 presents the principal components analysis showing the contribution (factor scores) of each character

Table 1. Means of nine characters measured in sixteen rice varieties (2002).

S/No	Variety	Tiller	Pan/m^2	Height (cm)	Flw DAS	Sterility	Leaf Blast	Stem-borer	AfRGM	Yield (t ha^{-1})
1	WAT 1176-B-FKR-B-B	167	41	113	90	2	2	2	13	2293
2	WAT 1184-B-FKR-B-B	104	56	118	86	1	1	1	23	2133
3	WAT 1191-B-FKR-B-B	173	40	88	88	1	3	1	16	2347
4	WAS 105-B-IDSA-B-WAS-2-1-FKR-B-B	161	46	90	90	1	2	0	16	2247
5	WAS 114-B-IDSA-B-WAS-5-1-FKR-B-B	118	44	97	87	1	2	1	18	2267
6	WAS 129-B-IDSA-B-WAS-1-1-FKR-B-B	142	62	97	85	1	1	1	17	1840
7	WAS 122-IDSA-1-WAS-6-1-FKR-B-B	129	37	94	84	1	1	1	20	933
8	WAS 122-IDSA-1-WAS-2-FKR-B-B	142	49	90	82	1	1	1	17	1520
9	WAS 122-IDSA-1-WAS-1-1-B-FKR-B-B	151	32	94	85	1	1	1	22	1280
10	WAS 161-B-6-4-FKR-B-B	124	57	76	87	4	2	2	23	933
11	WAS 161-B-9-3-FKR-B-B	141	37	73	86	5	2	1	17	1387
12	WAS 161-B-6-3-FKR-B-B	141	48	80	82	3	1	1	15	1413
13	WAS 163-B-5-3-FKR-B-B	141	24	85	89	4	2	1	19	1093
14	WAS 191-8-3-FKR-B-B	140	70	77	83	4	2	0	18	1387
15	WAS 191-9-3-FKR-B-B	114	60	88	80	3	1	0	15	1493
16	TOX 3055-10-1-1-1 (Check)	101	37	91	107	6	9	1	17	1707
	Mean	137	46	91	87	2	2	1	18	1642
	Significance	ns	ns	**	**	**	**	ns	ns	*

*, **; Significant at 5 and 1% probability levels, respectively.

Table 2. Correlation coefficients of nine traits used in characterizing sixteen rice varieties (2002).

Character	Tiller at 60 DAS	Pan m^2	Plant Height (cm)	Flw date	Sterility	Leaf Blast	Stemborer	AfRGM	Yield (t ha^{-1})
Tiller at 60 DAS	1.000								
Pan m^2	-0.226	1.000							
Plant height (cm)	-0.092	-0.036	1.000						
Flw date	-0.206	-0.395	0.134	1.000					
Sterility	-0.345	-0.066	-0.563*	0.450	1.000				
Leaf Blast	-0.294	-0.255	-0.075	0.933**	0.591*	1.000			
Stemborer	0.071	-0.301	0.236	0.209	0.030	0.063	1.000		
AfRGM	-0.415	-0.005	0.046	-0.061	-0.069	-0.140	0.196	1.000	
Yield (t ha^{-1})	0.245	0.086	0.577*	0.239	-0.423	0.142	-0.061	-0.428	1.000

*, **; Significant at 5 and 1% probability levels, respectively.

among the 16 varieties. The three principal components accounted for about 68.84% of total variance with the first and second principal component taking 29.84 and 22.79%, respectively. The relative discriminating power of the principal axes as indicated by the eigen values was highest (2.63) for axis 1 and lowest (1.53) for axis 3. The first principal component that accounted for the highest proportion (29.84%) of total variation was mostly correlated with flowering date, sterility and leaf blast. They have early cycle and mode-rately resistance to sterility and leaf blast. Characters that were mostly correlated with the second principal component were plant height and yield and third principal component were tiller and AfRGM.

Figure 1 presents the morphological dendrogram showing the minimum distance between cluster groups and genotypes were divided into three major groups.

Table 4 shows characteristics of morphological groups defined by topology. In group 1 tiller number at 60 days ranged from 104 to 173 m^2; panicle number ranged from 40 to 56; plant height ranged from 88 to 118 cm indicating that varieties are semi-dwarf to medium height type; flowering occurred between 86 to 90 while grain yield ranged from 2133 to 2347 kg ha^{-1}; sterility ranged from 1 to 3; leaf blast ranged from 1 to 3; stem-borer ranged from 0 to 3 and AfRGM ranged from 13 to 23

Table 3. Principal components analysis showing the contribution (factor scores) of each character among the sixteen rice varieties (2002).

Character	Prin1	Prin2	Prin3
Leaf blast	-0.933	0.252	-0.046
Flowering date (FLW)	-0.860	0.421	0.155
Sterility	-0.837	-0.341	-0.194
Yld (t ha^{-1})	0.159	0.886	0.033
Plant height (cm)	0.260	0.651	0.641
AfRGM	0.006	-0.506	0.705
Tiller at 60 DAS	0.331	0.379	-0.624
Panicle number (Pan m^{-2})	0.356	-0253	0.064
Stem Borer	-0.212	0.143	0.305
Eigen value	2.686	2.051	1.459
Variance (%)	29.840	22.794	16.206
Cumulative (%) variance	29.840	52.634	68.840

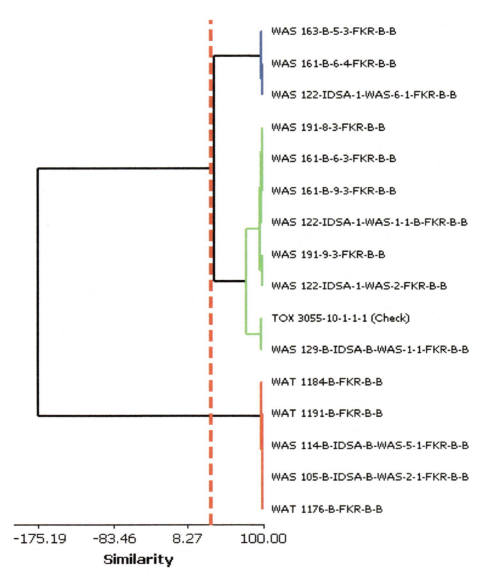

Figure 1. Morphological dendogram showing the minimum distance between cluster groups (2002).

Table 4. Characteristics of morphological groups defined by topology in 2002.

Character	Group 1		Group 2		Group 3	
	Min.	Max.	Min.	Max.	Min.	Max.
Tiller No. at 60 days	104	173	101	151	124	141
Panicle number (Pan m^{-2})	40	56	32	70	24	57
Plant height (Ht)	88	118	73	97	76	94
Flowering date (FLW)	86	90	80	107	84	89
Sterility	1	3	1	7	1	5
Leaf blast (LB)	1	3	1	9	1	3
Stemborer	0	3	0	1	1	3
AfRGM	13	23	15	22	19	23
Yield (Kg/ha)	2133	2347	1280	1840	933	1093

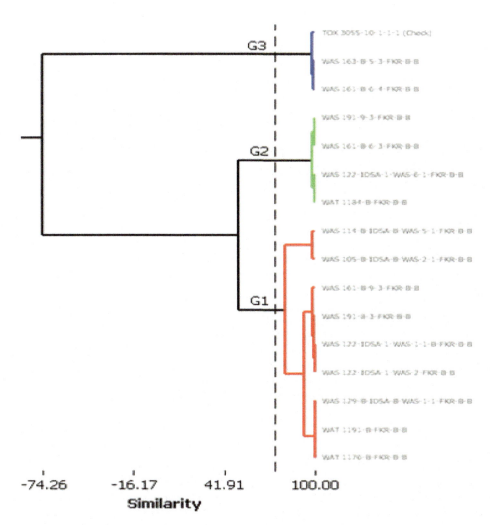

Figure 2. Morphological dendogram showing the minimum distance between cluster groups (2003).

while group 2 and 3 has its own distinct characteristics.

A plot of relationship between the 16 varieties as shown by the first and second principal components axes (Prin 1 and 2) (Figure 3) revealed that all the varieties were ordered into four distinct PCA clusters. Most of the varieties in group 2 were centrally distributed more towards the first principal axis and they have good yield. However, genotypes selected from group 1 were also

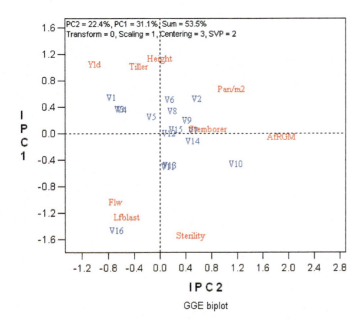

Figure 3. Plot of Prin 1 and Prin 2 showing the relationship between clusters of 16 Varieties in 2002.

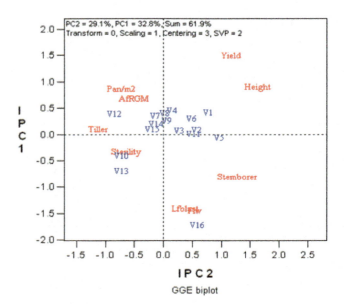

Figure 4. Plot of Prin 1 and Prin 2 showing the relationship between clusters of 16 varieties in 2003.

distributed towards the upper left side of first principal axis and they have good tillering ability and moderate height. Group three were distributed towards the lower right side and they are susceptible to AfRGM.

Figure 5 defined the genotypes that performed best in 2002 (which genotype won in which trait). The polygon is formed through connecting the best genotypes in each trait to other. Starting from the biplot origin, perpendicular lines are drawn to each side of the polygon, which divide the biplot into 4 sectors. The which-won-where pattern is examined as follows. The varietal number at the vertex of polygon in any sector is the genotype that produces the highest value in trait(s) that fall in that sector. Thus, genotypes (1, 3, 4 and 5) had good yield and tiller while genotype (1) produced the highest yields. The genotypes that had highest flowering date and leaf blast was genotype (16). While genotypes (2, 6 and 8) had high value in panicle/m^2.

2003 wet season

Table 5 presents the means of nine characters measured in 16 rice varieties. Analysis of variance showed highly significant differences in plant height, flowering date, sterility and leaf blast. There was no significant difference in panicle number, tillering numbe, yield, stem borer and AfRGM. However, 7 interspecifics had a higher number of tillers than the average (470 tillers). Out of the materials tested, seven varieties had height below an average value (111 cm). Thus, seven varieties were semi-dwarf while genotypes (1, 2 and 11) had medium height. Seven varieties had flowering days below

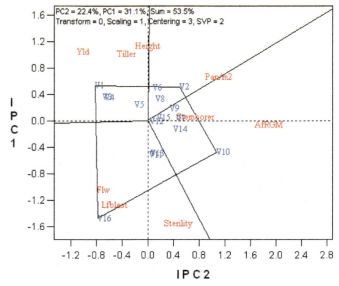

Figure 5. A polygon view of the GGE biplot of genotype x trait, showing which genotype won where or best for which trait (2002).

the average and this indicate that they are medium maturing varieties. Eight varieties had higher panicle number than average (265). The average yield recorded was 3214 kg/ha and all varieties were moderately resistant an tosterility except WAS 163-B-5-3-FKR-B-B that is genotype (13).

The correlation matrix showed that tiller was positively and significantly associated with Panicle/m^2 but negatively significant to plant height. Flowering date was positively and significantly associated with stem-borer

Table 5. Means of nine characters measured in twenty rice varieties (2003).

S/No	Variety	Tiller	Pan/m²	Height (cm)	Flw DAS	Sterility	Leaf blast	Stemborer	AfRGM	Yield (t ha⁻¹)
1	WAT 1176-B-FKR-B-B	437	273	139	85	1	2	2	9	3628
2	WAT 1184-B-FKR-B-B	448	253	130	88	1	1	1	4	2961
3	WAT 1191-B-FKR-B-B	471	264	105	88	1	2	2	9	3672
4	WAS 105-B-IDSA-B-WAS-2-1-FKR-B-B	436	295	111	89	2	1	1	10	4123
5	WAS 114-B-IDSA-B-WAS-5-1-FKR-B-B	386	213	115	86	1	4	2	7	3894
6	WAS 129-B-IDSA-B-WAS-1-1-FKR-B-B	427	259	117	85	1	2	1	7	3687
7	WAS 122-IDSA-1-WAS-6-1-FKR-B-B	451	268	111	88	1	0	1	16	2925
8	WAS 122-IDSA-1-WAS-2-FKR-B-B	497	268	113	85	1	0	1	9	3358
9	WAS 122-IDSA-1-WAS-1-1-B-FKR-B-B	533	209	114	87	1	0	1	11	3404
10	WAS 161-B-6-4-FKR-B-B	517	300	94	91	2	5	1	12	2325
11	WAS 161-B-9-3-FKR-B-B	423	200	128	86	2	4	0	7	3251
12	WAS 161-B-6-3-FKR-B-B	540	320	99	82	2	2	0	10	3002
13	WAS 163-B-5-3-FKR-B-B	503	255	96	88	4	6	1	9	2195
14	WAS 191-8-3-FKR-B-B	508	305	99	88	1	1	1	6	3491
15	WAS 191-9-3-FKR-B-B	544	288	102	86	1	1	2	9	3127
16	TOX 3055-10-1-1-1 (Check)	464	212	98	108	1	8	3	5	2070
	Mean	470	265	111	87	1	2	1	9	3214
	Significance	ns	ns	**	**	**	**	ns	ns	ns

*, **; Significant at 5 and 1% probability levels, respectively.

and highly significant to leaf blast. However, flowering date had negative but significant association with yield. While plant height was positively and significantly associated with yield and leaf blast was negatively associated with yield. Sterility was positively and significantly correlated to leaf blast (Table 6).

Table 7 presents the principal components analysis showing the contribution (factor scores) of each character among the 16 varieties. The three principal components accounted for about 73.26% of total variance with the first and second principal component taking 29.79 and 28.52%, respectively. The relative discriminating power of the principal axes as indicated by the eigen values was highest (2.68) for axis 1 and lowest (1.35) for axis 3. The first principal component that accounted for the highest proportion (29.79%) of total variation was mostly correlated with plant height, yield, plant height and tiller. They have medium height and high yield. Characters that were mostly correlated with the second principal component were leaf blast, panicle number, flowering date and AfRGM.

Figure 2 presents the morphological dendrogram showing the minimum distance between cluster groups. Accessions were divided into 3 major groups. Table 8 shows characteristics of morphological groups defined by topology as each of the three groups has its own distinct characteristics. A plot of relationship between the 16 varieties as shown by the first and second principal components axes (Prin 1 and 2) (Figure 4) revealed that all the varieties were ordered into four distinct PCA clusters. Most of the varieties in group 2 were centrally distributed more towards the first principal axis and they have good tiller and panicle number/m². However, genotypes selected from group 1 were also distributed towards the upper right side of first principal axis while group three were distributed towards the lower side.

Figure 6 defined the genotypes that performed best in 2003 (which genotype won in which trait). The polygon is formed through connecting the best genotypes in each trait to other. Starting from the biplot origin, perpendicular lines are drawn to each side of the polygon, which divide the biplot into 6 sectors. The which-won-where pattern is examined as follows. The varietal number at the vertex of polygon in any sector is the genotype that produces the highest value in trait(s) that fall in that sector. Thus, genotypes (1, 4, 6, 8 and 9) had good yield and moderately height while genotype (4) produced the highest yields. The genotypes (16) had long cycle and highest

Table 6. Correlation coefficients of nine traits used in characterizing sixteen rice varieties (2003).

Character	Tiller at 60 DAS	Pan m^{-2}	Plant height (cm)	Flw date	Sterility	Leaf blast	Stem-borer	AfRGM	Yield (t ha^{-1})
Tiller at 60 DAS	1.000								
Pan m^{-2}	0.462*	1.000							
Plant height (cm)	-0.605*	-0.363	1.000						
Flowering date	-0.060	-0.324	-0.333	1.000					
Sterility	0.184	0.112	-0.342	-0.070	1.000				
Leaf blast	-0.151	-0.343	-0.332	0.647**	0.453*	1.000			
Stemborer	-0.169	-0.242	-0.084	0.613*	-0.397	0.341	1.000		
AfRGM	0.270	0.303	-0.210	-0.258	0.135	-0.319	-0.239	1.000	
Yield (t ha^{-1})	-0.410	0.053	0.478*	-0.546*	-0.421	-0.619*	-0.082	-0.041	1.000

*, **; Significant at 5 and 1% probability levels, respectively.

Table 7. Principal components analysis showing the contribution (factor scores) of each character among the sixteen rice varieties (2003).

Character	Prin1	Prin2	Prin3
Yld (t ha^{-1})	0.850	-0.195	0.025
Plant height (cm)	0.806	0.196	-0.327
Sterility	-0.596	-0.227	-0.672
Leaf Blast	-0.569	0.685	-0.266
Tiller at 60 DAS	-0.560	-0.552	0.371
Panicle number (Pan m^{-2})	-0.181	-0.722	0.267
Flowering date (FLW)	-0.494	0.687	0.284
AfRGM	-0.130	-0.608	0.087
Stemborer	0.151	0.548	0.647
Eigen value	2.682	2.567	1.345
Variance (%)	29.795	28.518	14.950
Cumulative (%) variance	29.795	58.313	73.263

Table 8. Characteristics of morphological groups defined by topology in 2003.

	Group 1		Group 2		Group 3	
Character	Min.	Max.	Min.	Max.	Min.	Max.
Tiller No. at 60 days	386	533	448	544	464	517
Panicle number (Pan m-^{2})	200	305	253	320	212	300
Plant height (Ht)	99	139	99	130	94	98
Flowering date (FLW)	85	89	82	88	88	108
Sterility	1	3	1	3	1	5
Leaf Blast (LB)	0	5	0	3	5	9
Stemborer	0	3	0	3	1	3
AfRGM	6	11	4	16	5	12
Yield (Kg/ha)	3251	4123	2925	3127	2070	2325

value in leaf blast and stem-borer. Thus, interspecific performed well than the check.

Figure 7 represents the average tester coordination view, showing the performance of the genotypes across the years and their stability under valley fringe condition. Visualization of the mean and stability of genotypes is

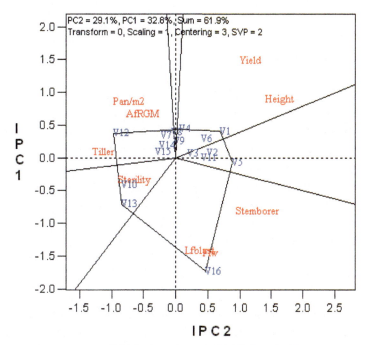

Figure 6. A polygon view of the GGE biplot of genotype x trait, showing which genotype won where or best for which trait (2003).

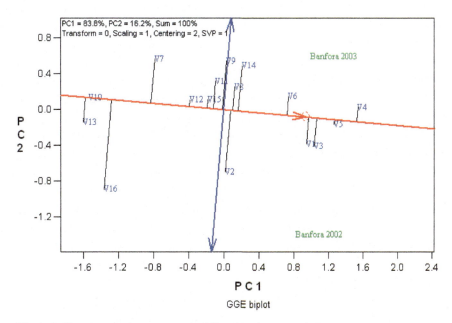

Figure 7. The mean performance vs. stability of the 16 rice varieties across the two seasons under valley fringe lowland condition.

achieved by drawing an average environment coordinate (AEC) on the genotype-focused biplot. First, an average environment, represented by the small circle, is defined by the mean PC1 and PC2 scores of the environments. The small circle near Banfora 2003 location suggests 2003 as the best in term of yield performance. The line connecting the biplot origin and the circle (Banfora) is referred to as the average-tester axis. Based on their mean performance, the genotypes are ranked along the average-tester axis with the arrow pointing towards genotype with greater value. Thus, the genotype is ranked along the AEC abscissa, with the arrow pointing to higher mean performance.

Based on this, Genotype (4) was clearly the highest-yielding, on average, in these environments, followed by (5, 3, 1, and 6) and the least in terms of performance was the genotype (13). The double arrow indicates that a greater projection onto the AEC ordinate, regardless of the direction, means greater instability. The bi-plot confirmed the conclusions drawn from Figures 5 and 6.

Trends in insects and diseases attacks according to varieties and seasons

In 2002, nine varieties including five intra-specifics and four interspecifics had value below means (18) recorded for AfRGM. Three varieties including one intra-specific and two interspecifics showed no stem borer attack (dead heart) compared to the mean obtained in the study. All intra or interspecific varieties rated below or equal to the average of 2 while TOX 3055-10-1-1-1(check) was rated high, thus, susceptible to leaf blast. In 2003, average score recorded for AfRGM was (9) and (1) for stemborer. Five varieties including one intra-specific and four interspecifics had values higher than the average recorded for AfRGM.

DISCUSSION

Variations did exist among the genotypes tested with respect to the traits that were evaluated. Total number of tillers, panicle number and flowering days were observed to greatly influence the yield among the genotypes that were evaluated. The yields recorded in 2003 were all higher than those recorded in 2002 and this was due to the differences observed between the genotypes in panicle numbers and insect attacks. The number of tillers produced which ascertains panicle number is the most important factor in high grain yield. However, this characteristic does not seem to be the causal factor in this study because WAS 191-9-3-FKR-B-B, that is genotype (15) which had highest number of tiller recorded 3127 kg/ha. Thus, the rate of fertile tillers was a factor that could justify this, because high tillering associated with high sterility rate reduces panicle number; meanwhile high tillering associated with low sterility rate increases panicle number.

Moreover, variations observed in the study may also due to difference in rainfall pattern in 2002 and 2003. The amount of rainfall recorded in 2002 and 2003 were 920.8 and 1226 mm, respectively. Therefore rainfall was an instrumental in the development of agro-morphological characters, and occurrence of insect attacks and diseases. The observation of these agro-morphological characters showed an average of 470 tillers and 265 panicles in 2003, which higher than those recorded in 2002. This potential development was also observed in yields. Regarding insect attacks no significant difference was observed between varieties over both years.

Since the cycle is one of the major concerns of African countries because of irregular rainfall pattern. Therefore, interspecific varieties showed cycles that were generally shorter than intra-specific varieties and check. Height was also an important factor in the lowland, because what was sought was a variety that was not too tall to avoid lodging, but not also too short to bear strong water levels. For this characteristic, genotypes (1, 5, 6, 7, 8, 9, 11, 12 and 14) were below average during the both years and no variety were susceptible to lodging during the both years.

Principal components analysis and hierarchical clustering generated from similarity or genetic distance matrices has provided an overall pattern of variation as well as the degree of relatedness among the genotypes. In addition, the principal component analysis, confirmed the contributions of the three traits to grain yield among the genotypes. The implication is that if selection is to be made between cluster groups for a future breeding exercise, panicle number, total number of tillers, days to heading, should be given high priorities. The GGE bi-plot generated several graphic bi-plots, strong genotype by environment interaction was confirmed. It also revealed the relationship among genotypes in terms of their responses and stability to the environments and traits. The results confirmed that the interspecific genotypes (WAS 122; WAS 161; 163; and 191 series) performed well across the locations and they were very stable. The interspecific varieties formed the most interesting group and have a better capacity for adaptation to the diversity of lowlands. They have acceptable yields, sometimes higher than those of intraspecific varieties and check.

Conclusion

In Burkina Faso the major concern of the national rice breeding and improvement program is to develop short or average height materials (lodging resistance) with short cycle, high yield potential and resistant or tolerant to various biotic and abiotic stresses. The results confirmed that the interspecific genotypes performed well across the seasons. Interestingly, interspecific varieties produced the greatest number of tillers and these observations confirm those made by Jones et al. (1996), who noted that interspecific *O. glaberrima* × *O. sativa* had a very high tillering capacity, which predisposed them to be more competitive with weeds.

In conclusion, the results obtained were quite encouraging and showed that, the varieties possess good agronomic traits that are well adapted to intensified lowland rice farming. The recent naming of some of these interspecific varieties as NERICA-L (New Rice for Africa Lowland) by WARDA has confirmed that they compare well with the traditional varieties. Based on the topology of the varieties, it is concluded that the interspecific crossings *O. glaberrima* × *O. sativa indica* can increase lowland rice biodiversity. Thus from this study, we now have a new set of interspecific lines that are adapted to lowland conditions and which the national research programs in Burkina Faso can use in various tests for satisfying farmers' needs.

REFERENCES

Adeyemi P, Vodouhe SR (1996). Amélioration de la productivité des variétés locales de *Oryza glaberrima* Steud. par des croisements intra et interspécifiques avec *Oryza sativa* Linn *in* Hybridations interspécifiques au Bénin. ADRAO, Bouaké (Côte d'Ivoire) pp. 159 - 175.

Akoroda MO (1983). Principal Components analysis and metroglyph of variation among Nigerian yellow yams. Euphytica 32: 565- 573.

Besançon G (1993). Le riz cultive d'origine africaine*Oryza glaberrima* Steud et les formes sauvages et adventices apparentées: Diversité, relations génétiques et domestication *O. breviliguta* A. Chev, et *O. Stapfii* A. Chev.. PhD Dissertation, University of Paris 11, France. p. 246.

Illy L (1997). La place de la riziculture irriguée dans le système de production agricole et animale au Burkina Faso. In Miézan KM, Wopereis MCS, Dingkuhn M, Deckers J and Randolph TF (eds.) Irrigated rice in the Sahel: Prospects for sustainable development. West Africa Rice Dev. Assoc. (WARDA), Bouaké, Ivory Coast pp. 131-135.

IRRI (1996). Standard Evaluation Systems for Rice. Manila Philippines p. 52.

Jones MP, Audebert A, Mande S, Aluko K (1996). Characterization and utilization of oryza glaberrima Steud. In upland rice breeding. In Proceedings Workshop Africa-Asia Joint Research on Interspecific Hybridisation between the African and Asian Rice Species *O.glaberrima* and *O.sativa,* WARDA, Bouaké, Côte d'Ivoire pp. 43-59.

Jones MP, Mande S, Aluko K (1997c). Diversity and potential of *Oryza glaberrima* Steud in upland rice breeding. Breed. Sci. 47: 395-398.

Jones MP, Dingkuhn M, Aluko GK, Semon M (1997a). Interspecific *Orza sativa* x *O.glaberrima* Steud. Progenies 92: 237-246

Jones MP, Dingkuhn M, Johnson DE, Fagade SO (1997b). Interspecific hybridization: progress and prospect, Bouaké, Côte d'Ivoire, WARDA. pp. 21-29.

Ogunbayo SA, Ojo DK, Guei RG, Oyelakin O, Sanni KA (2005). Phylogenetic diversity and relationship among forty rice accessions using Morphological and RAPDs techniques. Afr. J. Biotechnol. 4:1234 -1244.

Ogunbayo SA, Ojo DK, Popoola AR, Ariyo OJ, Sié M, Sanni KA, Nwilene FE, Somado EA, Guei RG, Tia DD, Oyelakin OO, Shittu A (2007). Genetic comparisons of landrace rice accessions by morphological and RAPDs techniques. Asian J. Plant Sci. 6(4): 653-666.

Pham JL (1992). Evaluation des ressources génétiques des riz cultivés en Afrique par hybridation intra et interspécifique. Thèse Docteur es sciences , Université de Paris XI ORSAY (France) p. 236.

Randolph TF (1997). Rice demand in the Sahel. In K.M. Miézan, M.C.S. p. 71-88.

SAS Institute Inc. SAS/STAT (1999). Guide for personal computer, version 8 edition, Cary, NC, SAS institute Inc. p. 1028.

Segdaa Z, Haefeleb SM, Wopereisc MCS, Sedogoa MP, Guinkod S (2005). Integrated soil, water and nutrient management for sustainable irrigated rice systems in Burkina Faso. In Synthesis of soil, water and nutrient management research in the Volta Basin Pp. 159-188.

Sie M (1999). Caractérisation des hybrides interspécifiques *(O. glaberrima* x *O. sativa)* pour leur adaptabilité à la riziculture de basfond. Formulaire de requête d'un financement spécial pour un projet d'un groupe d'action. p. 6.

Sneath PHA, Sokal RR (1973). The Principle and Practice of Numerical Classification. In: Kennedy D, Park RB (Eds.), Numerical Taxonomy. Freeman, San Francisco. p. 537.

Swofford DL, Olsen GJ, Waddell PJ, Hillis DM (1990). Phylogenetic Reconstruction. In: Molecular systematics. Hillis, D.M. and Moritz C. (Eds). Sinauer Associates, Sunderland, pp: 411-501.

WARDA (1996). Rice trends in sub-Saharan Africa. A synthesis of statistics on rice production, trade and consumption. WARDA, Bouaké, Ivory Coast. p. 8.

WARDA (2006). Annual Report for 2005. West Africa Rice Development Association. pp. 8-25.

Wopereis MCS, Donovan C, Nebié B, Guindo D, Ndiaye MK (1999). Soil fertility management in irrigated rice systems in the Sahel and Savanna regions of West Africa: Part I. Agronomic analysis. Field Crops Res. 61(2):125-145.

Yan W, Hunt LA, Sheng Q, Szlavnics Z (2000). Cultivar evaluation and mega-environment investigation based on the GGE biplot. Crop Sci. 40: 597-605.

Yan W (2001). GGE biplot—a Windows application for graphical analysis of multienvironment trial data and other types of two-way data. Agron. J. 93: 1111-1118.

Yan W (2002). Singular value partitioning in biplot analysis of multienvironment trial data. Agron. J. 94: 990-996.

Yan W, Rajcan I (2002). Biplot evaluation of test sites and trait relations of soybean in Ontario. Crop Sci. 42: 11-20.

Yan W, Kang MS (2003). GGE biplot analysis: A graphical tool for breeders, geneticists, and agronomists. CRC Press LLC. Boca Roton, Florida. p. 271.

Biochemical and molecular characterization of submergence tolerance in rice for crop improvement

Bishun D. Prasad[1,4]*, Ganesh Thapa[2,3], Samindra Baishya[2] and Sangita Sahni[4]

[1]Department of Agricultural Biotechnology, Assam Agricultural University, Jorhat, Assam, India.
[2]Department of Biochemistry, Assam Agricultural University, Jorhat, Assam, India.
[3]Department of Biotechnology, Indian Institute of Technology Guwahati, Assam, India.
[4]Depatment of Biology, the University of Western Ontario, ON, Canada N6A 5B7, Canada.

Submergence stress is considered as the third most important limitation of rice production contributing to its low productivity in lowland and rainfed ecosystem. Characterization of genotypes and using them in breeding programme is likely the best option to withstand submergence and stabilize productivity in these environments. However, progress in genotypes characterization has been slow but can substantially be enhanced by using potential of biochemical and molecular markers, to enhance and speedup progress through breeding. Survival percentage, alcohol dehydrogenase (ADH) activity and isozyme profiles were carried out in this investigation. The response to submergence appears to be complex and involves a number of enzymes, therefore, cultivars were also characterized by using other isozymes (Aspartate aminotransferase, AAT; Malate dehydrogenase, MDH; Esterase, EST and Peroxidase, POX) and molecular marker like random amplified polymorphic DNA (RAPD). Dendrogram constructed, using isozymes and RAPD data clustered the rice genotypes into three major groups, submergence tolerant, moderately tolerant and susceptible. The result obtained in this study will help plant breeders in breeding high yielding cultivars for lowland eco-systems, with submergence tolerant cultivars.

Key words: Rice, submergence stress, alcohol dehydrogenase (ADH), isozyme, random amplified polymorphic DNA (RAPD), FR13A.

INTRODUCTION

Submergence stress is a major constraint to rice production during the monsoon flooding season in the rainfed lowlands in south and southeast Asia, which causes annual losses of over US$1 billion and affects disproportionately the poorest farmers in the world (Dey and Upadhyaya, 1996; Xu et al., 2006). Out of 40 million ha in Asia grown under rainfed lowlands, about 15 million ha are frequently damaged by submergence stress (Huke and Huke, 1997). Recently, the extent of submergence stress has increased due to extreme weather events such as unexpected heavy rains that have inundated wider areas across many regions in Asia.

The onset of flooding leads to the condition of anaerobiosis or oxygen deprivation as gas diffusion from the atmosphere to water is nearly 10^4 times slower as compared to diffusion in air (Armstrong, 1979). To cope with the reduction in oxygen supply, plants have developed a number of metabolic and morphological adaptations that enable them to survive transient periods of complete or partial submergence (Kende et al., 1998; Drew et al., 2000; Bailey-Serres and Voesenek, 2008). Escape from hypoxia involves shoot elongation, development of aerenchyma and adventitious root formation (Drew et al., 2000; Sauter, 2000; Voesenek et al., 2006; Perata and Voesenek, 2007). Nonetheless, oxygen may ultimately become limiting, necessitating a switch from aerobic respiration to anaerobic fermentation, a key catalytic pathway for recycling NAD^+ to maintain glycolysis and substrate level phosphorylation in the absence of oxygen (Davies, 1980). Enzymes associated with anaerobic fermentation such as alcohol dehydrogenase (ADH) have been considered important in submergence tolerance (Setter, 1992).

The adverse effects of flooding or submergence are

---

*Correspondence author. E-mail: bprasad2@uwo.ca.

Table 1. Description of rice cultivars used in this study.

Cultivars	Description
FR13A	Native rice cultivar, source of submergence tolerance gene
Jalashree	Pankaj X FR13A
Jalkunwari	Pankaj X FR13A
Bahadur	Pankaj X Mahsuri
Ranjit	Pankaj X Mahsuri
Luit	Hira X Annada
Keteki	Bahadur X Savitri
Lachit	CRM13-3241 X Kalinga 2
Chilarai	IR24 X CR44-118-1
Mahsuri	(Taichung 63 X Mayung Eboss) X Mayung Eboss.

multiple and complex which leads to death of most rice cultivars under complete submergence for 1 to 2 weeks (Xu et al., 2006; Perata and Voesenek, 2007). However, some cultivars, such as FR13A, a native rice cultivar from eastern India, can survive up to 2 weeks of complete submergence (Jung et al., 2010). But, these tolerant varieties lack many of the desirable traits of the widely grown varieties, referred to as "mega varieties" that are popular in major rice-growing areas of Asia, because of their high yield and grain quality (Mackill et al., 2006). Considerable progress has been made in breeding submergence tolerant cultivars with mega varieties for lowland eco-systems. Therefore, characterization of genotypes and using them in breeding programme is likely the best option to withstand submergence and stabilize productivity in these environments. Biochemical methods are valuable in differentiating and identifying the different genotypes and the development of molecular marker technology in recent years has revolutionized the whole concept of plant breeding.

However, information about these physiological and biochemical parameters in relation to rainfed lowland cultivars in eastern India with different durations of submergence is quite limited. The study aimed to screen rice genotypes for submergence by using biochemical (ADH activity and isozymes) and molecular marker (RAPD; Random Amplified Polymorphic DNA) technique.

In addition, we aim to establishing the relationships between survivals of genotypes under study with the effects of submergence (up to 15 days) stress using ADH activity and its isozymic profiling. The results obtained in this study will help plant breeders in breeding high yielding cultivars for lowland eco-systems, with submergence tolerant genotypes.

MATERIALS AND METHODS

Screening of rice varieties

Ten rice genotypes were collected from RARS, Titabor, Assam, India for the study (Table 1). Ten days old seedlings grown in earthen pots containing soil were completely submerged in a concrete tank in the greenhouse (28 and 23°C, day and night, respectively) at 0.75 m depth for 10 days. A control set for all the genotypes was also maintained. Scoring was carried out 10 days after total desubmergence and submergence tolerance defined as the ability of a seedling to produce new, growing leaves as described by Singh et al. (2001). The whole experiment was laid out under complete randomized design and replicated three times.

Alcohol dehydrogenase activity and isozyme analysis

Activity and isozymic forms of ADH for both submerged and control plants were studied at 3-day intervals for 15 days of submergence. The leaf tissues were ground in prechilled pestle and mortar with 0.1 M Tris-HCl, pH 7.4 and 10 mM dithiothreitol (DTT) at 4°C. The supernatant obtained after centrifugation (15,000 rpm for 15 min at 4°C) was assayed for the total soluble protein (Lowry et al., 1951) and ADH activity was determined by the Racker method as modified by Stafford and Vennesland (1953). The same extract (60 µg protein) was also used to analyze isozymic forms of ADH (EC 1.1.1.1).

Soluble protein extraction for isozyme studies

Shoots of 7 days old etiolated seedlings were used for aspartate amino transferase (AAT, EC 2.6.1.1), esterase (EST, EC 3.1.1.11), malate dehydrogenase (MDH, EC 1.1.1.37) and peroxidase (POX, EC 1.11.1.7) isozymes studies. For analysis of isozymes except for POX, the sample was ground in prechilled pestle and mortar with 50 mM Tris-Cl buffer (pH 7.6) containing 5 mM β-ME and 5 mM EDTA in the ratio of 1:2 (w/v). For POX, 50 mM Tris-Cl buffer (pH 7.6) alone was used for extraction. The ground mixture was then centrifuged at 15,000 rpm for 15 min at 4°C. The supernatant obtained was immediately used for native polyacrylamide gel electrophoresis (PAGE) (60 µg protein was loaded in each lane) and gels were stained for AAT, EST, MDH, and POX. The relative mobility (Rm) value of each band was computed and stained bands were scored in the format of binary data sets as presence (1) or absent (0).

DNA extraction and RAPD analysis

Genomic DNA of rice cultivars were extracted from 7 day old etiolated seedlings (Dellaporta et al., 1983). RAPD analysis was conducted by using 15 decamer arbitrary primers obtained from Operon Technologies, California. RAPD amplification was performed in 25 µl volume containing 1X Taq DNA Polymerase

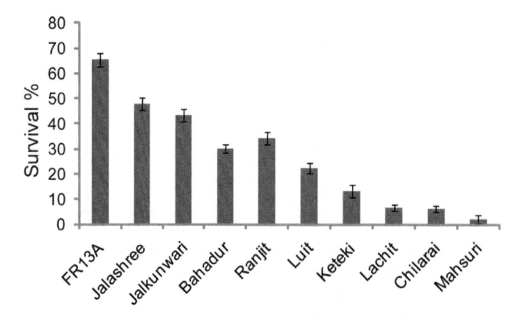

Figure 1. The effect of complete submergence on the survival of rainfed, lowland rice cultivars.

buffer, 200 µM dNTPs mixture, 0.5 µM primer, 25 ng of template DNA and 1 U of Taq DNA polymerase (Bangalore Genei, Bangalore, India) in a thermal cycler (Gene Amp® 2400, Applied Biosystems, USA). PCR amplification was carried out with initial denaturation at 94°C for 5 min, followed by 40 cycles of denaturation at 94°C for 1 min, annealing at 36°C for 1 min and extension at 72°C for 1 min. The 40th cycle was followed by final extension step at 72°C for 5 min. The amplified products were separated by electrophoresis on 1.5% agarose gel and visualized by ethydium bromide (EtBr) staining.

Phylogenetic analysis

Bands were scored as present (1) and absent (0) for isozyme and RAPD profile. An index of genetic similarity using Jaccard's coefficient was calculated. Cluster analyses were performed using UPGMA method and dendrograms were generated using "SPSS for MS Windows Release 16.0".

Data analysis

All the data were analysed statistically with the computer software SPSS and subjected to ANOVA. ADH activity means at each time point under submergence treatment of different rice genotypes were compared to Mahsuri by using least significance difference (LSD) test.

RESULTS

Survival percent

The survival percentage of survived plants is presented in Figure 1. The survival ranged from 1.5 to 65%. The highest survival percentage was recorded in FR13A and the lowest in Mahsuri.

ADH activity during submergence

The ADH enzyme activity in different rice cultivars, as influenced by the different submergence durations, is presented in Figure 2. ADH activity in FR13A, ranged from 0.0076 to 0.328779 units/mg protein during submergence period. ADH activity increased sharply after 6 days of submergence followed by steady increment between 9 and 12 days and then increased sharply from day 12 to reach the maximum (0.328779 units/mg protein) at 15 days of submergence in FR13A. Jalashree (0.0133 to 0.3383 units/mg of protein) and Jalkunwari (0.0064 to 0.3384 units/mg protein) follow almost similar pattern of ADH activity as in FR13A and their ADH activity also reached their maximum at 15 days of submergence. The ADH activity in Bahadur (0.0039 to 0.0488 units/mg of protein) and Ranjit (0.0042 to 0.0384 units/mg of protein) reached maximum at 12 days of submergence followed by a decline at 15 days of submergence. Luit (0.0028 to 0.0582 units/mg of protein), Keteki (0.0033 to 0.0533 units/mg protein), Lachit (0.0012 to 0.0411 units/mg protein), Chilarai (0.0033 to 0.0374 units/mg protein) and Mahsuri (0.0017 to 0.0316 units/mg protein) follow similar pattern of ADH activity. Their ADH activity increased and reached their maximum at 12 days of submergence followed by a sharp decline at 15 days of submergence.

Alcohol dehydrogenase (ADH) isozyme analysis under submergence condition

To investigate the profile of ADH isozyme under different submergence conditions, extract of leaf tissues were

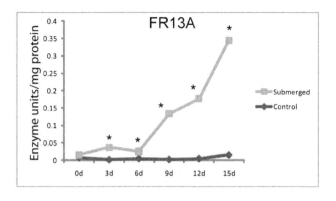
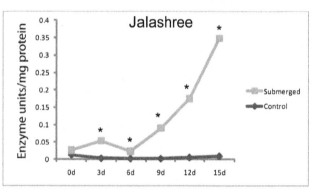
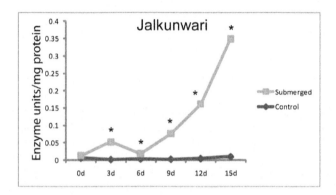
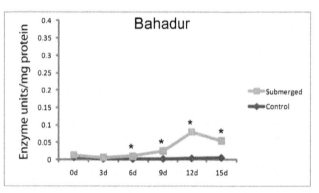
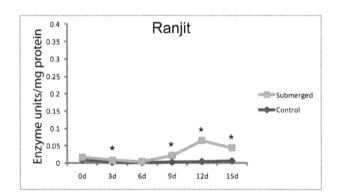
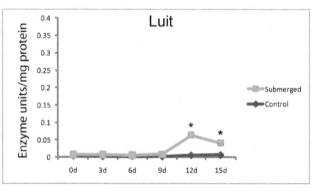
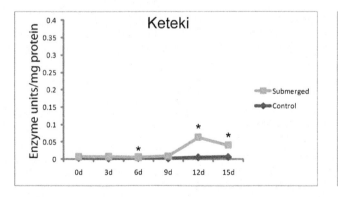
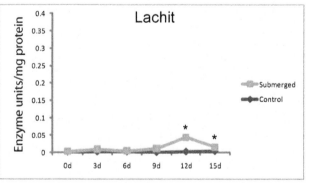

Figure 2. ADH activity of different rice genotypes under stress and control condition. ADH activity was studied at 3-day (d) intervals for 15 days of submergence. ADH activity means at each time point under submergence treatment of different rice genotypes were compared to Mahsuri by using least significance difference (LSD) test. P values are indicated by asterisk (*p< 0.05). Error bars have been omitted for clarity of graph.

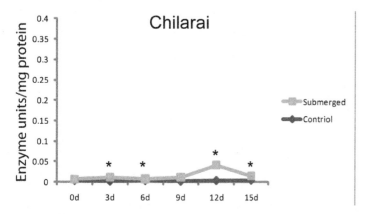
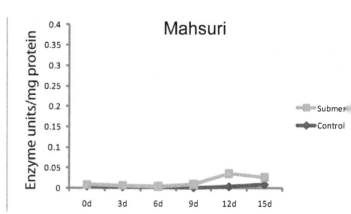

Figure 2. Contd.

subjected to native PAGE. Electrophoretic banding pattern of ADH revealed variable banding profile at different submergence durations (Figure 3). Isozyme pattern of ADH at 0 day of submergence showed only one intense band in all rice genotypes investigated. At 3 days of submergence two clear isozymes were observed in FR13A, Jalshree, and Jalkunwari. Second, isozyme has just started to appear in Bahadur and Ranjit, whereas only one isozyme was observed in other genotypes. At 6 and 9 days of submergence, two monomorphic bands were observed in all the genotypes studied. At 12 days of submergence, three isozymes were observed only in FR13A, Jalshree, and Jalkunwari, whereas only two isozymes were identified in all other rice genotypes. At 15 days of submergence, three monomorphic bands were observed in all rice genotypes studied.

Other isozyme analysis

The isozymic profile of AAT, EST, MDH and POX from 7 days old etiolated seedlings were analyzed in order to differentiate among rice genotypes. Two monomorphic bands were observed in all genotypes for AAT and MDH (Figure 4). EST pattern in rice genotypes yielded 5 bands in total and out of 5 bands, 2 were found to be polymorphic bands (Figure 4). An EST isozyme at Rm 0.80 was present only in FR13A, Jalshree, Jalkunwari, Ranjit and Bahadur, whereas isozyme at Rm 0.85 was found to be common in Luit, Keteki, Lachit, Chilarai, and Mahsuri. Three other EST isozymes were found common in all rice genotypes studied. POX yielded 7 monomorphic and 2 polymorphic bands. Isozymes with Rm 0.42 and 0.90 were present in FR13A, Jalshree, Jalkunwari, Bahadur and Ranjit. However, isozyme with Rm 0.48 was observed in only Mahsuri. The banding patterns of EST and POX isozyme were found to be polymorphic and clearly differentiated rice genotypes used.

Random amplified polymorphic DNA (RAPD) analysis

The banding pattern with all the 15 RAPD primers generated a total of 147 bands, with an average of about 9.8 bands per primer. Out of these amplified fragments 62 were polymorphic. The level of polymorphism was different with different primers among different genotypes. Four primers viz., OPD-06, OPH-07, OPN-04 and OPS-03 (Figure 5) clearly discriminated rice genotypes investigated.

Clustering of genotypes according to their level of submergence tolerance

In order to represent the relationships among genotypes, a cluster analysis (UPGMA) was used to generate dendrogram based on isozyme and RAPD analysis. Isozyme analysis showed two major clusters, FR13A, Jalashree and Jalkunwari were grouped together in one cluster, whereas all other genotypes except Ranjit and Bahadur were grouped together in another cluster. Ranjit and Bahadur were grouped closer to cluster having FR13A, Jalashree and Jalkunwari (Figure 6A). Dendrogram obtained from the RAPD data also showed two major clusters, FR13A, Jalashree and Jalkunwari were grouped together in one cluster, whereas all other genotypes except Bahadur were grouped together in the remaining cluster (Figure 6B).

DISCUSSION

Submergence due to flash flood is the key factor limiting yield of lowland rice which adversely affects grain yield of rice crop (Mohanty et al., 2000). On the whole, improved submergence tolerance is an important trait for rice growing in rainfed lowland areas. Therefore, efforts are being directed towards improving submergence tolerance character without affecting grain yield.

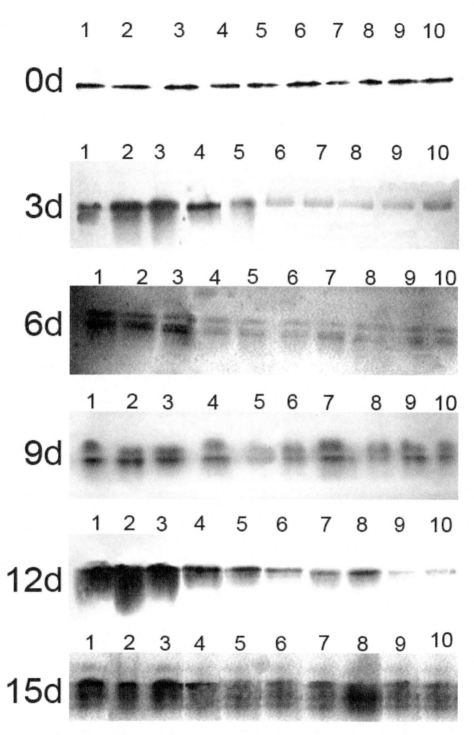

Figure 3. Isozyme profile analyses of ADH at 3-d intervals for 15 days under submergence treatment. Lane 1_FR13A, Lane 2_Jalashree, Lane 3_Jalkunwari, Lane 4_Bahadur, Lane 5_Ranjit, Lane 6_Luit, Lane 7_Keteki, Lane 8_Lachit, Lane 9_Chilarai, and Lane 10_Mahsuri.

The present investigation mainly attempts to characterize different rice genotypes on the basis of their submergence tolerance using biochemical and molecular techniques.

Screening of rice genotypes for submergence tolerance

Identification of submergence tolerant genotypes has

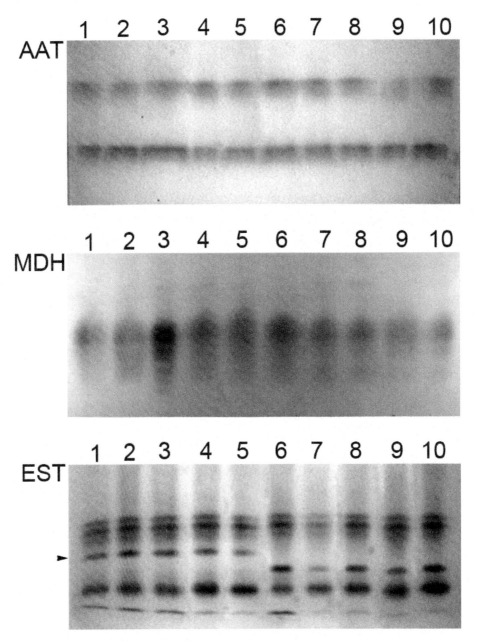

Figure 4. Isozyme profile analyses of AAT, MDH and EST of 7 days old etiolated seedlings. Lane 1_FR13A, Lane 2_Jalashree, Lane 3_Jalkunwari, Lane 4_Bahadur, Lane 5_Ranjit, Lane 6_Luit, Lane 7_Keteki, Lane 8_Lachit, Lane 9_Chilarai, and Lane 10_Mahsuri.

required the use of stress-specific screens, because direct evaluation of tolerance is not as simple as it might seem. Therefore, in the present investigation, 10-day-old seedlings were submerged for 10 days at a depth of 0.75 m water to check their ability to withstand submergence.

The highest survival percentage was observed in FR13A, while the lowest was recorded in Mahsuri. This may be because of high level of ADH activity in FR13A whereas, least ADH activity was observed in Mahsuri throughout the period of submergence.

These results were in consistence with previous report (Singh et al. 2001) where FR 13 A showed the least damage and the best survival, whereas, the poorest survival was observed in Mahsuri at the end of the submergence treatment.

ADH activity during submergence

Anaerobic fermentation is one of the major metabolic adaptations that plants assume when they are submerged or faced with lack of oxygen (Sachs et al.,

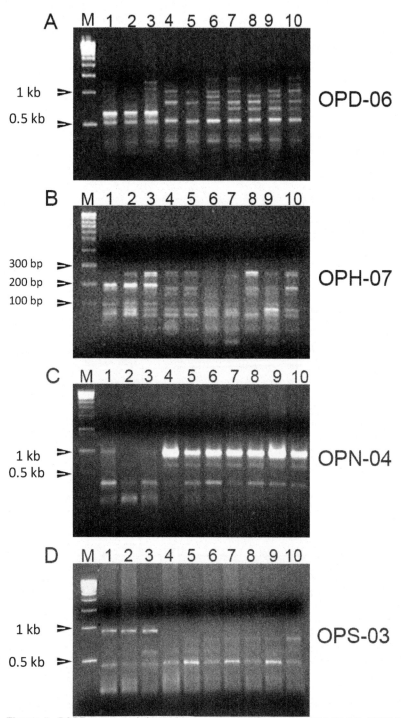

Figure 5. RAPD pattern of rice cultivars using OPD-06 (A), OPH-07 (B), OPN-04 (C), and OPS-03 (D) primers. Lane M_DNA ladder, Lane 1_FR13A, Lane 2_Jalashree, Lane 3_Jalkunwari, Lane 4_Bahadur, Lane 5_Ranjit, Lane 6_Luit, Lane 7_Keteki, Lane 8_Lachit, Lane 9_Chilarai, and Lane 10_Mahsuri.

1996). Higher levels of ADH, an enzyme associated with anaerobic fermentation and ethanol production during anaerobiosis have been reported for flood tolerant plants (Tripepi and Mitchell, 1984, Kato-Noguchi and Kugimiya, 2003). In all rice genotypes investigated, the ADH activity under submerged condition increased as the duration of submergence increased. A sharp increase in the ADH activity was observed after 9 days of submergence. In FR13A, Jalashree, Jalkunwari and Bahadur, the increase in the activity continued till 15 days of submergence.

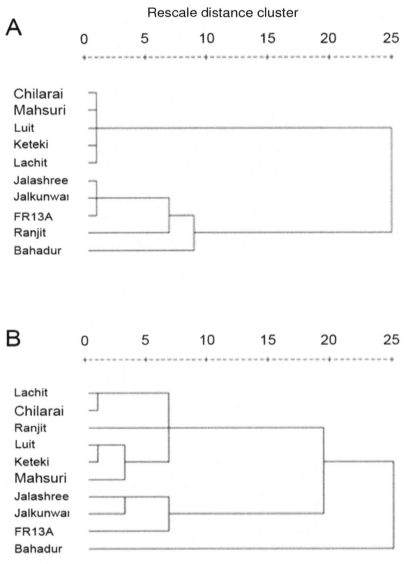

Figure 6. Cluster analysis dendrogram of rice varieties based on (A) Isozymes, and (B) RAPD data.

This finding is similar to that of Bertani et al. (1980) who observed that ADH undergoes a rise in activity upon imposition of anaerobiosis which then tends towards a plateau. Increased level of ADH activity during O_2 deficiency and its level measured at any point of time has been function of rate of transcription and on its transcript stability (Ferl et al., 1980; Dennis et al., 1984, 1985; Rowland and Strommer, 1986). The decline in the level of ADH activity in the susceptible genotypes after initial induction might be attributed to the decline in the energy metabolism under anoxia condition.

ADH isozyme in submerged rice genotypes

Increased rice ADH enzyme activity under submergence condition prompted us to see the change in isozymic profiles in different under submergence stress. The number of ADH isozyme observed in rice genotypes varied from 1 to 3. Plant ADH enzymes were considered to be dimers and the two subunits of ADH are encoded by two unlinked genes (Gottlieb, 1982; Sachs and Ho, 1986). The products of these genes dimerize randomly to yield three electrophoretically distinct isozymes: ADH1-ADH1 homodimer, ADH1-ADH2 heterodimer and ADH2-ADH2 homodimer (Gottlieb, 1982; Newman and VanToai, 1991).

In this study, early appearance of 2nd and 3rd isozyme at 3 and 12 days of submergence in FR13A, Jalshree, and Jalkunwari indicated that the increase in ADH activity lead to increase in enzyme synthesis. Increased rice ADH enzyme activity has been shown by a change in isozymic profiles of the ADH protein (Rivoal et al., 1989; Xie and Wu, 1989). Hence, the results obtained in our study strengthen the hypothesis that change in isozymic profiles may be used to distinguish submergence tolerant and susceptible rice genotypes.

Other isozymes

Response to submergence appears to be complex and involves a number of enzymes. Hence, we investigated AAT, EST, MDH, and POX to see the differences, if any, between the cultivars under study. Isozyme profile of AAT and MDH were found to be monomorphic, whereas EST and POX pattern were polymorphic and able to differentiated cultivars under study. Such differences may not be a mere coincidence, as EST and POX has been previously shown to be a useful induction of submergence and other abiotic stress tolerance in rice (Mandal et al., 2004; Zhang et al., 1988).

The differences in the isozyme binding pattern were due to variation in the amino acid content of the molecule, which in turn was dependent on the sequence of nucleotides in DNA (Micales et al., 1986). Different bands obtained indicate different electrophoretic mobilities of the isozymes, which were coded by different alleles or separate genetic loci.

Random amplified polymorphic DNA (RAPD) analysis

Genetic markers such as morphological markers and biochemical markers were more prone to environment effect and limited by small number of loci (Tanksley et al., 1989). The use of PCR based assays having advantage of being quick, easy to use and refractory to many environmental influences can complement traditional and biochemical methods of cultivars characterization. The RAPD technique for detecting genetic variation among cultivars and identifying germplasm is well established (Cao and Oard, 1997; Gorji et al., 2010).

In this study, four primers viz.; OPD-06, OPH-07, OPN-04 and OPS-03 were found to clearly discriminate rice genotypes used in this study. OPH-07 primer was found to be linked with Sub 1 locus of chromosome number 9 and produced bands associated with either tolerant or susceptible F_3 families in rice (Xu and Mackill, 1996). Dendrogram generated based on RAPD analysis, FR13A, Jalashree and Jalkunwari were grouped together in one cluster, whereas all other genotypes except Bahadur were grouped together in the remaining cluster.

Conclusion

With the identification of physiological traits, enzyme activity, isozymes, and DNA markers associated with submergence tolerance, the prospects for breeding suitable rice cultivars for rainfed lowlands have been improved. The characterization of genotypes based on survival percentage, ADH activity, isozymes, and RAPD analysis for its survival under submergence has again been well established in our study. Dendrogram generated by isozymes and RAPD analyses were able to discriminated genotypes under study. The genotypes grouped in each cluster showed similar pattern of ADH activity and its isozymic profile and were well correlated with survival percentage under different period of submergence. The rice genotypes used in the present study can be grouped into three categories:

(I) Submergence tolerant (FR13A, Jalashree and Jalkunwari),
(II) Moderately submergence tolerant (Bahadur and Ranjit),
(III) Submergence susceptible (Luit, Keteki, Lachit, chilarai and Mahsuri).

Thus, the results of this study clearly indicate the utility of biochemical and molecular markers for the characterization of rice genotypes for submergence tolerance. Therefore, such studies are useful in identifying and characterizing submergence tolerant genotypes in rice and information obtained can be of use for breeders.

REFERENCES

Armstrong W (1979). In: Advances in Botanical Research. Woolhouse, H. W. (ed.) Acad. Press, New York, 7: 225-332.

Bailey-Serres J, Voesenek LACJ (2008). Flooding stress: acclimations and genetic diversity. Annu. Rev. Plant Biol., 59: 313-339.

Bertani A, Brambilla I, Manegus F (1980). Effects of anaerobiosis on rice seedlings: Growth, metabolic rate, and fate of fermentation products. J. Exp. Bot., 31: 325-331.

Cao D, Oard JH (1997). Pedigree and RAPD based DNA analysis of commercial US rice cultivars. Crop Sci., 37: 1630-1635.

Davies DD (1980). Anaerobic metabolism and the production of organic acids. In: Davies DD, ed. The biochemistry of plants. Vol 2. New York: Academic Press, 581-611.

Dellaporta SL, Wood J, Hicks JB (1983). Maize DNA minipreps. Maize Genet. Crop Newsl., 57: 26-29.

Dennis ES, Gerlach WL, Pryor AJ, Bennetzen JL, Inglis A, Llewellyn D, Sachs MM, Ferl RJ, Peacock WJ (1984). Molecular analysis of the alcohol dehydrogenase (Adh1) gene of maize. Nucleic Acids Res., 12: 3983-4000.

Dennis ES, Sachs MM, Gerlach WL, Finnegan EJ, Peacock WJ (1985). Molecular analysis of the alcohol dehydrogenase 2 (Adh2) gene of maize. Nucleic Acids Res., 13: 727-743.

Dey MM, Upadhyaya HK (1996). Yield loss due to drought, cold and submergence in Asia. In: Everson RE, Herdt RW, Hossain M, eds. Rice research in Asia: progress and priorities. Manila, The Philippines: IRRI, pp. 291-303.

Drew MC, He CJ, Morgan PW (2000). Programmed cell death and aerenchyma formation in roots. Trends Plant Sci., 5: 123-127.

Ferl RJ, Brennan M, Schwartz D (1980). In vitro translation of maize ADH: evidence for the anaerobic induction of mRNA. Biochem. Genet., 18: 681-691.

Gorji AH, Davish F, Esmaeilzadehmoghadam M, Azizi F (2010). Application RAPD Technique for Recognition Genotypes Tolerant to Drought in some of Bread Wheat. Asian J. Biotechnol., 2: 159-168.

Gottlieb LP (1982). Conservation and duplication of isozymes in plants. Science, 216: 373-380.

Huke RE, Huke EH (1997). Rice area by type of culture-in South, Southeast, and East Asia. A revised and updated data base. IRRI, Los Baños, Philippines.

Jung KH, Seo YS, Walia H, Cao P, Fukao T, Canlas PE, Amonpant F, Bailey-Serres J, Ronald PC (2010). The submergence tolerance regulator Sub1A mediates stress-responsive expression of AP2/ERF transcription factors. Plant Physiol., 152: 1674-1692.

Kato-Noguchi H, Kugimiya T (2003). Preferential Induction of Alcohol Dehydrogenase in Coleoptiles of Rice Seedlings Germinated in Submergence Condition. Biol. Plant, 46: 153-155.

Kende H, van der Knaap E, Cho HT (1998). Deepwater rice: a model plant to study stem elongation. Plant Physiol., 118: 1105-1110.

Lowry OH, Rosenbought NJ, Ferr AL, Randall RJ (1951). Protein measurement with the Folin-phenols reagent. J. Biol. Chem., 193: 265-275.

Mackill DJ, Collard BCY, Neeraja CN, Rodriguez RM, Heuer S, Ismail AM (2006). QTLs in rice breeding: examples for abiotic stresses. In: Brar DS, Mackill DJ, Hardy B (eds) Rice genetics 5: proceedings of the international rice genetics symposium. International Rice Research Institute, Manila, pp. 155-167.

Mandal S, Mandal N, Hazra K, Mukherjee SK, Das PK (2004). Submergence tolerant androgenic rice dihaploids derived from intra- and interspecific crosses: I. Seed Storage Protein and Isozyme Pattern Through Gel Electrophoresis. Plant Tissue Cult., 14: 25-35.

Micales JA, Bonde MR, Peterson GL (1986). The use of isozyme analysis in fungal taxonomy and genetics. Mycotaxonomy, 27: 407-449.

Mohanty HK, Mallik S, Grover A (2000). Prospects of improving flooding tolerance in lowland rice varieties by conventional breeding and genetic engineering. Curr. Sci., 78(2): 132-140.

Newman KD, VanToai TT (1991). Developmental regulation and organ-specific expression of soybean alcohol dehydrogenase. Crop Sci., 31: 1253.

Perata P, Voesenek LA (2007). Submergence tolerance in rice requires Sub1A, an ethylene-response-factor-like gene. Trends Plant Sci., 12: 43-46.

Rivoal J, Ricard B, Pradet A (1989). Glycolytic and fermentative enzyme induction during anaerobiosis in rice seedlings. Plant Physiol. Biochem., 27: 43-52.

Rowland LJ, Strommer JN (1986). Anaerobic treatment of maize roots affects transcription of Adh1 and transcript stability. Mol. Cell Biol., 6: 3368-3372.

Sachs MM, Ho T-HD (1986). Alteration of gene expression during environmental stress in plants. Annu. Rev. Plant Physiol., 32: 363-376.

Sachs MM, Subbaiah CC, Saab IN (1996). Anaerobic gene expression and flooding tolerance in maize. J. Exp. Bot., 47: 1-15.

Sauter M (2000). Rice in deep water: how to take heed against a sea of troubles. Naturwissenschaften, 87: 289-303.

Setter TL (1992). Important physiological mechanisms of submergence tolerance in rice. In:Kuo, G.C. (ed.) Adaptation of food crops to temperature and water stress. Proc. Int. Symp., Taiwan. 13-18th August, 1992. pp. 220-230.

Singh HP, Singh BB, Ram PC (2001). Submergence tolerance of rainfed lowland rice: search for physiological marker traits. J. Plant Physiol., 158: 883-889.

Stafford HA, Vennesland B (1953). Alcohol dehydrogenase of wheat germ. Arch. Biochem. Biophys., 44: 404-414.

Tanksley SD, Young ND, Paterson AH, Bonierbale MW (1989). RFLP mapping in plant breeding: New tools for old science. Biotechnology, 7: 257.

Tripepi RR, Mitchell CA (1984). Metabolic response of river birch and european birch roots to hypoxia. Plant physiol., 76: 31-35.

Voesenek LA, Colmer TD, Pierik R, Millenaar FF, Peeters AJ (2006). How plants cope with complete submergence. New Phytol., 170: 213-226.

Xie Y, Wu R (1989). Rice alcohol dehydrogenase genes: anaerobic induction, organ specific expression and characterization of cDNA clones. Plant Mol. Biol., 13: 53-58.

Xu K, Mackill DJ (1996). RAPD and RFLP mapping of a submergence tolerance locus in rice. pp. 244. Rice Genetic Newsletter, Vol. 12.

Xu K, Xu X, Fukao T, Canlas P, Maghirang-Rodriguez R, Heuer S, Ismail AM, Bailey-Serres J, Ronald PC, Mackill DJ (2006). Sub1A is an ethylene-response-factor-like gene that confers submergence tolerance to rice. Nature, 442: 705-708.

Zhang ZH, Zheng ZL, Gao HY, Cao HX (1988). Breeding, evaluation and utilization of anther cultured varieties "Zin Xion" and "Hua Han Zoo" in rice (Oryza sativa L.). In : IRRI and Academia Sinica (eds.) Genetic manipulations in crops, Cassel Tycooly, Philadelphia, pp. 36-37.

Participatory selection and characterization of quality protein maize (QPM) varieties in Savanna agro-ecological region of DR-Congo

K. Mbuya[1,2], K. K. Nkongolo[4]*, A. Kalonji-Mbuyi[1,3] and R. Kizungu[1,2]

[1]University of Kinshasa, B.P 117 Kinshasa 11, RD–Congo.
[2]Institut National pour l' Etude et la Recherche Agronomiques (INERA), B.P. 2037, Kinshasa 1, RD-, Congo.
[3]Nuclear Research Center, Kinshasa, RD-Congo.
[4]Department of Biological Sciences, Laurentian University, Sudbury, Ontario, Canada.

Maize (*Zea mays* L.) is a major cereal crop for human nutrition in the Democratic Republic of Congo (DR- Congo). Prevailing normal maize is deficient in two essential amino acids, lysine and tryptophan. Participatory variety selection was applied to select diversified quality protein maize (QPM) varieties that possess farmers' preferred plant and grain traits. The varieties were planted with and without chemical fertilization. Selection was based primarily on agronomic traits such as time to maturity, plant and ear aspect, disease and insect resistance, yield and yield components as well as flour quality. There were significant differences among QPM varieties for several agronomic traits. The use of participatory approach in agricultural research allowed selection of one QPM, (QPMSRSYNTH), and one normal improved maize (AK9331-DMR-ESR-Y) for their yield advantage over currently released normal maize varieties in more than one criterion. The adoption of these newly introduced varieties is expected to be high since they were selected based on farmer's preference.

Key words: Quality protein maize, participatory varietal selection, DR-Congo.

INTRODUCTION

Several hundred million people in developing countries rely on maize as their main staple food. However conventional maize (corn) has two significant flaws; it lacks the full range of amino acids, namely lysine and tryptophan, needed to produce proteins, and has its niacin (vitamin B_3) bound in an indigestible complex. In addition diets high in corn produce a condition known as wet-malnutrition that leads to 'Kwashiorkor' caused by a chronic lack of protein in the diet (Hugo, 2000; Olakojo et al., 2007; Upadhyay et al., 2009).

Quality protein maize developed by the international maize and wheat improvement center (CIMMYT) in the late 1990's produces 70 to 100% more of lysine and tryptophan and yields 10% more grain than the most modern varieties of tropical maize (Bjarnason and Vasal, 1992; Vasal, 2000). These two amino acids allow the body to manufacture complete proteins, thereby eliminating wet-malnutrition. In addition tryptophan can be converted in the body to Niacin, which theoretically reduces the incidence of Pellagra. QPM offers 90% of the nutritional value of skim milk, the standard for adequate nutrition value (National Research Council, 1988; Hugo, 2000; Olakojo et al., 2007; Upadhyay et al., 2009).

QPM varieties have yielded positive results in China, Mexico, and Central America for yield and reduction of wet malnutrition. In Africa, 17 countries have introduced and promoted QPM. These include South Africa, Burkina Faso, Cameroon, Ivory Coast, Ethiopia, Ghana, Guinea, Kenya, Malawi, Mali, Mozambique, Nigeria, Ouganda, Senegal, Tanzania, Togo and Zimbabwe (Bressani, 1991; Olakojo et al., 2007; Upadhyay et al., 2009). The introduction of QPM in Democratic Republic - Congo is limited and localized in "National Institute for Agronomic Research and Studies (INERA) in Gandajika.

*Corresponding author. E-mail: knkongolo@laurentian.ca.

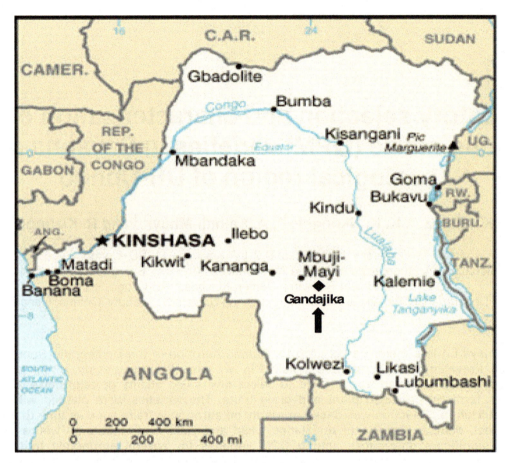

Figure 1. Map (in white) of DR- Congo. The arrow indicates the site location (Gandajika).

The low adoption rate of many of the crop varieties, especially in subsistence and small scale farming systems, has been attributed to the release of station-bred varieties that were evaluated and managed by researchers under conditions that are most favorable to crop growth without regard to local constraints and farmers (Gyawali et al., 2007; Weltzien et al., 2003). More recently, plant breeding strategies that make use of, and maintain crop diversity have been advocated by some researchers as one way of improving crop yields, productivity, stability, and adoption rate (Witcombe et al., 1996; Mekbib, 2006). Participatory plant breeding (PPB) or participatory variety selection (PVS) is one of such strategy that aims at strengthening cooperation between researches, especially breeders, and farmers in evaluating plant germplasm and establishing plant breeding goals that take into account farmers' knowledge and gender factors. Witcombe et al. (1996) were able to distinguish between works involving segregating or stable lines using participatory approach. They refer to work involving farmers in evaluating stable lines as "participatory variety selection" (PVS) while work with still-segregating material are referred to as PPB. The strategy is premised on the observation that the agronomic, socio-economic, and socio-cultural requirements of smallholder farmers and consumers are too diverse to be filled by a limited number of genotypes. Efforts to develop agriculture in a manner that will benefit the poor must fully address gender equality and the empowerment of women. Sex-disaggregated data, gender analysis, and women's participation in decision-making are necessities in agricultural planning and implementation, including development of new varieties.

The objective of this study was to use participatory variety selection (PVS) strategy to evaluate agronomic performance and flour quality of QPM among male and female farmers in a maize growing region of Gandajika, DR-Congo.

MATERIALS AND METHODS

Thirteen open pollinated (OP) quality protein maize varieties and ten normal maize varieties were evaluated for agronomic characteristics and disease reaction at two sites in Gandajika which falls within the savanna agro-ecological region. Gandajika is in Eastern Kasai of DR-Congo with latitude 6° 45' S, longitude 23° 57' E and altitude 780 m (Figure 1). Three of the normal varieties were hybrids and seven were OP. Five varieties adapted to local conditions were used as checks. They include four genetically improved and released varieties (Salongo 2, Mus 1, Kasaï, and GPS 5) along with a local farmer variety. The environmental

Table 1. Temperature, Relative humidity, and rainfalls during the 2008 to 2009 growing conditions in Gandajika (DR-Congo).

Month	Temperature (°C)			Relative humidity (%)			Rainfalls (mm)
	Max.	Min.	Mean	Max.	Min.	Mean	
October	33.5	21.2	27.4	75.1	52.1	63.6	155.9
November	32.8	21.6	27.2	81.4	64.3	72.8	164.0
December	31.6	22.2	26.9	86.7	67.0	76.8	125.1
January	32.5	21.2	26.8	80.6	71.4	76.0	115.8

Source: INERA Gandajika.

Table 2. Origin, year of introduction, and types of the 24 varieties evaluated with farmers in Gandajika, DR-Congo.

Varieties	Origin/Provider	Year of introduction	Type	category
ECA-QVE 3	CIMMYT-Kenya	2008	QPM	OP*
ECA-QVE 4	CIMMYT-Kenya	2008	QPM	OP
ECA-QVE 6	CIMMYT-Kenya	2008	QPM	OP
ECA-POPE1	CIMMYT-Kenya	2008	QPM	OP
POOL15QC7-SRC1-F2	CIMMYT-Kenya	2008	QPM	OP
DMR-ESR-W	IITA-Ibadan	1994	NORMAL	OP
AK9331-DMR-ESR-Y	IITA-Ibadan	1994	NORMAL	OP
QPMSRSYNTH	CIMMYT-Kenya	2008	QPM	OP
SUSUMA	CIMMYT-Kenya	2008	QPM	OP
OBATANPA-SRC1-F3	CIMMYT-Kenya	2008	QPM	OP
POOL15QPM-S	CIMMYT-Kenya	2008	QPM	OP
S99TLWQHG-AB	CIMMYT-Kenya	2008	QPM	OP
SOOTLWQ-AB	CIMMYT-Kenya	2008	QPM	OP
SOOTLWQ-B	CIMMYT-Kenya	2008	QPM	OP
LONGE 5	NARI-Ouganda	2008	QPM	OP
OPAQUE-2	CIMMYT -Kenya	2007	O-2**	OP
SALONGO-2	INERA-Gandajika	1976	NORMAL	OP
KASAI-1	INERA-Gandajika	1976	NORMAL	OP
MUS-1	INERA-Gandajika	1996	NORMAL	OP
GPS-5	INEAC-Gandajika	-	NORMAL	OP
LOCALE	Farmers	-	NORMAL	OP
PAN 67	South Africa	2008	NORMAL	Hybrid
PAN 77	South Africa	2008	NORMAL	Hybrid
MH 18	South Africa	2008	NORMAL	Hybrid

*Open pollinated varieties, **opaque-2 gene, NARI : Namulonge agriculture research institute, CIMMYT: International maize and wheat improvement center, INERA : 'Institut National pour l'Etude et la Recherche Agronomiques' (in french), INEAC: 'Institut National d'Etude Agronomique au Congo Belge' (in french), IITA: international institute of tropical agriculture.

conditions (temperature, relative humidity, and total monthly rainfall) recorded during the growing season at the evaluation site are presented in Table 1.

The source and year of introduction of each variety to DRC are described in Table 2. Selection of farmer for this study was based on willingness to participate as well as their knowledge of maize crop and its production. Care was taken for gender considerations, therefore both men and women were involved. Staff members of the National Institute for Agronomic Studies and Research or 'Institut National pour l'étude et la recherche agronomique (INERA)' and local NGOs participated in the site and farmer selection. Participating farmers (23 men and 16 women) were from Gandajika and five surrounding villages (Kalunga, Mpembanzeu, Muyembi, Mpiana and Bena Kayumba). They were briefed on their role as well as the objectives and expectations of project activities.

The community plot at each site was ploughed and ridged at a spacing of 0.75 × 0.5 m. Gross plot size (experimental unit) was 5 m long and 1.5 m wide. Two seeds were planted per stand and later thinned to one two weeks after seedling emergence to provide a uniform plant population of 53,333 plant / ha. Manual weeding

Table 3. Plant aspect, reaction to downy mildew and maize streak virus (MSV), days to flowering and plant height of 13 QPM, 7 normal open pollinated maize varieties and 3 normal maize hybrids evaluated under chemical fertilization in community plots in Gandajika, DR-Congo.

Varieties	Plant aspect	Reaction to downy mildew	Reaction to maize streak virus	Days to 50% male flowering	Days to 50% female flowering	Plant height
	1-5	1-5	1-5	Days	Days	cm
ECA-QVE 3	1.5	1.0	1.5	53.3	56.0	141.7
ECA-QVE 4	1.5	1.0	1.8	51.3	54.7	132.3
ECA-QVE 6	1.5	1.0	1.3	52.7	56.0	135.0
ECA-POPE1	1.5	1.2	1.3	51.3	55.7	138.0
POOL15QC7-SRC1-F2	1.5	1.0	1.3	52.0	55.3	145.3
DMR-ESR-W	1.7	1.0	1.0	52.7	56.0	132.7
AK9331-DMR-ESR-Y	1.3	1.0	1.3	56.0	58.7	149.7
QPMSRSYNTH	1.3	1.0	2.0	56.7	59.3	162.7
SUSUMA	1.7	1.0	1.8	57.0	59.7	158.7
OBATANPA-SRC1-F3	1.5	1.2	1.3	57.3	60.7	167.0
POOL15QPM-S	1.5	1.0	1.8	53.0	57.0	132.3
S99TLWQHG-AB	4.0	1.0	4.2	60.7	63.3	137.7
SOOTLWQ-AB	4.0	1.0	3.8	59.3	62.7	132.7
SOOTLWQ-B	4.0	1.0	3.5	59.0	61.3	156.7
LONGE 5	1.5	1.0	1.2	59.7	59.7	152.3
OPAQUE-2	1.3	1.0	1.0	63.7	63.7	144.7
SALONGO-2	1.3	1.0	1.8	65.3	65.3	171.0
KASAI-1	2.0	1.0	2.2	63.7	63.7	165.3
MUS-1	1.3	1.0	1.3	59.7	59.7	157.7
GPS-5	1.5	1.0	1.7	64.7	64.7	165.0
LOCALE	1.8	1.0	1.5	63.7	63.7	181.0
PAN 67	1.2	1.2	1.5	60.7	60.7	163.7
PAN 77	1.8	1.0	1.5	61.0	61.0	167.7
MH 18	3.0	1.0	3.0	60.3	60.3	152.7
LSD (0.05)	0.7	ns	0.8	1.5	1.5	16.9
CV (%)	23.3	10.2	25.7	1.6	1.6	6.8

was carried out as to keep the field clean. The trial was a completely randomized block design (RCBD) with three replicates. At each site, two trials were conducted, one with chemical fertilizer application and the second with no fertilizer application. The fertilizer treatment consisted of the application of 140 kg/ha of NPK (17-17-17) at 14 days after planting followed by 20 Kg/ha of nitrogen (urea fertilizer) application 25 to 30 days later.

Data collection was based on farmer and breeder criteria. The following data were collected from the two middle rows of each plot: plant aspect, days to 50% flowering, plant heights, husk tip cover, and cob aspect using ratings 1 to 5, where 1 = excellent, 2 = very good, 3 = good, 4 = fair and 5 = poor. All varieties were observed for the natural symptoms of two main local diseases, maize streak virus (MSV) and Downy mildew (caused by *Perosclerospora (sclerospora) sorghi*). Severity of each of the two diseases was scored using a rating 1- 5 where, 1 = no symptom, 2 = slight infection, 3 = moderate infection, 4 = severe infection, and 5 = very severe. Other yield related characters included ear number, kernel rows, kernels per row, and grain filling. At harvest, ears from the middle two rows were harvested together, shelled and grain yield per plot was determined at 14% moisture content from which grain yield per ha was estimated. Grain yield was adjusted to 83% shelling recovery from the de-husked cob weight per plot. The key criteria for male and females farmers include grain yield, big cob size, small cob rachis, easiness for shelling, and corn flour quality for local dishes (fufu).

Collected data were subjected to analysis of variance (ANOVA) using Genstat discovery version 3 and the least significant differences among means were calculated to identify differences among specific treatments.

RESULTS

Data related to plant development are summarized in Tables 3 and 4 for both trials with and without fertilization. In the first trial (with chemical fertilization), days to 50% flowering varied from 51.3 to 62.3 for male flowering and 54.7 to 65.3 for female flowering. Plant aspect rating varied between 1 and 4. ECA-QVE3, ECA-QVE6, POOL15QC7-SRC1-F2, DMR-ESR-W, AK9331- DMR-ESR-Y, QPMSRSYNTH and MUS-1 were overall

Table 4. Average agronomic performance of 13 QPM varieties, 7 normal open pollinated varieties and 3 normal hybrids evaluated under chemical fertilization (NPK and urea) in community plots in Gandajika, DR-Congo.

Varieties	Cob aspect* 1-5	Kernel rows per cob Number	Cob size/ diameter (cm)	Cob rachis size (cm)	Cob length (cm)	Adjusted grain yield* (t/ha)	Flour quality* 1-5
ECA-QVE 3	1.8	14.7	4.3	2.6	15.1	3.1	1.5
ECA-QVE 4	1.7	14.7	4.3	2.4	13.7	3.8	1.5
ECA-QVE 6	1.7	14.0	4.3	2.5	13.5	3.2	1.5
ECA-POPE1	1.7	14.6	4.3	2.4	13.8	3.7	2.0
POOL15QC7-SRC1-F2	1.7	14.7	4.3	2.5	15.3	3.8	2.0
DMR-ESR-W	1.7	13.3	3.8	2.1	13.7	2.9	1.5
AK9331-DMR-ESR-Y	1.5	14.0	4.3	2.4	15.0	4.8	1.5
QPMSRSYNTH	1.7	14	4.3	2.5	16.6	4.2	1.5
SUSUMA	1.5	14.7	4.6	2.4	15.3	3.7	1.0
OBATANPA-SRC1-F3	1.7	14.0	4.5	2.4	15.4	3.3	1.0
POOL15QPM-S	2.2	14.0	4.5	2.5	14.1	2.6	1.5
S99TLWQHG-AB	2.0	16.0	4.4	2.4	15.3	-	-
SOOTLWQ-AB	2.5	14.7	4.4	2.5	15.3	-	-
SOOTLWQ-B	1.7	15.3	4.4	2.5	15.7	-	-
LONGE 5	1.7	14.0	4.3	2.7	14.8	3.1	1.0
OPAQUE-2	3.0	13.3	4.4	2.4	17.5	2.5	2.5
SALONGO-2	1.8	14.7	4.4	2.5	15.0	3.5	1.5
KASAI-1	1.8	16	4.6	2.5	15.2	3.4	1.5
MUS-1	1.8	14.7	4.3	2.7	15.1	3.8	1.0
GPS-5	2.2	12	3.8	2.2	15.1	2.2	1.5
LOCALE	1.8	10.7	3.7	1.4	14.0	3.5	1.5
PAN 67	1.8	12	4.5	2.2	16.2	3.9	2.0
PAN 77	2.0	13.3	4.5	2.7	16.7	4.1	2.0
MH 18	2.2	14.0	4.3	2.3	15.4	3.5	1.5
LSD (0.05)	0.6	1.4	0.3	0.3	1.1	0.4	-
CV (%)	20.1	6.1	4.5	6.9	4.3	7.5	-

1= best and 5 = worst.

preferred by farmers for their plant aspect (Table 3), whereas among the QPM varieties, S99TLWQHG-AB, SOOTLWQ-AB, and SOOTLWQ-B received the lowest score of 4 and were not selected by farmers. Other QPM entries were not significantly different from released normal varieties and the local farmer variety. Among the hybrids, MH18 showed a moderate score of 3 (Table 3). All the varieties showed good husk cover. Plant heights vary from 116.7 to 167.3 cm with the local farmer variety, Salongo-2, GPS-5, Kasai-1, Obatanpa-SRC1-F3, and the hybrid PAN 77 producing the tallest plants (Table 3). In general, early maturing varieties were shorter than late maturing.

There were no significant differences among the varieties under the trial conditions for incidence and severity of downy mildew. All the varieties showed resistance to this disease with ratings varying from 1 to 1.2 (Table 3). However there were significant differences among the varieties for maize streak virus ratings. S99TLWQHG-AB, SOOTLWQ-AB, and SOOTLWQ-B, and MH18 hybrids were considered susceptible under experimental conditions with disease rating ranging from 3.0 to 4.2. Kasaï 1 (score 2.2) was moderately resistant while the remaining 18 varieties were resistant (scores ranged between 1 and 2.0).

Result of yield related characters revealed that cob aspect rating ranged from 1.5 to 3. Most varieties produced good to very good ears (score 1.5 to 2.2) (Table 4). Cob length varied from 13.5 to 17.5 cm with OPAQUE 2 recording the longest cob, while the shortest cobs were observed in ECAQVE-6. However, QPMRSYNTH had the longest cob among the QPM varieties.

There were significant differences among varieties for number of kernel rows per cob. Kasaï 1 (an improved check) and S99TLWQHG-AB had the highest number of kernel rows (16 under chemical fertilization and 15.3 for Kasai 1 in trials without chemical fertilization) while the local farmer variety had the lowest (10.7 in plots with chemical fertilization and 11.3 in plots without fertilization).

Table 5. Characteristics (plant aspect, reaction to Downy Mildew and Maize Streak Virus) days to flowering and plant height of 13 QPM varieties, 7 normal open pollinated maize varieties and 3 normal maize hybrids evaluated under no-chemical fertilization in community plots in Gandajika, DR-Congo.

Varieties	Plant aspect	Reaction to downy mildew	Reaction to maize streak virus	Days to 50% male flowering	Days to 50% female flowering	Plant height
	1-5	1-5	1-5	Days	Days	cm
ECA-QVE 3	1.8	1.3	1.0	53.7	56.3	136.0
ECA-QVE 4	2.2	1.0	1.3	53.0	57.3	128.0
ECA-QVE 6	2.2	1.3	1.7	53.3	58.3	123.3
ECA-POPE1	2.0	1.3	1.3	53.3	57.7	119.7
POOL15QC7-SRC1-F2	2.2	1.0	1.8	53.0	57.3	121.0
DMR-ESR-W	1.8	1.0	1.0	53.3	58.3	124.3
AK9331-DMR-ESR-Y	1.5	1.0	1.5	58.7	62.0	125.7
QPMSRSYNTH	1.8	1.0	2.2	60.0	64.3	133.7
SUSUMA	1.7	1.2	1.5	59.3	63.0	140.3
OBATANPA-SRC1-F3	2.0	2.0	1.7	59.7	63.7	159.0
POOL15QPM-S	2.0	1.2	2.0	53.7	58.3	134.3
S99TLWQHG-AB	4.0	1.0	4.0	61.7	66.0	134.0
SOOTLWQ-AB	4.0	1.0	3.7	63.0	66.0	119.0
SOOTLWQ-B	4.0	1.0	3.7	62.0	65.0	135.3
LONGE 5	1.7	1.0	1.2	59.3	62.3	131.3
OPAQUE-2	2.2	1.0	1.0	63.7	69.7	116.7
SALONGO-2	1.5	1.0	2.2	62.7	68.7	138.7
KASAI-1	2.7	1.0	2.8	64.3	71.3	128.3
MUS-1	1.7	1.0	1.2	62.7	65.3	132.0
GPS-5	2.0	1.0	1.7	62.0	69.7	167.3
LOCALE	1.7	1.0	2.2	62.7	67.3	153.3
PAN 67	1.5	1.2	1.2	57.3	61.7	155.0
PAN 77	1.5	1.0	1.5	61.7	66.0	156.0
MH 18	2.0	1.3	2.3	59.7	63.7	131.3
LSD (0.05)	0.6	0.3	0.6	1.9	2.6	16.5
CV (%)	18.0	16.3	20.6	2.0	2.5	7.4

All the QPM varieties had relatively high number of kernel rows per cob varying from 14 to 15.3 on average. These values were significantly higher than for the local variety but similar to data recorded for most normal varieties. Cob diameter varied from 3.7 to 4.6 cm. The local farmer variety, GPS-5, and DMR-ESR-W had significantly smaller cobs than most of QPM and improved normal varieties. Cob rachis size (diameter) varied from 1.4 to 2.9 cm (Table 4). The smallest size of cob rachis was observed in the local farmer's variety.

There were significant differences among the mean grain yield of entries. It varied from 2.2 to 4.8 kg ha^{-1}. Only one QPM variety (QPMSRSYNTH) and one improved normal variety from IITA, AK9331-DMR-ESR-Y yielded more than the currently released varieties (Mus 1, Kasaï 1, GPS 5, Salongo 2) (Table 4). The same trend was observed in trial without fertilizer application. The flour quality score of OPAQUE 2 for making local dish (fufu) ranges from 1 to 1.5 and was considered excellent and very good (Table 4). The taste and palatability of local dish (fufu) made with the selected varieties was considered good by all the participants.

The data of trials without chemical fertilizer conducted in the same sites as the first one were summarized in Tables 5 and 6. Generally, the farmer's rankings of the varieties were similar to those described for trials with fertilizer application. However grain yield was 21% higher when NPK and urea were applied when compared to trials without fertilizer. In both trials, the normal improved variety AK9331-DMR-ESR-Y had the highest grain yield (4.8 T/ha and 3.9 T/ha, for fertilized and non fertilized trials, respectively) followed by the QPMSRSYNTH with 4.2 T/ha and 3.4 T/ha, for fertilized and non fertilized trials, respectively. The grain yield advantages of AK9331-DMR-ESR-Y over the improved normal varieties Salongo 2, Mus 1, and Kasaï were 25, 39 and 62% respectively, in trials without fertilization. Moreover, under the same condition the grain yield for this variety (AK9331-DMR-ESR-Y) was 77% higher than the local farmer variety. Average grain yield for AK9331-

Table 6. Average agronomic performance of 13 QPM varieties, 7 normal open pollinated varieties and 3 normal hybrids evaluated without chemical fertilization (NPK and urea) in community plots in Gandajika, DR-Congo.

Varieties	Cob aspect 1-5	Kernel rows / Cob Number	Cob size/diameter (cm)	Cob rachis size (cm)	Cob length (cm)	Adjusted grain yield (t/ha)
ECA-QVE 3	1.7	14.7	4.3	2.6	13.3	2.3
ECA-QVE 4	2.0	13.3	4.0	2.5	12.5	2.7
ECA-QVE 6	2.0	14.7	4.1	2.5	12.4	2.9
ECA-POPE1	2.0	14.0	4.1	2.6	12.0	2.4
POOL15QC7-SRC1-F2	2.0	14.7	4.0	2.6	12.4	2.9
DMR-ESR-W	1.5	13.3	3.7	2.3	12.7	2.5
AK9331-DMR-ESR-Y	1.7	14.0	4.2	2.5	13.7	3.9
QPMSRSYNTH	1.8	13.3	4.0	2.5	13.7	3.4
SUSUMA	1.7	14.0	4.3	2.4	14.3	3.2
OBATANPA-SRC1-F3	1.7	14.7	4.0	2.5	12.8	3.0
POOL15QPM-S	1.8	14.0	4.1	2.5	12.5	2.3
S99TLWQHG-AB	1.8	14.7	4.1	2.5	14.0	2.9
SOOTLWQ-AB	1.8	14.7	4.2	2.5	14.0	-
SOOTLWQ-B	1.8	14.0	4.1	2.5	13.8	-
LONGE 5	1.8	14.0	4.3	2.5	13.4	-
OPAQUE-2	3.0	13.3	4.1	2.4	13.9	1.7
SALONGO-2	1.5	14.7	4.3	2.5	14.3	3.1
KASAI-1	1.7	15.3	4.2	2.4	13.7	2.4
MUS-1	1.7	14.0	4.1	2.5	13.8	2.8
GPS-5	1.7	12.0	3.7	2.5	13.6	2.1
LOCALE	2.7	11.3	3.7	1.4	12.4	2.2
PAN 67	1.7	12.0	4.2	2.2	13.2	3.1
PAN 77	1.5	14.0	4.7	2.6	14.9	3.4
MH 18	1.8	12.7	4.2	2.3	14.0	2.9
LSD (0.05)	0.5	1.6	0.4	0.3	1.1	0.4
CV (%)	15.1	7.2	5.4	6.6	5.0	5.1

DMR-ESR-Y variety was 37, 26 and 41% higher than Salongo 2, Mus 1, and Kasai 1, respectively under chemical fertilization regimen. The yield for AK9331-DMR-ESR-Y variety was 37% higher than for the local farmer variety in trials with chemical fertilization. The yield advantages of QPMSRSYNTH over currently released and local varieties vary from 10 to 62% in trials without fertilization. These values ranged from 10 to 24% under chemical fertilization. The agronomic performance of AK9331-DMR-ESR-Y and QPMSRSYNTH over the released normal improved maize varieties was confirmed in subsequent field observations during the 2009 to 2010 growing season.

DISCUSSION

The exotic QPM varieties tested under local conditions in Gandajika (DR-Congo) showed significant variation for grain yield, days to flowering, plant height, and plant and cob aspects. Two varieties, AK9331-DMR-ESR-Y and QPMSRSYNTH stood out of the collection for their agronomic performance in the two sites. The agronomic performance of other QPM and normal varieties were equal or slightly better than the genetically improved normal varieties currently released or the local farmer variety. The QPM variety Obatanpa developed in Ghana in 1992 and widely grown in many African countries (Sallah et al., 2003) had similar scores with currently released normal varieties for most yield components. The low incidence of downy mildew and maize streak virus symptoms in QPM plots was surprising because all the QPM varieties showed some level of susceptibility to MSV and downy mildew during pre-screenings at the Gandajika research station. This low incidence is likely a result of late planting of the trials during the 2008 to 2009 growing season that resulted in poorer conditions for pathogen inoculum growth. It has been reported that quality protein maize varieties are more vulnerable to diseases than normal maize varieties (National Research Council, 1988; Akande and Lamidi, 2006).

The two elite varieties (AK9331-DMR-ESR-Y and

QPMSRSYNTH) produced 23% more grain yield in fertilized plots than in unfertilized ones under the evaluation conditions. Considering that it costs 50% more per ha to use chemical fertilizers, it is recommended that the selected varieties could be grown without chemical fertilizer. The scores between male and female farmers in this study were very close, however, slight difference observed was an indication that male and female farmers have particular preferences for certain traits. The females, ranked flour qualities for making local dishes second to grain yield among the important criteria while males considered disease resistance second to grain yield during the selection. Farmers' knowledge about the crops in their areas has a high potential of strengthening PVS/PPB programs. It would serve well to empower scientists' knowledge for designing site-, crop- and farmer–specific activity.

One of the advantages of using PPB or PVS is that selected varieties can be released in a particular community without a formal release. Decentralization is often intertwined with participation in PPB or PVS programs in ways that makes it difficult to separate out the effects of these two distinct phenomena (Weltzien et al., 2003). The decision to decentralize can be based on the extent of G X E (Genetics X Environment) interactions and the target region. In the present study there was no significant G X E when most criteria were considered. Likewise no significant differences in ranking of accessions from location to location were observed. The two varieties, AK9331-DMR-ESR-Y and QPMSRSYNTH did score first and second place, respectively, at both locations. If decentralization is deemed to be beneficial, organizational issues come into play in determining how to best structure a decentralized program. Particular models of farmer participation may be especially appropriate for highly decentralized programs. But the degree and nature of participation can and should be considered separately from decentralization. The present project was decentralized since it targeted only two locations within an agricultural region and no multi-location evaluation of the corn varieties was conducted.

The two elite varieties are being considered for release locally in Gandajika since they have good culinary qualities and farmer preferences varied little among the two sites within the region. It is expected that they will be released across the corn growing regions in DR-Congo when multi-location evaluation is completed. There is abundance evidence supporting the fact that when the knowledge that farmers have is integrated in breeding scheme, it will increase the adoption rate of new varieties. This study clearly confirms previous reports that collaborative breeding and selection with farmers is extremely useful for decentralizing breeding programs (Gyawali et al., 2007).

ACKNOWLEDGEMENTS

This research was conducted through a partnership between Laurentian University (Ontario, Canada), University of Kinshasa (DR-Congo), and Caritas Congo. The authors are grateful to the Canadian International Development agency (CIDA) for financial support and the Association of Universities and Colleges of Canada (AUCC) for managing the partnership program. We would like to express our sincere thanks to Caritas –Congo, Caritas, Mbuji Mayi, and INERA Gandajika for logistical arrangement. Technical assistance of Sister Nicole Ntumba is appreciated. We thank Dr. M. Mehes – Smith and Dr. Clulow for reviewing the manuscript.

REFERENCES

Akande SR, Lamidi GO (2006). Performance of quality protein maize varieties and disease reaction in the derived-savana agro-ecology of South – West Nigeria. Afr. J. Biotech., (5): 1744-1748.

Bjarnason M, Vasal SK (1992). Breeding quality protein maize (QPM). In Janick 9 ed.). Plant breeding review, Vol. 9. John Wiley & son Inc. New York, USA.

Bressani R (1991). Quality Protein maize, In: Proceedings of the international Symposium on Quality Protein Maize (Eds Larkins, B.A. and Mertz, E.T.). EMBRAPA-CNPMS' sete Lagaos, Brazil, pp. 41-63.

Gyawali S, Sunwar S, Subedi M, Tripathi M, Joshi KD, Witcombe JR (2007). Collaborative breeding with farmers can be effective. Field Crop Res., (101): 88-95.

Hugo C (2000): Quality protein maize: Improved nutrition and livelihood for the poor. Maize Research Highlights 1999-2000. CIMMY, p. 27.

Mekbib F (2006). Farmer and formal breeding of sorghum (*Sorghum bicolor* (L.) Moench) and the implications for integrated plant breeding. Euphytica, 152 (2): 163-176.

National Research Council (1988). Quality Protein Maize. National Academy Press. Washington D.C. USA, pp. 41-54.

Olakojo SA, Omueti O, Ajomale K, Ogunbodede BA (2007). Development of Quality Protein maize: Biochemical and Agronomic evaluation. Trop. Subtrop. Agroecosyst., (7): 97-104.

Sallah PYK, Asiedu EA, Obeng-Antwi K, Twumasi-Afriyie S, Ahenkora K, Okai DB, Osel SA, Akuamoah-Boaten A, Haaq W, Dzah BD (2003). Overview of quality protein maize development and promotion in Ghana. Paper presented at the training Workshop on Quality Protein Maize Development and Seed Delivery System held at Crops research Institute, Kumasi, Ghana, August 4-15, p. 21.

Upadhyay SR, Gurung DB, Paudel DC, Koirala KB, Sah SN, Prasad RC, Pokhrel BB, Dhakal R (2009). Evaluation of Quality Protein Maize (QPM) Genotypes under rainfed mid hill environments of Nepal. Nepal J. Sci. Technol., 10: 9-14.

Vasal SK (2000). The quality protein maize story. Food Nutr. Bull., 21: 445-450.

Weltzien E, Smith ME, Meitzner LS, Sperling L (2003). Technical and institutional issues in participatory plant breeding- from the perspective of formal plant breeding. A global analysis of issues, results, and current experiences. CGIAR systemwide program on participatory research and gender analysis for technology development and institutional innovation. CIAT. PPB monograph No. 1. p. 208

Witcombe JR, Joshi A, Joshi KD, Sthait BR (1996). Farmer participatory crop Improvement I: Varietal selection and breeding methods and their impact on Biodiversity. Exper. Agric., 32: 445-460.

Effects of Gibberellic acid and 2,4-dichlorophenoxyacetic acid spray on fruit yield and quality of tomato (*Lycopersicon esculentum* Mill.)

Dandena Gelmesa[1]*, Bekele Abebie[2] and Lemma Desalegn[3]

[1]Haramaya University P.O. Box 138, Dire Dawa, Ethiopia.
[2]Adama University, P.O. Box., 1888, Adama Ethiopia.
[3]Ethiopian Institute of Agricultural Research-Melkassa Agricultural Research Center, P. O. Box 436 Nazreth Ethiopia.

Experiment was conducted at Melkassa Agricultural Research Center, centeral rift valley of Ethiopia from September 2008 to January 2009 with the objective to determine the effects of different concentrations and combinations of 2,4-dichlorophenoxyacetic acid (2,4-D) and gibberellic acid (GA_3) spray on fruit yield and quality of tomato. The experiment consisted of two tomato varieties-one processing (Roma VF) and one fresh market (Fetan), three levels of 2,4-dichlorophenoxyacetic acid (2,4-D) (0, 5 and 10 mg l^{-1}) and four levels of gibberellic acid (GA_3) (0, 10, 15 and 20 mg l^{-1}) arranged in 2 × 3 × 4 factorial combinations, in randomized completed block design with three replications. The result showed increase in fruit length from 5.44 to 6.72 cm at 10 mg l^{-1} 2,4-D combined with 10 mg l^{-1} GA_3 above the control, increased fruit weight by 13% due to 2,4-D and reduced fruit weight in single or combined application of GA_3 with 2,4-D. Fruit pericarp thickness was increased by about 50% due to 2,4-D and GA_3 application above the control. Titratable acidity, total soluble solids and lycopene content were also increased due to combined application of 2,4-D and GA_3 spray. Lower fruit pH is another quality attributes of tomato affected by 2,4-D application while that of GA_3 has no effect. Final fruit yield were significantly improved above the control even though both varieties responded differently. For Roma VF, GA_3 at concentration of 10 and 15 mg l^{-1} resulted in maximum fruit yield of 69.50 and 67.92 ton ha^{-1}, respectively in the absence of 2,4-D. For Fetan, maximum marketable fruit yield of 74.39 and 74.20 ton ha^{-1} was obtained from treatment combinations of 10 + 15 and 5 + 0 2,4-D and GA_3, respectively. Hence, yield increment of about 35% for Roma VF and 18% for Fetan were produced at 10 mg l^{-1} GA_3 and 10 + 15 mg l^{-1} 2,4-D and GA_3, respectively over the control. Significant increase in fruit size and weight due to 2,4-D and increased fruit number due to GA_3 spray contributed to increased fruit yield. The results indicated that both PGRs are important in tomato production to boost yield and improve fruit quality under unfavorable climatic conditions of high temperature. Therefore, it is important to further investigate application methods and concentrations of the PGRs under concern in different growing conditions on different tomato cultivars.

Key words: Gibberellic acid, 2,4-dichlorophenoxy acetic acid, tomato, *Lycopersicon esculentum* Mill, fruit yield, quality.

INTRODUCTION

In Ethiopia, tomato (*Lycopersicon esculentum* Mill.) is an important cash crop and widely cultivated both under irrigation and rain fed throughout the year (Lemma, 2002). Tomato has a significant role in human nutrition because of its rich source of lycopene, minerals and vitamins such as ascorbic acid and β-carotene which are anti-oxidants and promote good health (Wilcox et al., 2003). The general dietary deficiencies of vitamins in Ethiopian

---

*Corresponding author. dandenagalmesa@gmail.com.

population could be alleviated by a liberal consumption of many vegetables including tomato (Fekadu et al., 2004). Considering the significant value of the crop emphasis has been given by the national vegetable crops research program in the country to improve yield and quality of fresh market and processing tomato in order to satisfy the demands of both local and export markets (Lemma, 2002). A number of improved varieties and other agronomic packages have been recommended to the users to overcome the low productivity and quality of tomato in the country. However, the average national yield still remains very low which is around 7 ton/ha (CSA, 2009) and less than 50% of the current world average yield of about 27 ton/ha (FAOSTAT, 2007). Increasing temperature, viral diseases and salinity are the major limiting factors in sustaining and increasing tomato productivity (Fekadu and Dandena, 2006). Lack of adabtive cultivars and poor fruit setting of existing varieties especially during the hot/dry season where the demand for tomato is very high is one challenge farmers are facing in tomato production even though there is potential land for cultivation. Breeding for heat tolerance in tomato crop has been difficult due to many factors like moderate heritability inheritance being complex or the cultivars becoming lower in yield (George et al., 1984). For good fruit set and better yield, pollination, germination of pollen grains, pollen tubes growth, fertilization and fruit initiation must take place successfully (Kinet and Peet, 1997). The author further explained that, high relative humidity of the air, low light intensity and extreme low and high temperature, and improper mineral nutrition seems to be involved in the control of those phenomena and result in low fruit set and quality.

Induction of artificial parthenocarpy through application of PGRs enables fertilization-independent fruit development that can reduce yield fluctuation in crops like tomato, pepper and likes (Heuvelink and Korner, 2001). This could be possible by application of certain PGRs like auxin and GA_3 that bring the possibility of tomato production under adverse environmental conditions. Gemici et al. (2006) reported that application of synthetic auxin and gibberellins (GAs) are effective in increasing both yield and quality of tomato. Those PGRs are used extensively in tomato to enhance yield by improving fruit set, size and number (Batlang, 2008; Serrani et al., 2007a) and could have practical application for tomato growers. Tomato fruit setting was promoted by gibberellic acid (GA_3) at low concentration (Sasaki et al., 2005; Khan et al., 2006) and reduced pre-harvest fruit drop with increased number of fruits per plant and yield was observed due to Naphthalene Acetic Acid (NAA) or β-NAA spray (Alam and Khan, 2002). Additionally, the report of Anwar (2010) indicated that application of 2, 4-D at 5 mg l^{-1} significantly improved growth attributes and fruit yield of tomato plant but those attributes decreased beyond this concentration. Furthermore, Bensen and Zeevaart (1990) reported that GA_3 is more effective on tomato stem growth at concentration of 10 ppm (10 mg l^{-1}) or below.

However, information regarding the practical use of 2,4-D and GA_3 their combined application, rate and concentration in crop production in general and tomato in particular is lacking in Ethiopia under condition where tomato production is affected due to adverse environmental conditions. On the other hand, most report indicated that synthetic auxin like 2,4-D has herbicidal or ephinastic effect (Pandolfini et al., 2002) which lead to flower bud abscission, poor fruit set, fruit defects and puffiness beyond certain concentrations. Contrary to this, GA_3 seems to have opposite effect to 2,4-D and promote vegetative growth and reproductive organ formation (Gemici et al., 2000) with extended flowering, maturity period and less fruit size formation (Graham and Ballesteros, 2006). If 2,4-D combined with GA_3 it may have not express such effects and resulted in intermediate result for better fruit setting, yield and quality (Serrani et al., 2007a). Hence, coordinated action of the two PGRs may be important to overcome the side effects of their single application and enhance fruit yield and quality in addition to the possibility of tomato production under adverse conditions in Ethiopia. This study was therefore initiated to investigate the best dose of the PGRs under consideration in single or combined application that improve tomato fruit yield and quality.

MATERIALS AND METHODS

The experiment was conducted at Melkassa Agricultural Research Center, Ethiopia which is located at 8°24′ N latitude, 39°21′ E longitude and at an altitude of 1550 m above sea level, in the middle rift valley of Ethiopia. The center is characterized by low and erratic rainfall with mean annual rainfall of 796 mm with peaks in July and August. The dominant soil type of the center is Andosol of volcanic origin with pH that ranges from 7 to 8.2. The mean annual temperature is 21.2°C with a minimum of 14°C and maximum of 28.4°C (MARC, 2008).

Two improved tomato varieties Roma VF and Fetan were obtained from Melkassa Agricultural Research Center and used as a test material. Roma VF is processing type, compact and strong stem with determinate growth habit while Fetan is fresh market type with strong stem having determinate growth habit. Commercial tissue culture grade of GA_3 ($C_{19}H_{22}O_6$) powder with 95% purity and 2,4-D ($C_8H_6Cl_2O_3$- CSA No 94-75) powder (salt formulation) were obtained from Sigma Chemical Co Ltd, Germany used for the experiment.

The two tomato varieties with three levels of 2,4-D (0, 5 and 10 mg l^{-1}) and four levels of GA_3 (0, 10, 15 and 20 mg l^{-1}) are arranged in a 2 × 3 × 4 factorial combinations, in RCB design with three replications. Seedlings were raised in mid September in an open nursery bed and transplanted to the experimental field after 35 days at a spacing of 30 cm between plants on ridges having 100 cm width. A net plot size of 12 m^2 (4 × 3 m) having 40 plants/plot was used. A total of 20 plants per plot were considered for data collection from the two middle rows. The field was irrigated using furrow irrigation when rainfall was not sufficient for plant growth. Urea as a source of nitrogen fertilizer was applied at a rate of 46 kg/ha in split form, half at transplanting and half at first flowering as a side dress. Phosphorous fertilizer was applied at a rate of 40

Table 1. Interaction effects of 2,4-D and GA₃ on fruit set percentage of tomato plants grown at Melkassa, Ethiopia.

2,4-D (mg l^{-1})	GA$_3$ (mg l^{-1})				Mean
	0	10	15	20	
0	54.75f	71.10b	75.43a	70.50b	67.95
5	60.86e	66.31c	66.06cd	63.49de	64.18
10	60.79e	63.68de	63.27de	62.04e	62.44
Mean	58.80	67.03	68.25	65.34	
CV (%)			9.94		

Means followed by the same letter within the table are not significantly different from each other according to DMRT at 5% probability level.

Table 2. Interaction effects of variety, 2,4-D and GA₃ on marketable fruit number per plant of tomato plants grown at Melkassa, Ethiopia.

2,4-D (mg l^{-1})	GA$_3$ (mg l^{-1}) Roma VF				Mean	GA$_3$ (mg l^{-1}) Fetan				Mean
	0	10	15	20		0	10	15	20	
0	41ef	62a	55b	55b	53	28c	35g	33gh	33g	32
5	47cd	49c	48c	40f	46	21j	21j	20j	22j	21
10	39f	34g	44de	30hi	37	19j	14k	15k	20j	17
Mean	42	48	49	42		23	24	23	25	
CV (%)					10.97					

Means followed by the same letter within the table are not significantly different from each other according to DMRT at 5% probability level.

kg/ha all at transplanting using di-ammonium phosphate as a side dress. Weeding, cultivation and pest control were done following previous recommendations (Lemma, 2002).

The required weight of the PGRs was taken using electronic sensitive balance (model BOSCH SAE200) and a stock solution was prepared by dissolving in 1 ml of 97% ethanol. Latter the stock solution was diluted in distilled water (dH₂O) to prepare the working solutions, just before application. Tween-20 at the rate of 0.05% (v/v) was added before spray and mixed well to act as cohesive agent. The solution was poured into hand-held atomizer sprayer and was directly sprayed on the plants at early flowering (42 days after transplanting). Spraying was performed early in the morning to avoid rapid drying of the spray solution, due to transpiration.

Data were collected from randomly selected plants in the two middle rows except for fruit yield where the two middle rows were considered. The collected data includes fruit set percentage marketable fruit number per plant, fruit yield (marketable, unmarketable and total) (ton/ha), fruit length (cm), pericarp thickness (mm), total soluble solids (°Brix), pH, titratable acidity and lycopene content (mg/100 g). The data was analyzed using analysis of variance (ANOVA) by SAS (2002) software and mean separation was carried out by DMRT at 5% probability level.

RESULTS

Fruit set and number per plant

Percent fruit set was significantly affected by PGRs in both in single and combined application. Fruit set was increased from 54.75 to 75.43 (about 21%) when GA₃ concentration increased from 0 to 15 mg l^{-1} while the concentration of 2,4-D was kept at 0 mg l^{-1} and decreased in the presence of 2,4-D with or without GA₃ (Table 1). Hence, increased concentration of 2,4-D application resulted in low fruit set percentage compared to GA₃ and control treatment. On the other hand, GA₃ both in single and combined application with 2,4-D improved fruit set even though more increments was observed in increased concentration of GA₃ alone. Variation in percent fruit set observed subsequently lead to the variation in marketable fruit number (MFN) per plant (Table 2) due to the effects of PGRs under concern. The result indicated that combined and single application of 2,4-D and GA₃ significantly (P < 0.001) affected MFN per plant of both varieties. For instance, increased concentration of 2,4-D from 0 to 5 mg l^{-1} resulted in decreased MFN per plant from 42.79 to 33.42 by about 22% and further increase of 2,4-D concentration to 10 mg l^{-1} decreased MFN per plant to 26.92 by about 37% (data not shown). The data also indicated that MFN per plant significantly reduced by half from 62 to 30 for Roma VF and from 35.33 to 15 for Fetan depending on increased 2,4-D of concentration. However, MFN per plant was increased significantly when GA₃ concentration increased having an opposite effect with that of GA₃ in MFN per

Table 3. Interaction effects of variety, 2,4-D and GA$_3$ spray on marketable and total fruit yield of tomato plants grown at Melkassa, Ethiopia.

2,4-D (mg l^{-1})	GA$_3$ (mg l^{-1})	Marketable yield (ton/ha)		Total yield (ton/ha)	
		Roma VF	Fetan	Roma VF	Fetan
0	0	51.20d	62.90cd	59.68d	73.44c
	10	69.50a	66.78c	83.05a	83.08ab
	15	67.91ab	71.23b	79.96ab	83.34ab
	20	57.08cd	72.51ab	65.72cd	75.04c
5	0	57.14cd	74.20ab	64.24cd	85.23a
	10	62.54bc	72.10ab	72.61bc	84.14a
	15	66.40ab	73.67ab	75.44b	84.54a
	20	59.96c	66.47c	66.09cd	75.34c
10	0	55.59c	62.44d	74.31b	72.52c
	10	62.27c	73.93ab	67.87c	84.68a
	15	66.52ab	74.39a	76.98ab	84.52a
	20	56.18cd	66.73c	64.33cd	77.22bc
Mean		65.82		75.56	
CV (%)		7.54		8.33	

Means followed by the same letter within the column are not significantly different from each other according to DMRT at 5% probability level.

plant. Hence, the more the plants were exposed to high dose of 2,4-D spray resulted in low percent fruit set and succeeding MFN per plant due to high rate of flower bud abscission and subsequent fruit drop.

Fruit yield (total, marketable and unmarketable)

Significantly highest total and marketable fruit yield of 83.05 and 69.50 ton ha^{-1} were obtained at 0 mg l^{-1} of 2,4-D and 10 mg l^{-1} GA$_3$ and 0 and 15 mg l^{-1} 2,4-D and GA$_3$ combination, respectively in Roma VF (Table 3). Variety Fetan gave highest marketable fruit yield of 74.39 and 74.20 ton ha^{-1} at treatment combinations of 10 + 15 and 5 + 0 mg l^{-1} 2,4-D and GA$_3$, respectively. Lowest marketable and total fruit yield was obtained from control treatments for variety Roma VF and from 10 mg l^{-1} 2,4-D and control for variety Fetan. Hence, both PGRs at higher concentration in single and/or combined application resulted in lower fruit yield compared to the single or combined application at lower concentration.

Fruit length and percarp thickness

Significant variation in fruit length between the two varieties ($P < 0.001$), 2,4-D ($P < 0.001$), GA$_3$ ($P < 0.05$) and their interactions ($P < 0.05$) were observed. In variety Roma VF, interaction between 2,4-D and GA$_3$ significantly increased fruit length in almost all combinations tested except at higher level of GA$_3$ compared to the control (Table 4). Maximum fruit length (6.72 cm) was recorded when the concentrations of both PGRs were at 10 mg l^{-1} in variety Roma VF and significantly started to decline for all concentration of 2,4-D when concentration of GA$_3$ was beyond 15 mg l^{-1}. Lowest fruit length of 5.34 cm and 5.44 cm were recorded from Fetan and Roma VF, respectively from the control treatmets. Hence, the interaction effects of 2,4-D and GA$_3$ indicated the possibility of increasing tomato fruit length with a combined application of the two PGRs than when each of them were applied independently. In addition to fruit length, fruit pericarp thickness (PT) was also significantly affected by 2,4-D ($P < 0.01$), GA$_3$ ($P < 0.05$) and interaction of 2,4-D with variety ($P < 0.05$) and 2,4-D with GA$_3$ ($P < 0.01$). The interaction effect of varieties and 2,4-D showed an increase in fruit PT for Roma VF from 4.06 to 5.03 mm and Fetan from 4.79 to 5.29 mm when the concentrations of 2,4-D increased to 10 mg l^{-1} (Table 5). Similarly, 2,4-D and GA$_3$ in combined applications showed increasing trend in fruit PT than the control (Table 6) even though it is significant only at 5 mg l^{-1} 2,4-D with 50.64% increment above the control.

Total soluble solids and titratable acidity

Fruit total soluble solids (TSS) (°Brix) was significantly affected due to 2,4-D alone and when combined with GA$_3$ while other treatment effects were not significant. The

Table 4. Interaction effects of variety, 2,4-D and GA$_3$ on fruit length (cm) of tomato plants of tomato plants grown at Melkassa, Ethiopia.

2,4-D (mg l^{-1})	GA$_3$ (mg l^{-1})							
	Roma VF				Fetan			
	0	10	15	20	0	10	15	20
0	5.44fg	6.02a-g	6.20a-e	5.88c-g	5.34g	5.49e-g	5.69d-g	5.47e-g
5	6.22a-e	6.49a-c	6.23a-e	6.29a-d	5.62d-g	5.97b-g	5.69d-g	5.67d-g
10	6.66a-c	6.72a	6.36a-d	5.96b-g	5.49e-g	5.70d-g	5.63d-g	5.86c-g
Mean	6.11	6.41	6.26	6.04	5.48	5.72	5.67	5.67
CV (%)	3.74							

Means followed by the same letter in the table are not significantly different from each other according to DMRT at 5% probability level.

Table 5. Interaction effects of variety and 2,4-D on fruit pericarp thickness (mm) of tomato plants grown at Melkassa, Ethiopia.

Variety	2,4-D (mg l^{-1})			Mean
	0	5	10	
Roma VF	4.06b	5.11ab	5.03ab	4.90
Fetan	4.79ab	4.87ab	5.29a	4.98
Mean	4.72	4.95	5.16	
CV (%)		8.93		

Means followed by the same letter in the table are not significantly different from each other according to DMRT at 5% probability level.

data (Table 6) indicated that, increased in TSS content up to 10% above the control at 10 + 0 and 10 + 20 mg l^{-1} 2,4-D and GA$_3$ treatment combinations were observed. Variation on fruit TA content due to varietal difference and combined application of the PGRs with increasing trends either due to increased concentration in single or combined application of the two PGRs under concern were another atirbute of the PGRs observed in this experiment. However, highest TA values of the two varieties were attained at different concentration and combination of the PGRs under concern. Hence, highest TA value of 0.76 was observed for variety Roma VF at 10 + 20 mg l^{-1} 2,4-D and GA$_3$ and for variety Fetan at 10 + 0 mg l^{-1} 2,4-D and GA$_3$, respectively (Figure 1).

Lycopene content

Lycopene content of the varieties were significantly affected by GA$_3$ in both single and combined application with 2,4-D (Tables 6 and 7). Variety and GA$_3$ interaction indicated that lowest lycopene content of 12.75 mg/100 g was obtained from the control (0 mg l^{-1} GA$_3$) and the highest (13.72 mg/100 g) from 10 mg l^{-1} GA$_3$ for Roma VF. In the case of Fetan, lowest lycopene contents (12.76 mg/100 g) was obtained at 10 mg l^{-1} and the highest (15.93 mg/100 g) at 20 mg l^{-1} GA$_3$ (Table 7). Combined application of GA$_3$ and 2,4-D also indicated that highest lycopene content of 16.03 and 15.33 mg/100 g were obtained at 20 + 0 and 5 + 10 mg l^{-1} GA$_3$ and 2,4-D respectively while the lowest (12.33 mg/100 g) lycopene content was from the control treatment.

DISCUSSION

Fruit set and number per plant

The present result indicated that, 2,4-D beyond certain concentration leads to flower bud abscission and fruit drop due to its herbicidal effect. Our result also support the findings of Gimici et al. (2006) who suggested that high concentrations of 2,4-D at 10 mg l^{-1} produced fewer fruits than with 4-CPA. The more the plants were exposed to high dose of 2,4-D spray resulted in lower number of fruits per plant due to increased rate of flower bud abscission. However, the report by Khan et al. (2006) indicated the significant role of GA$_3$ in tomato plant to increase fruit set that leads to larger number of fruits per plant and increased fruit size and final yield. An increased MFN per plant due to GA$_3$ application observed in this study also hold the finding of Khan et al. (2006) who stated that GA$_3$ at 10^{-8}, 10^{-6} and 10^{-4} molar proved to induce higher number of fruit per plant than the untreated

Table 6. Interaction effects of 2,4-D and GA$_3$ on fruit pericarp thickness (PT), total soluble solids (TSS) and lycopene contents of tomato plants grown at Melkassa, Ethiopia.

2,4-D (mg l^{-1})	GA$_3$ (mg l^{-1})	PT (mm)	TSS (°Brix)	Lycopene (mg/100 g)
0	0	3.93b	4.59b	12.33c
	10	4.93b	4.79b	12.46c
	15	5.13b	4.88b	13.01bc
	20	4.97b	4.82b	16.03a
5	0	5.92a	5.03a	14.25b
	10	4.81b	4.92ab	15.33a
	15	4.96b	4.83b	13.69bc
	20	5.10b	4.90ab	12.83bc
10	0	5.34ab	5.27a	13.30bc
	10	5.10b	5.03a	12.76bc
	15	5.37ab	4.95ab	14.28b
	20	4.83b	5.05a	14.60b
Mean		5.03	4.94	13.74
CV (%)		8.93	4.34	15.31

Means followed by the same letter in the table are not significantly different from each other according to DMRT at 5% probability level.

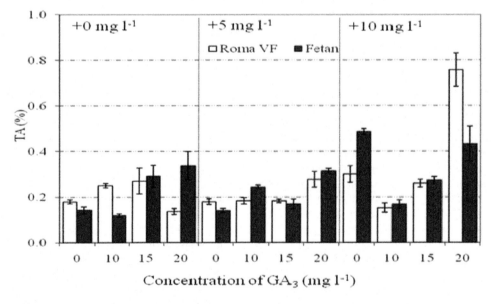

Figure 1. Interaction effects of variety, 2,4-D and GA$_3$ on fruit titratable acidity (TA) (%) of tomato plants grown at Melkassa, Ethiopia.

control. On the other hand, application of GAs can cause fruit set and growth of some fruits, in case where auxin may have no effect (Taiz and Zeiger, 2002). The significant effect of GA$_3$ in tomato plant further explained via its role in synthesis of protein including various enzymes, increased rate of shoot elongation and photosynthetic capacity leading to total leaf area and leaf dry weight (Ballantyine, 1995; Mostafa and Saleh, 2006).

Fruit yield (total, marketable and unmarketable)

Subsequent improvement in fruit yield as a result of higher percent of fruit set and MFN per plant due to 2,4-D

Table 7. Interaction effects of variety and GA$_3$ on fruit lycopene content (mg/100 g) of tomato plants grown at Melkassa, Ethiopia.

Variety	GA$_3$ (mg l^{-1})				Mean
	0	10	15	20	
Roma VF	12.76e	13.72c	12.99e	13.04de	13.13
Fetan	14.55ab	12.59e	14.32b	15.93a	14.35
Mean	13.66	13.16	13.66	14.49	
CV (%)			15.31		

Means followed by the same letter in the table are not significantly different from each other according to DMRT at 5% probability level.

and GA$_3$ was observed. Previous research result indicated increased MFN per plant, reduced fruit drop, increased fruit weight due to GA$_3$ spray (Naeem et al., 2001) which could result in increased fruit yield. Increased fruit size and weight due to 2,4-D and fruit number per plant due to GA$_3$ contributed to the overall increased fruit yield. However, 2,4-D at 10 mg l^{-1} resulted in reduced fruit set percentage and deformed and inferior fruits which contributed to low yield. On the other hand, increased concentration of GA$_3$ resulted in less unmarketable fruit size which reduces the final yield. The findings of Gimici et al. (2006) also indicated that high concentrations of 2,4-D at 10 mg l^{-1} decreased yield in tomato plant due to its herbicidal effect beyond certain concentration which cause flower bud abscission and fruit drop. However, combined application of GA$_3$ with 2,4-D can increase fruit yield to some extent when the concentration of 2,4-D has negative effect. Reduced concentration of 2,4-D at or below 5 mg l^{-1} seems to increase yield and other fruit quality attributes especially in variety Fetan by decreasing fruit drop compared to higher concentration beyond 5 mg l^{-1}. The current finding is also in consistent with the findings of Anwar (2010) and Pudir and Yadav (2001) that indicated improvement in tomato fruit yield at low concentration (≤ 5 mg l$^{-1)}$ of 2,4-D. The effect of 2,4-D and GA$_3$ in increasing fruit size was also another factor which could increase fruit yield.

Fruit length and percarp thickness

The interaction effect of 2,4-D and GA$_3$ indicated that fruit length was maximum for both levels of 5 and 10 mg l^{-1} 2,4-D with 10 mg l^{-1} of GA$_3$ but significantly reduced when the concentration of GA$_3$ increased. Gillaspy et al. (1993) indicated that after cell division and during cell expansion, which is associated with maximum fruit growth. The increase in cell volume due to expansion may contribute to the final size of the fruits as observed in this study. The effects of GA$_3$ resulted in smaller fruit size compared to Cycocel treated plants reported by Graham and Ballesteros (2006) also justify that fruit development (length) induced by 2,4-D may be regulated by GA$_3$ beyond certain concentrations. According to Rasul et al. (2008), 2,4-D at 25, 50 and 100 mg l^{-1} produced longer fruits as compared to Fulmet and CPPU in Teasle Gourd which indicating that 2,4-D is the most responsive auxin. According to Graham and Ballesteros (2006), cycocel treated plants born larger fruits while GA$_3$ treated ones bear smaller fruits even less than the control one. The effect of 2,4-D in our study indicated longer fruits with bigger size is also in agreement with the work of Khan et al. (2006). The authors reported an increase in fruit size due to 2,4-D application which could be due to stimulation of parthenocarpic fruit growth that resulted in increased fruit weight. Another possible reason that can be ascribed for the reduction in fruit size of tomato fruits at increased GA$_3$ concentration could stimulate shoot growth and suppressing the growth of developing fruit lets due to competition for assimilates results in decreased fruit width, size and number (Bakrim et al., 2007). Similar to our observation, Gimici et al. (2006) also reported that 2,4-D resulted in increased tomato fruit size, fresh and dry weight when used at recommended concentration. Similarly, Serrani et al. (2007a) reported that, tomato fruits induced by GA$_3$ and 2,4-D had thicker pericarp than pollinated fruits throughout its development, and more in response to 2,4-D than GA$_3$. The authors observed that the combined application of GA$_3$ and 2,4-D had an intermediate effect on pericarp thickness and number of layers compared to separate application. Therefore, this increase in fruit pericarp thickness may have an advantage to help the fruit skin become more resistant to mechanical bruising, pest attack and better shelf life due to reduced rate of water loss.

Total soluble solids (TSS) and titratable acidity (TA)

The significant increased in fruit TSS and TA content of tomato fruits observed due to 2,4-D and GA$_3$ at indicated treatment combinations have be maintained the previous reports. Increased In most reports it was generalized that parthenocarpic tomato fruits were shown to have higher levels of soluble solids and sugar, but lower level of acid compared with seeded fruits (Ho and Hewitt, 1986).

According to the authors, the rate of assimilate export from the leaves; rate of import by fruits, and the fruit carbon metabolism are factors that finally influence TSS of tomato fruit. The role of GA_3 in increasing tomato fruit TSS was reported by many authors. For instance, Graham and Ballesteros (2006) reported that GA_3 increased proteins, soluble carbohydrates, ascorbic acid, starch and β-carotene in the tomato. Higher sugar content in tomato fruits was obtained from plants treated with 50 mg l^{-1} GA_3 (Kataoka et al., 2009). In general, TSS has been of major interest to the processing industries that manufacture concentrated tomato products (Ram, 2005) and for fresh market consumption (Ho, 1998). It is believed that increased TSS content of fruits could give more finished product per ton of raw tomato fruit and thus, require less energy to produce a certain quantity of concentrated product. Hence, the use of 2,4-D and GA_3 spray for tomato production is one option to improve TSS content of tomato fruit. Tomato pH is dependent on several factors, including cultivar, maturity stage, cultural practices as well as growing location and seasonal variations (Gould, 1992) but achievement of low fruit pH (data not shown) and high TA value by spray of 2,4-D in our study could be a useful investigation. Comparable to the present result, significant increase in TA content due to application of PGRs was reported due to increased formation of organic acids in the tissues most likely to accelerate activities within the Krebs cycle (Graham and Ballesteros, 2006). Thakur et al. (1996) indicated that acidity of tomato fruits was reduced when the whole plant was sprayed with GA_3 and 2,4-D; however, the ascorbic acid content increased with higher concentrations of 2,4-D and Para-chlorophenoxy acetic acid. On the other hand, reduced pH value of tomato fruit is a desirable quality and essential factor accounting for flavor. Thus processors typically add citric acid to tomato juice to ensure low pH values. Thus, inverse relationship between decreased pH and increased TA content is the desirable fruit quality parameter (Erdal et al., 2007; Fontes et al., 2000) to reduce the risk of microbial spoilage and requires moderate conditions for processing and enzyme inactivation.

Lycopene content

The role of PGRs in increasing tomato flavor, color intensity and lycopene contents was realized through its enhanced up take or utilization of certain plant nutrieits like nitrogen, phosphorous and potassium (Oded and Uzi, 2003). For instance, Khan et al. (2006) observed an increase in leaf phosphorous, nitrogen, and potassium content in addition to increased lycopene content of tomato fruit treated with GA_3. Low potassium inhibits lycopen synthesis and delays the development of a full red color in tomato ripening (Oded and Uzi, 2003). This seems that GA_3 application on tomato enhanced potassium uptake which was responsible for lycopen synthesis. Similarly, Afaf et al. (2007) indicated that GA_3 application increased phosphorous accumulation in leaves and stems of tomato plants that was also responsible for required lycopene content in the fruit.

Conclusions

This study clearly indicates that the effective concentration of these PGRs to improve tomato fruit yield and quality depends on the chemical nature of the PGRs used and the tomato cultivar. In the case of 2,4-D improvement of fruit yield and quality seems to be at or lower concentration of 5 mg l^{-1} especially in the case of variety Fetan but GA_3 at both 10 and 15 mg l^{-1} indicated better result for both varieties in yield and quality. In general, it is important to continue the experiment in different growing seasons and conditions on different types of tomatoes to confirm the present result. Method of application should be considered as spray flower only seems to be more important than applying on the whole plant in case of 2,4-D to avoid the herbicidal effect on the leaf part.

ACKNOWLEGMENTS

The authors would like to thank Melkassa Agricultural Research Center and Detscher Akedemischer Austauschdienst (DAAD) for material and financial support, respectively during this research work.

REFERENCES

Afaf TMK, Abd El-hameid AM, El-Greadly NHM (2007). A comparison study on the effect of some treatments on earliness, yield and quality of globe artichoke (*Cynara scolymus* L.). Res. J. Agric. Biol. Sci., 3: 695-700.

Alam SM, Khan, MA (2002). Fruit yield of tomato as affected by NAA spray. Asian J. Plant Sci., 1(1): 24.

Anwar W, Aziz T, Naveed F, Sahi ST (2010). Short Communication. Foliar applied 2, 4-dichlorophenoxy acetic acid improved tomato growth and yield. Soil Environ., 29(1): 77- 81.

Ballantyine DJ (1995). Cultivar, photoperiod and gibberellin influence shoot elongation and photosynthetic capacity of Hardy Azaleas. Hortic. Sci., 2: 257-259.

Bakrim A, Lamhamdi M, Sayah F, Chibi F (2007). Effects of plant hormones and 20-hydroxyecdysone on tomato (*Lycopersicon esculentum*) seed germination and seedlings growth. Afr. J. Biotechnol., 6: 2792-2802.

Batlang U (2008). Benzyladenine plus gibberellins (GA_{4+7}) increase fruit size and yield in greenhouse-grown hot pepper (*Capsicum annuum* L.). J. Biol. Sci., 8(3): 659-662.

Bensen RJ, Zeevaart JAD (1990). Comparison of *ent-kaurene synthetase* A and B activities in cell-free extracts from young tomato fruits of wild-type and *gih-1, gih-2,* and *gih-3* tomato plants. J. Plant Growth Regul., 9: 237-242.

Central Statistical Agency CSA (2009). Agricultural sample survey 2008/2007. Report on area and production of crops (Private peasant holdings, main season). Stat. Bull., Addis Ababa Ethiopia, 01-446.

Erdal I, Erteke A, Senyigit U, Koyuncu MA (2007). Combined effects of

irrigation and nitrogen on some quality parameters of processing tomato. World J. Agric. Sci., 3: 57-62.
FAOSTAT (2007). FAOSTAT on-line. Rome: United Nations Food and Agriculture Organization. http://faostat.fao.org/default.aspx?lang=e.
Fekadu M, Ravishanker H, Lemma D (2004). Relationship between yield and plant traits of tomato genotypes under dry land condition of central Ethiopia. J. Veg. Crop Res., 60: 45-54.
Fekadu M, Dandena G (2006). Review of the status of vegetable crops production and marketing in Ethiopia. Uganda J. Agric. Sci., 12(2): 26-30.
Fontes PCR, Sampalo RA, Finger FL (2000). Fruit size, mineral composition and quality of trickle-irrigated tomatoes as affected by potassium rates. Pesq. Agropec. Bras. Brasilia, 35: 21-25.
Gemici M, Türkyilmaz B, Tan K. (2006). Effect of 2,4-D and 4-CPA on yield and quality of the tomato, *Lycopersicon esculentum* Mill. JFS. 29: 24-32.
Gemici M, Güve A, Yürekli AK (2000). Effect of some growth regulators and commercial preparations on the chlorophyll content and mineral nutrition of *Lycopersicon esculentum* Mill. Turk. J. Bot., 24: 215-219.
George WL, Scott JW, Splittstoesser WE (1984). Parthenocarpy in tomato. Hortic. Rev., 6: 65-84
Gould WA (1992). Tomato Production, Processing and Technology, 3rd edition. CTI Publishers, Baltimore, MD.
Graham HD, Ballesteros M (2006). Effect of plant growth regulators on plant nutrients. J. Food Sci. Article first published online: 25 AUG 2006. DOI: 10.1111/j.1365-2621.1980.tb04086.x.
Heuvelink E, Korner O (2001). Parthenocarpic Fruit Growth Reduces Yield Fluctuation and Blossom-end Rot in. Sweet Pepper. Ann. Bot., 88: 69-74.
Ho LC (1998). Improving Tomato Fruit Quality by Cultivation. In KE Cockshull, D, Gray, GB Seymour and B. Thomas (eds.). Genetic and Environmental Manipulation of Horticultural Crops. CABI Publishing, pp. 17-29.
Ho LL, Hewitt JD (1986). Fruit Development. In: JG Atherton and J. Rudich (eds). The Tomato Crop. A Scientific Basis for Improvement. Chapman and Hall, pp. 201-233.
Kataoka K, Yashiro Y, Habu T, Sunamoto K, Kitajima A (2009). The addition of gibberellic acid to auxin solutions increases sugar accumulation and sink strength in developing auxin-induced parthenocarpic tomato fruits. Scientia Horticulturae, 123(2): 228-233.
Khan MMA, Gautam AC, Mohammad F, Siddiqui MH, Naeem M, Khan MN (2006). Effect of gibberellic acid spray on performance of tomato. Turk. J. Biol., 30: 11-16.
Kinet JM, Peet MM (1997). Tomato. In H.C Wien (*ed.*). The Physiology of Vegetable Crops. CAB International, pp. 207-248.
Lemma D (2002). Tomatoes. Research Experience and Production Prospects. Res. Report, Ethiopian Agricultural Research Organization Addis Ababa, Ethiopia, 43: 48.

Melkassa Agricultural Research Center MARC (2008). Ethiopian Institute of Agricultural Research, Center Profile, Melkassa, Ethiopia. p. 12.
Mostafa EAM, Saleh MMS (2006). Influence of spraying with gibberellic acid on behavior of Anna Apple trees. J. Appl. Sci. Res., 2: 477-483.
Naeem N, Ishtiaq M, Khan P, Mohammad N, Khan J, Jamiher B (2001). Effect of gibberellic acid on growth and yield of tomato Cv. Roma. J. Biol. Sci., 1(6): 448-450.
Oded A, Uzi K (2003). Enhanced performance of processing tomatoes by potassium nitrate-based nutrition. Acta Hortic. (ISHS), 613: 81-87. www.actahort.org/books/613/613_8.htm.
Pandolfini T, Rotino GL, Camerini S, Defez R, Spena A (2002). Optimization of transgene action at the post-transcriptional level: High quality parthenocarpic fruits in industrial tomatoes. BMC Biol., 2: 1-11.
Pudir JPS, Yadav PK (2001). Note on effect of GA_3 NAA and 2, 4-D on growth, yield and quality of tomato var. Punjab Chuhara. Curr. Agric., 25: 137-138.
Ram HH (2005). Vegetable Breeding Principles and Practices, 2nd edition. Kalyani Publishers, Ludhiana-New Delhi. p. 653.
Rasul MG, Mian MAK, Cho Y, Ozaki Y, Okubo H (2008). Application of plant growth regulators on the parthenocarpic fruit development in Teasle Gourd (Kakrol, *Momordica dioica* Roxb.). J. Fac. Agric., 53: 39-42.
Statistical Analysis System SAS (2002). SAS institute version 9.00 Cary, NC, USA.
Sasaki H, Yano T, Yamasaki A (2005). Reduction of high temperature inhibition in tomato fruit set by plant growth regulators. JARQ., 39: 135-138.
Serrani JC, Fos M, Atare´s A, Garcı´a-Martı´nez JL (2007a). Effect of gibberellin and auxin on parthenocarpic fruit growth induction in the cv Micro-Tom of tomato. J. Plant Growth Regul., 26: 211-221.
Taiz L, Zeiger E (2002). Plant Physiology, 3rd edition. Sinaure Associates, Inc. USA, p. 690.
Thakur BR, Singh RK, Nelson P (1996). Quality attributes of processed tomato products: A review. Food Res. Int., 12: 375-401.
Wilcox J, Catignani G, Lazarus C (2003). Tomatoes and cardiovascular health. Crit. Rev. Food Sci. Nutr., 43(1):1–18.

Studies on effectiveness and efficiency of gamma rays, EMS and their combination in soybean [*Glycine max* (L.) Merrill.]

Mudasir Hafiz Khan[1] and Sunil Dutt Tyagi[2]

[1]Department of Agriculture, Srinagar (J&K), India.
[2]Department of Plant Breeding and Genetics, Kisan (P. G.), College, Simbhaoli, Ghaziabad, (U. P.), India.

Mutagenic effectiveness and efficiency of gamma rays, EMS and combined treatments was studied in terms of M_2 (progenies) lethality and chlorophyll mutations in two cultivars of soybean (Pusa-16 and PK-1042). In general the frequencies of chlorophyll mutations were high in gamma rays and combined treatments. Four types of mutants viz., albina, xantha, chlorine and viridis were observed in the study. Gamma rays were found to be more effective to induce chlorophyll mutations in both cultivars. PK-1042 cultivar exhibited higher mutagenic efficiency as compared to Pusa-16 in EMS and gamma rays treatment.

Key words: Effectiveness, efficiency, gamma rays, EMS, soybean.

INTRODUCTION

Mutagenic effectiveness is a measure of the frequency of mutations induced by unit dose of a mutagen, while mutagenic efficiency gives an idea of the proportion of mutations in relation to other associated undesirable biological effects such as gross chromosomal aberrations, lethality and sterility induced by the mutagen (Wagner and Foster, 1965). The usefulness of any mutagen in plant breeding depends not only on its mutagenic effectiveness but also on its mutagenic efficiency. Chlorophyll mutations are used to evaluate the genetic effects of various mutagens. Gaul (1964) reported the appearance of greater number of viridis type which is attributed to the involvement of polygene in chlorophyll formation. The present study was undertaken to gather information on efficiency, effectiveness and chlorophyll mutations on the consequences of induction of physical (Gamma rays) and Chemical (EMS) mutagens in Soybean.

MATERIALS AND METHODS

Dry seeds (9 - 12% moisture) of two cultivars Pusa-16 and PK-1042 of soybean were treated with ethyl methane sulphonate (EMS) (0.1, 0.2 and 0.3% concentration) and Gamma rays (15, 30 and 45 Kr) at NRL, IARI, New Delhi. 100 irradiated seeds (Gamma ray) were subjected to 0.2% EMS for eight hours. The treated material along side with two controls (untreated) were immediately sown in a single unreplicated plots with 4 rows at a distance of 30 and 10 cm between rows and plants, respectively at Research Farm of Kisan (P.G.) College, Simbhaoli, UP, India. The seeds treated with chemical mutagen were washed in running water before sowing. The data was recorded from 20 randomly selected plants from each treatment.

Survival of plants was recorded at the time of maturity in the field and was expressed as percentage of control. Selfed seeds of the individual M_1 plants were harvested separately and were grown as individual M_2 families in separate lines in Modified Single Seed Bulk Design. The treated and the control material were screened for the frequency of chlorophyll mutations in the M_2 generation. Mutagenic frequency was estimated as percentage of segregating M_1 plant progenies (Gaul, 1964). The mutagenic efficiency and effectiveness were determined by the methods suggested by Konzak et al. (1965).

RESULTS

Chlorophyll mutation

The chlorophyll mutation frequencies were calculated as per cent M_1 plants (M_2 progenies) and the data is presented in Table 1. It may be noticed from the data that in cultivar PK-1042, the mutation frequency increased with increase in the dose of gamma rays, while in case of

*Corresponding author. E-mail: kmudasirhafiz@yahoo.com.

Table 1. Frequency of chlorophyll mutations in EMS and Gamma rays treated M_2 generations of soybean cultivars.

Cultivars	Dose	Number of M_2 seedlings		Mutation frequency/100 M_2 seedlings	Spectrum (relative %) of Chlorophyll mutants				
		Tested	Chlorophyll mutants		Albina	Xantha	Chlorina	Viridis
	Control								
PK-1042	0	600	-	-	-	-	-	-
Pusa-16	0	580	-	-	-	-	-	-
	Gamma rays (kR)								
PK-1042	15	500	9	1.80	0.00	0.00	55.56	44.44
	30	420	13	3.09	7.69	23.07	30.76	38.46
	45	400	15	3.75	13.33	33.34	20.00	33.33
Pusa-16	15	535	15	2.80	6.66	20.00	33.34	40.00
	30	495	17	3.43	0.00	29.41	41.17	29.41
	45	480	13	2.70	23.07	0.00	15.38	61.53
	EMS (%)								
PK-1042	0.1	600	9	1.50	0.00	33.33	22.22	44.45
	0.2	574	15	2.61	6.66	33.33	26.67	33.33
	0.3	525	12	2.28	16.67	16.67	41.67	25.00
Pusa-16	0.1	520	5	0.96	0.00	25.00	50.00	25.00
	0.2	515	9	1.74	11.11	22.22	33.34	33.33
	0.3	485	16	3.29	0.00	31.25	25.00	43.75
	Combination [EMS + Gamma rays (%)]								
PK-1042	15 kR + 0.2	485	11	2.26	0.00	36.36	27.27	36.36
	30 kR + 0.2	320	15	4.68	20.00	0.00	33.34	46.46
	45 kR + 0.2	315	13	4.12	7.69	38.46	23.07	46.15
Pusa-16	15 kR + 0.2	435	12	2.75	16.66	41.66	25.00	16.67
	30 kR + 0.2	410	15	3.65	20.00	6.67	33.33	40.00
	45 kR + 0.2	385	19	4.93	10.52	26.31	36.84	26.31

EMS and combined treatments, the mutation frequencies exhibited addition up to the intermediate dose levels and then showed a decline with further increase in the dose of mutagen. In Pusa-16, the situation was somewhat different. An increase in the mutation frequencies at intermediate level and a gradual decrease at higher doses, whereas EMS and combined treatment showed slight increase in the mutation frequency with the increase in the dose/conc. of mutagen.

Spectrum and frequency of chlorophyll

The data of frequencies and spectra of different chlorophyll mutants in the two soybean cultivars that is PK-1042 and Pusa-16 are presented in Table 1. The details for individual mutation types are as follows: -

a) Albina: In PK-1042, out of nine mutagenic treatments the *albina* mutants were observed in six treatments. Among these treatments, the highest frequency (20.00%) of mutants was observed at 30 + 0.2% combination treatment. In other treatments, the values ranged from 0.00 to 16.67%. In Pusa-16, the *albina* mutants were observed in six treatments. The highest frequency (23.07%) was observed at 45 Kr gamma rays and 30 + 0.2% (20.00) combination treatment. Other treatments showed a range of 0.00 to 16.66%.

b) Xantha: The frequency of *xantha* chlorophyll mutants showed a slight increase in PK-1042 as compared to Pusa-16 in EMS and combined treatments. The frequencies of *xantha* mutants showed a decrease at intermediate dose levels and then increase with the increase in dose of mutagen, whereas in Pusa-16 it showed a decreasing trend. With gamma rays treatments in PK-1042, an increase in the frequency of xantha mutants was noticed with the increase in the doses of mutagen, while in Pusa-16 an increase upto intermediate level and thereafter a gradual decrease (0.00%) at higher dose (45 Kr) was noticed.

c) Chlorina: Dissimilar to the *xantha* mutants, the frequency of the *chlorina* mutants were high in Pusa-16 then PK-1042. In Pusa-16, the frequen-cies of *chlorina* mutants showed an increase with the increase in dose of mutagens. In PK-1042, on the other hand, the increase in the dose of gamma rays was associated with the decrease in the frequency of *chlorina* mutants, whereas EMS and combined treatments showed increase at the intermediate levels. In Pusa-16 as compared to other treatments, the frequencies of *chlorina* mutants were high following combined treatments. The increase in the dose was associated with the increase in the frequencies of mutants. 45 Kr gamma rays and combination treatment (45Kr + 0.2%) showed the highest frequency of *chlorina* mutants in Pusa-16 among all the treatments in both cultivars.

d) Viridis: The frequencies of the *viridis* mutants were high as compared to the other types of chlorophyll mutants. In both the cultivars, the frequencies of the *viridis* showed an increase at the intermediate level of combination treatment. Contrary to this, an increase in the dose of gammarays was associated with the decrease in the frequency of mutants, whereas it showed a reverse trend in case of EMS treatment.

The frequencies of different chlorophyll mutant types in both the cultivars were found in the following order: - *Viridis* > *Chlorina* > *Xantha* > *Albina*.

Mutagenic effectiveness and efficiency

The effectiveness of different mutagens and that of different treatments of mutagens were calculated and the data are presented in Table 2. The mutagenic effectiveness was as high as 1.6310 in gamma rays treatments. Whereas the highest effectiveness of EMS treatments was 1.3291 in the cultivar Pusa-16. EMS was 19 to 60 times more effective than gamma rays, while it was less effective that is 0.40 to 5 times in case of PK-1042. In both cultivars mutagenic effectiveness decreased with the increase in the dose of the mutagens except in EMS treatment of Pusa-16 where the increase (1.3291) was observed at higher concentration (0.3%). Thus, the order of mutagens based on effectiveness was: - EMS > Gamma rays.

The mutagenic efficiency was calculated on the basis of lethality and the data are presented in Table 2. The highest (0.0297) and lowest (0.092) mutagenic efficiency was exhibited by Pusa-16 (15 Kr + 0.2%) and Pk-1042 and Pusa-16 (0.3%), respectively. In general, PK-1042 exhibited higher mutagenic efficiency as compared to Pusa-16 in EMS and gamma rays treatment. However, Pusa-16 showed higher frequency in combined treatment. The mutagenic efficiency decreased with the increase in the dose of the mutagen in both cultivars the differences in number of genes controlling chlorophyll development in the two cultivars. It may also be noted that the frequencies of chlorophyll mutants were higher in gamma rays and combined treatments. This is in conformity with the previous reports of Prasad and Das (1980). In 50% of the treatments (Gamma rays, PK-1042; EMS, Pusa-16 and Combined treatment, Pusa-16) an increasing trend in the frequency of mutants with increase in the dose of mutagen was observed but the increase was not linear. The non-linearity in the frequency of mutants and dose of mutagens may to some extent, be attributed to the irregular segregation ratios of chlorophyll mutants leading to deficits of recessives in the progenies.

Albina, xantha, chlorina and viridis were the different types of chlorophyll mutants found in the present study. Xantha and Chlorina types of chlorophyll mutants in soybean were earlier reported by Geetha and Vaidyanathan (2000). In general, viridis occurred in maximum proportion than other types of chlorophyll mutants in gamma rays and combined treatments.

Since chlorophyll mutations are most conspicuous and easily detectable as they have been extensively used to find out sensitivity of crop plants to mutagens and to elucidate effectiveness and efficiency of mutagens (Gaul, 1960). It may be noted from the Table 2 that relative to the EMS, gamma rays were more effective in inducing chlorophyll mutation in both the cultivars. Similar results have also been reported in Lathyrus by Nerkar (1977). In the present study, the efficiency of EMS, Gamma rays and combined treatments decreased considerably except Pusa-16 (combined treatment) with the increase in the dose of mutagens in both cultivars. The decrease in efficiency may be attributed to the failure in proportionate increase in dose of mutagens (Table 2).

Similar findings have also been reported by Dixit and Dubey (1986). It is obvious that the higher efficiency

Table 2. Effectiveness and efficiency of various EMS and Gamma rays treatments in M_2 generation in two cultivars of soybean.

Dose	Mutation frequency/100 M_2 seedlings		Chlorophyll mutant frequency as (per ment) M_2 seedlings		Mutagenic effectiveness		Mutagenic efficiency	
	PK-1042	Pusa-16	PK-1042	Pusa-16	PK-1042	Pusa-16	PK-1042	Pusa-16
Control	0.00	0.00	0.00	0.00	0.00	0.00	0.00	0.00
Gamma rays (kR)								
15	1.80	2.80	1.80	2.80	0.1200	0.1866	0.181	0.203
30	3.09	3.43	3.09	3.43	0.1030	0.1143	0.253	0.137
45	3.75	2.70	3.75	2.70	0.0833	0.0600	0.098	0.108
EMS (%)								
0.1	1.50	0.96	1.50	0.96	1.8750	1.2000	0.166	0.168
0.2	2.61	1.74	2.61	1.74	1.6310	1.0875	0.169	0.133
0.3	2.28	3.29	2.28	3.29	0.9500	1.3291	0.092	0.092
Combination								
15 kR + 0.2(%)	2.26	2.75	2.26	2.75	---	---	0.208	0.297
30 kR + 0.2(%)	4.68	3.65	4.48	3.65	---	---	0.172	0.243
45 kR + 0.2(%)	4.12	4.93	4.12	4.93	---	---	0.106	0.232

at lower and intermediate doses of mutagens may be due to the fact that the biological damage (lethality and sterility)increased with the dose at a r ate greater than the frequency of mutations (Konzak et al.,1965).

REFERENCES

Dixit P, Dubey DK (1977). Mutagenic efficiency of gamma rays, NMU and their combinations in Lentil (Lens culnaris Medik.) var. T. 36. Indian J. Genet. Plant Breed. 46(3): 501-505.

Gaul H (1960). Critical analysis of the methods for determining the mutation frequency after seed treatment with mutagens. Genet. Agraria. 12: 297-318.

Gaul H (1964). Mutation in Plant Breeding. Radiation Bot. 4: 155-232.

Geetha K, Vaidyanathan P (2000). Studies on induction of chlorophyll mutation in soybean through physical and chemical mutagens. Agricultural Science Digest 20(1): 33-35.

Konzak CF, Nilan RA, Wagner J, Foster RJ (1965). Efficient chemical mutagenesis. In the use of Induced Mutations in Plant Breeding. Radiation Bot. (Suppl.) 5: 49-70.

Nerkar YS (1977). Mutagenic effectiveness and efficiency of gamma rays, EMS and NMU in Lathyrus sativus. Indian J. Genet. Plant Breed. 37(1): 137-141.

Prasad AB, Das AK (1980). Studies of induced chlorophyll mutations in Lathyrus sativus L. Cytologia 45: 335-341.

Indigenous pest and disease management practices in traditional farming systems in north east India. A review

Gopal Kumar Niroula Chhetry[1] and Lassaad Belbahri[2]*

[1]Manipur University, Department of Life Sciences, Imphal 795003, India.
[2]Agronomy Department, University of Applied Sciences Western Switzerland technology, architecture and landscape, 150 Route de Presinge, 1254 Jussy, Switzerland.

Traditional farming system is an ecologically based age-old farming system developed by ancient farmers through generations of their interaction with nature and natural resources for food, fodder and fiber. Indigenous knowledge is the knowledge of the indigenous people inhabiting different geographical regions of the world with their own language, culture, tradition, belief, folklore, rites and rituals. This report is an attempt to document some of the indigenous practices followed by traditional farmers for the management of pests of certain common crops grown in traditional farms in general and north east India in particular.

Key words: Agricultural farming, indigenous traditional knowledge, pest and disease management, traditional ecological knowledge.

INTRODUCTION

Traditional farming system is an ecologically based age-old farming system developed by ancient farmers through generations of their interaction with nature and natural resources for food, fodder and fiber. It is an indigenous method of cultivating crops using self reliance locally available resources without external inputs. Indigenous farming system, once prevalent all over the world is now almost vanished from the developed countries and confined to some tribal dominated regions of developing countries occupying more than half of arable land (Thurston, 1992). Wherever they are, the common feature of traditional farming is the presence of spatial and temporal heterogeneity often with complex plant age structure, mixed cropping, multiple, host-pathogen interactions and use of simple tools to plow and harvest the crop. India is rich in traditional farming systems because of diversity in agro-ecological habitats inhabited by diverse ethnic groups who have been practicing age-old farming in location specific situations since generations. It is a community based farming system that brings the local people closer and respects their environment.

Tools and techniques used by them are either unknown or least understood. However, with the advent of modern agriculture, traditional farming in India is largely confined to hilly regions that too in peripheral states where green revolution could not influence the traditional farmers. Crops and pest scenario of crops in traditional farming system is by and large same as that of the conventional agriculture but degree of severity caused by various pests may differ. Management of crops in general and pest management approaches in particular are different among traditional farmers practicing traditional farming systems in different regions of the country because of difference in indigenous knowledge they accrued over generations of their close contact with the nature for evolving sustainable and locally adapted agriculture system.

Traditional farming system in north east of India is complex and unique of its own as the land use system in this regions is dominated by slash and burn agriculture where management of crops and pests are carried out by integrating indigenous knowledge and traditional ecological knowledge of the communities. This twin knowledge have been recognized by the world scientific communities and scientists are showing keen interest in traditional agriculture because indigenous farmers and their system may be of great help to remedy the deficiencies of pest management in modern agriculture (Sofia et al. 2006). Attempt is being made here to document some of the indigenous practices followed by tra-

*Corresponding author. E-Mail: lassaad.belbahri@hesge.ch.

ditional farmers for the management of pests of certain common crops grown in traditional farms in general and north east India in particular.

About indigenous/traditional knowledge

Indigenous knowledge is the knowledge of the indigenous people inhabiting different geographical regions of the world with their own language, culture, tradition, belief, folklore, rites and rituals. In course of their close interactions with nature and natural resources, they are to make a certain decisions as to the solutions of their problems they encounter in their day to day life while managing the land and environmental resources for survival. Compelling situations motivate them to generate knowledge out of necessities. Therefore, indigenous knowledge so developed is based on necessities, extinct, curiosity and observations of ethnic groups to mitigate the immediate situations. Hence, indigenous knowledge is used in the decision making process as to how, when and where to act depending on the situations.

This knowledge has been tested using the thumb rule of trial and error methods over a period of time through generations and validated to make the established knowledge for the purpose for which it is designed for use of the ethnic group who propounded for the first time. Thus, indigenous knowledge varies from tribe to tribe and usually passes on to the nest generations though the words of mouth, actions or even practices, usually by the elders of the family/society. In fact, these ethnic bases cumulative knowledge took generations of time to penetrate into the social fabrics of inter and intra- ethnic groups because of communications gap and orthodox nature of the society. Eventually, this local knowledge in course of time, gets socially accepted and validated which finally inters into the social life and subsequently become the Indigenous traditional knowledge (ITK) of the society as a whole. Thus, socially validated indigenous knowledge which shares common values and gains popularity with human element attached to it particularly the culture of the ethnic groups may be known by the name Indigenous traditional knowledge. However, now days it is known to people synonymously by various modern names such as indigenous technical knowledge, technical know how of indigenous people, peoples knowledge etc. (Singh and Sureja, 2008).

Indigenous traditional knowledge is distinct from international knowledge which is derived through hypothetico-inductive process such as knowledge generated in universities, research organizations, private and public research institutions through research activities. As such, scientists in recent years are keen to learn indigenous knowledge in its various dimensions as to how indigenous people view, perceive and interact with their environment and mobilize their cumulated knowledge for designing appropriate actions. Although many people, so to say self styled modern educationists have discarded this knowledge branding them as outdated primitive knowledge, yet its importance and hidden principles, particularly in the field of agricultural and environmental issues have been recognized by international bodies such as the united nations conference on environment and development, 1992; international union for conservation of nature and natural resources, 1980 and world convention on environment and development, 1987 (Kanoujia, 2004). As such, scientist in this knowledge base economy who are in research of new ideas and innovations expect that indigenous knowledge may hold significant message which may be of use to remedy the deficiencies in modern agricultural and environment related issues (Berkes et al., 2000).

Despite its potentiality in addressing the environment issues including agriculture, indigenous traditional knowledge could not develop to its fullest extent primarily for fast technological development and secondly due to crude and outdated nature of the knowledge, that too in the process of degeneration with the passing away of elders and finally due to the exposure of the society to their modern innovations. As the elderly persons in any ethnic group hold the key and custodians of the traditional knowledge, they are the masters to make correct decisions to overcome the adverse situations of their immediate environment. Being a local need based knowledge, indigenous traditional knowledge system is based on local resources without any external inputs. Eco-friendly manipulation of their immediate environment, judicious application of plant and animal products either in raw or simple processed forms are important components of indigenous knowledge system. Land being the basis of survival of mankind, indigenous knowledge system of indigenous people revolved mostly about the ecological management of their land and environment in harmony with nature. Here lies the importance and significance of indigenous traditional knowledge and therefore it should be protected, preserved and documented in the interest of future generations before it is lost.

Indigenous disease and pest management practices in traditional farming system

In recent years there is a resurgence of interest in reviving the age old farming system through scientific approach which is known by modern man as organic farming, because of hazardous effect of excessive chemicals in agricultural system, environment and human health. Further, the negative impact of green revolution such as loss of genetic diversity of indigenous crops because most farmers confined production in mono cropping of selected crops (maize, rice, wheat) ignoring indigenous crops, stagnation in crop production even with the increase dose of chemical fertilizers, development of pesticide resistant pests, emergence of new pests which were either non existent or present as minor pest degradation of arable land to the extent of unfit for cultivation without the application of synthetic chemical fertilizer and overall ill effects of chemical based products to human health etc are some of the challenges which invoke to look for alternative farming system that minimizes the use of chemicals (fertilizers or pesticides). Indiscriminate use of pesticides that results killing of natural enemies of pests, beneficial soil micro organisms and its residual effect in food chain necessitates the development of integrated pest management concept to minimize and judicious application of pesticides as last resort if other eco friendly alternative methods fail to contain the harmful agricultural pests. However, the success story of IPM is limited to few crops only and as such, search of viable alternatives for the management of pest is an ongoing process.

India is a large country with diverse agro-climatic habitats, arable land, crops and cropping pattern suitable for growing location specific crops being cultivated by hundreds of ethnic farming communities with their own indigenous technologies. Therefore, crop cultivation practices vary not only from one agro-climatic zone to another but also from one ethnic group to another. These ethnic farming communities are the store house of indigenous knowledge of technical knows how regarding the overall management of indigenous crops. The uniqueness of this knowledge is that it is environmentally benign, ecologically protective, socially acceptable, economically viable and sustainable. In appreciation of technical know how knowledge of ethnic groups, attempt is being made to collect, document and understand the rationale behind the traditional practices in general and pest management in particular which are disappearing at fast rate under the influence of high tech modern agriculture. While doing so, it not only preserves the age old agricultural heritage of the country and identity of Indian farmers but also promotes the scientific development of traditional practices with honour and dignity of indigenous farmers for sustainable agriculture because it holds the potential eco friendly message for pest management.

North eastern region of India bestowed with natural beauties and biodiversity resources is rich in eco friendly indigenous knowledge maintained and nourished by hundreds of ethnic tribes (more than 120 ST) and communities. Despite diversity in ethnicity, they practice a common but unique farming system locally called jhum around which their culture and tradition revolves .In this farming system farmers grow as many as 35 crops (Ramakrishnan, 2004) in mixed, multiple or polyculture form of which rice is the staple food crop that dominates the system in Manipur, Nagaland, Tripura, Arunachal Pradesh but potato in Meghalaya. Whatsoever be the cropping pattern, the cultivation practices are eco friendly and tuned to the need of the people. Irrespective of ethnic groups practicing jhum, interesting features of the system is that it has inbuilt pest and disease management mechanisms as reflected in their cultural practices such as mixed /multiple cropping , zero tillage , clean cultivation, slash and burning , green manuring, sequential cropping and harvesting, fallowing, flooding etc.

Use of plants and animal parts and products are the important components of indigenous knowledge in the management of pest and diseases of crops in jhum system. Indigenous farmers of the region also possessed rich traditional ecological knowledge such as growing location specific nitrogen fixing trees such as *Alnus nepalensis*, *Flemingia vestita* sparsely for enrichment of soil fertility keeping tree boles /trunk and erecting /pegging wooden structures amidst jhum/ terrace fields for facilitating perching of birds which prey on harmful crop pest, recycling of jhum based waste products for the management of crops etc.

Traditional cultivation practices, although it was jhum only in the past but due to population explosion, jhum cycle has been reduced due to population explosion, Jhum cycle has been reduced to 2 - 3 years, some of which have been converted to terrace wet farm while others to sedentary form of agriculture. Thus, jhum system of cultivation have undergone changes to terrace form to settled agriculture forms which one finds in mixed forms at different locations in N.E. Some of the indigenous traditional knowledge (ITK) being used by these traditional farmers in all the 3 system of crop cultivation have been explained with rationale which may be of use for integration in the existing IPM concept.

Traditional belief - a myth or reality

Indigenous people have rich store house of traditional beliefs, folklore, rituals and rites which may not hold any truth and have any practical value but expected to hold some message and therefore need in depth observation in the light of empirical sciences to discover some of these beliefs as sound agricultural practices. These beliefs include that seeds collected and thrashed on new moon day (Amawasia) for sowing in the next season are usually not infested by pest and pathogens, plant diseases are caused by halo around the sun, sowing seeds should be sprinkle first with gold water etc. North eastern region of India being the home of rich culture and nature friendly traditional knowledge, these beliefs have been shaped and nurtured by hundreds of ethnic groups scattered throughout the regions.

Conclusion

Formal pest and disease management knowledge and ecological knowledge derived through hypothetico-deductive method whereas indigenous pest management knowledge and traditional ecological knowledge derived through long experiences and perceptions accumulated by traditional farmers during the course of their interacttions with the nature and natural resources need to be effectively integrated and should not be viewed in isolation.

Although indigenous pest and disease management knowledge fitted well in the age-old land use system, yet need thorough validation in view of changing agricultural scenario from traditional to integrated farming system through inorganic and organic farming methods. Some of this validated Indigenous traditional knowledge (ITK) may be incorporated as an integral component to dovetail neatly into the IPM concept for evolving better pests and disease management strategies in any of these farming systems.

This systematic approach not only protects this fast disappearing ITK under the influence of modern agriculture but also preserve the indigenous pests and disease management identity of farming communities of this country. Therefore, this rich heritage of the county should be harnesses, preserved, documented and developed as modern science such as indigenous integrated pest management before they are lost.

REFERENCES

Berkes F, Colding J, Folke C (2000). Rediscovery of traditional ecological knowledge as adaptive management. Ecological Applications (10 (5): 1251-1262.

Chhetry, GKN (2008). Personal observations.

Ramakrishnan PS (2004). Sustainable development of jhum with conservation of biodiversity in North-East India: Where do we stand? Man and Society 1(1): 59-85.

Singh RK, Sureja AK (2008). Indigenous knowledge and sustainable agricultural resources management under rainfed agro-ecosystem. Ind. J. Trad. Knowledge 7 (4): 642-654.

Sofia PK, Prasad R, Vijay VK (2006). Organic farming- Tradition reinvented. Ind. J. Trad. Knowledge 5: 139-142.

Thurston HD (1992). Sustainable practices for plant disease management in traditional farming systems. West views Press, Boulder.

APPENDIX

S/No.	Name of indigenous traditional knowledge (ITK)	Details of ITK and rationale	Reference
1	Management of seed health free from pest and diseases in jhum, terrace, sedentary and home garden traditional farming system.	Good seed health is managed for sowing in the next season through: (i) Collection of healthy seeds before general harvest (ii) hanging over fire furnace/ kitchen for constant smoking (iii) mixing with ashes of fire wood (iv) smoking well dried healthy seeds with edible and non- edible oils (v) Mixing with neem seed powder etc and storing the same in a seed bin. (vi) using aromatic plants such as citronella grass, lemon grass, peels of pomelo etc against maize weevil by mixing /placing these plants over maize grains granary. Rationale: Unsuitable environment is created to inhibit the growth and proliferation of pest and other microorganisms. Aromatic plants act as repellent or fumigant e .g; Leguminous seeds ,vegetable seeds, maize etc. (tribals and non-tribals of N.E)	Chhetry, 2008
2	Methods of keeping seeds free from pest and diseases for use in traditional farming system	Seeds of maize and leguminous crops are often kept intact along with their outer husk and hang over the kitchen/ furnace. Here, maize cobs are tied up in bunches of 10 - 12 cobs by folding their next to outermost husk and hang over the wooden beams of kitchen and sometimes roof beam in the periphery of the house. Rationale: Open air mixed with smoke seemed to inhibit the pest and pathogen as well as the entry of this pest take time through hard husk of maize and beans. (Tribal and non- tribal, N.E.)	Chhetry, 2008
3	Storage method of paddy in traditional granary for keeping away pest and pathogens.	Specially prepared bamboo granaries plastered with mixture of fresh cowdung and mustard oil cakes are in use for storing paddy on the top of which branches with leaves of *Zanthoxylum acanthopodium* are placed to keep away pest mostly white butterflies -a common pest-of the stored paddy grains. Paddy granaries are either placed near the kitchen or vicinity to kitchen in a separate house. Rationale: Well plastering of bamboo crevices inhibits the entry of pest and pathogen .It may also be possible that oil cakes emits unfavourable odour to pest and diseases. Further, *Zanthoxylum acanthopodium*-a plant of carminative properties emits unpleasant smell that inhibits the white flies.	Chhetry, 2008

4	Foliar application of plant and animal products for the management of pest and diseases.	Following traditional methods are used for the management of pest and diseases of crops: (i) foliar application of wood ashes in the wee hours of the day keeps away aphids pod borers and diseases from plants (mostly vegetables) (ii) Dusting finely ground tobacco leaves keep the aphid pest and diseases away from plants. (iii) Hookah water is very much effective for controlling pest and diseases of major and minor crops such as blast of rice, pod borers, sucking bugs of vegetables etc. (iv) Dusting with saw dust is also sometimes used but their effect is not encouraging. (v) Fish and meat wash water application is also a mild deterrent in keeping away of pest due to unpleasant environment for the proliferation of pest and pathogens. Rationale: Thin film of ash coat with dew inhibits the attack of pest and pathogens. Ash also acts as a nutrient when it gets washed due to rain. Tobacco leaves and hookah water which contains nicotine prevents the foliages from pest and pathogens. (Well practiced traditionally in the N.E.)	Chhetry, 2008	
5	Management of fungal diseases and insect pest of upland paddy.	Pest and diseases of paddy are controlled/managed using the following traditional methods: (i) By spreading leaves of *Artemisia vulgaris, Croton caudatus, Munromia wallichi, Adhatoda vessica* etc. (ii) By erecting or pegging branches of *Cymbopogon Khasianum, Saccharum spantaneum* which inhibits stem borer of paddy. Rationale: Leaves of these medicinal plants on decomposition release substances /molecules which inhibit the pest and pathogen of paddy in jhum land (Manipur, Meghalaya, Sikkim)	Chhetry, 2008	

6	Management of diseases and pest of rice through plant products	Pest and diseases are also managed by (i) Pomace (wine residue) Here, well fermented wine pomace usually made up of millets are placed at the source of irrigation canal of terrace rice fields which slowly spread over the rice field and inhibits the growth of pests such as leaf folder and blast of rice . (ii) Oak tree bark are also grounded and placed over the source of irrigation canal which inhibits the insect pests of rice such as brown plant hoppers. Rationale: Unpleasant odour of pomace may be the reason behind inhibiting the fungal disease and leaf folder in particular.	Chhetry, 2008
7	Management of blast and chara problem in terrace as well as in settled wet land paddy.	Traditional farmers use paddy husk before five months to contain the blast of rice at 0.3 to 0.5 ton/h for effective control of blast disease of rice. Paddy husk also makes clay/ loamy soil porous for better aeration of plants /tillers. Chara, a green alga infested field water is drained off first and paddy husk were applied to get rid of chara problem in the field .This method is also effective for controlling blast of paddy. Rationale: After draining of water, chara get settled on the ground which when paddy husk is applied suppress the chara and get decomposed which becomes nutrients of the plant on irrigation of field again. Chara do not have the chance to come up and suck the nutrient meant for paddy again (Mao, Maram, Ukhrul, Manipur).	Chhetry, 2008
8	Management of thatch grass - a nuisance weed common in jhum rice field.	Thatch grass (*Imperita__cylindrica*) is a common weed in jhum fields often not only inhibits the healthy growth of paddy but also acts as alternate host of diseases and perches of brown plant hopper of paddy .This menace is usually overcome by cultivating cassava (*Manihot esculenta*), *Sesamum indicum*, *Glycine_ max* and *Cajanus cajan-* plants densely. This technique is very effective. Rationale: The ramification of the root system of these plants make the soil loose and it may also possible that root exudates of the plants inhibit the growth of thatch grass. (Manipur, Meghalaya, Arunachal Pradesh and Assam)	Chhetry, 2008

9	Control of nematodes in ginger chillies, tomato and turmeric by intercropping Chrysanthemum coronarium.	Nematodes of turmeric, tomato, chillies and ginger are controlled by either intercropping with *Chrysanthemum coronarium*, *Tagetes erecta*, or growing *Tagetes erecta* as border crops. This is a very effective method and often farmers incorporate leaves of these trap crops into the soil to enhance effectiveness and nutrients enrichment of crops. Rationale: The sharp smell of trap crops may be the reason in the inhibition of nematodes.	Chhetry, 2008
10	Cultural practices in the management of crops on jhum system	Traditional farmers manage the pest and diseases of paddy (major crop)and vegetables through cultural practices as the system has inbuilt mechanisms for the control of pest and diseases that include: (i) Burning of slashed debries which kills the resident pests and pathogens from the system because many plants and grasses serves as the alternate host of crops .This clean cultivation practices enable farmers to harvest crops less infected by pest and pathogen . (ii) Zero tillage practices (often seeds are sown by dibbling methods) enables the natural growth of nodulated frankia found in socially valued alder trees and undisturbed mycorrhizal root of slashed plants that promotes the healthy growth of crop plants. (iii) Mulching through the removal of unwanted weeds soon after the establishment of paddy .The decomposed mulch may inhibit the pathogen propagules and also provide nutrients to crop plants. It also protects the soil from erosion in jhum slopes. Thus, mulching has multipurpose use in jhum system. (iv)Mixed cultivation of rice with sparsely grown maize, legume crops, job's tear (*Coix lacryma jobi. L)*, shorghum and ground vegetables, protects the diseases and pest of rice probably due to the physical barriers of intercrops in the movement of air borne propagules, augmenting microclimate and humidity etc Maize and sorghum not only provide food but also acts as perch for birds to feed on insects and pest of paddy in jhum field. Further, maize plants also serve to locate the burrows of rodents that destroy paddy crops.	Chhetry, 2008

11	Biological management of pest and diseases in jhum cultivated crops	Biological management of pest on jhum cultivated crops include : (i) Keeping tree boles / trunk and partially cut alder trees amidst jhum field which facilitate perching of diverse birds for selective prey of pest such as brown grasshopper of paddy, pod borers of pigeon pea etc. (ii) Growing of intercrops such as maize and sorghum long duration pigeon pea etc. with rice. These crops not only yield food but the main purpose of growing them sparsely in the jhum field is to facilitate perching of birds which pick up diverse pest of rice and other ground vegetable crops. (iii) Growing job's tear which is a medicinal plant that not only serve as alternate food to rodents but the root exudates of this medicinal plant inhibit the root pathogens of the soil. Root exudates of other slashed plants in jhum too probably make significant contribution in the inhibition of root pathogens of jhum crops.	Chhetry, 2008
12	Traditional BUN cultivation for the management of pest pathogen in jhum.	Elaborate preparation of land into ridges for soil conservation and furrows for channeling water which one often finds in high and mid elevation *jhum* land where fallen *pinus* needles, twigs and slashed bushes are burnt by covering with a thin layer of soil drag from furrows. In this technique biological material and soil get burnt slowly by suffocating the whole ridges with smoke. Thus, the whole system becomes sterile making the soil free from soil based pest and pathogen and therefore crops such as potato and ginger raised by employing this technique are often healthy.	Chhetry, 2008
13	Management of diseases and pest of rice in terrace cultivation	Shortening of jhum cycle led to the terracing of land for paddy cultivation. Disease and pest of terrace cultivated paddy is overcome through clean cultivation in the following way: Terraced paddy field bund is well plastered using clay soil of the field to avoid the growth of grass which serves as the alternate host of disease and habitats for number of paddy pest. Sloppy walls of terrace fields too are equally cleaned leaving no room for the growth of grasses. Further, wooden perches are erected for biological management of pest through carnivorous birds. (Mao, Maram, Manipur)	Chhetry, 2008

14	Biological management of pest in sedentary wetland system by establishing perches and conserving sacred grooves	Carnivorous birds such as Egrets and others which prey on leaf folder larvae. Stem borer, grass hoppers etc. of rice in the sedentary wet land rice are facilitated by erecting suitable perches in sufficient numbers. Sacred grooves conserved by community also facilitate perching of birds which effectively manage the unwanted pest from rice field. Further, the excreta of these birds acts as fumigant to the growth of pathogens and hence less pest pathogens are found in wet land rice system wherever there are sufficient number of perches for the birds vis-à-vis excreta.	Chhetry, 2008
15	Cultural practices for insect, pest control in wet valley land paddy and sedentary dry land farming system.	Farmers often plough their field repeatedly soon after harvest for exposing soil inhabiting insect pests, arthopods, nematode etc. to harsh weather and to facilitate natural predators . Insects such as grasshopper, crickets and borers lay their eggs in the upper layer of soil in paddy fields eventually exposed during the course of repeated operation and their eggs either desiccate or preyed by the Egrets and their natural predators. This indigenous repeated ploughing technique for getting rid of soil borne insects which damage paddy crops is very effective in wet land system and dry sedentary farming system. (Manipur)	Chhetry, 2008
16	Control of rodents by smokes through burning of paddy husk and dry chillies.	Rodents cause heavy loss to paddy in jhum fields. To control this menace burrows of rats are stuffed with smoke by burning paddy husk and land race dry chillies variety. Complete control over rodents depends on the number of burrows plugged by smokes Rationale: The suffocating pungent smokes promptly affect the respiration systems of rats and killed. (North East)	Chhetry, 2008
17	Traditional eco-technologies for the management of disease in traditional land use system	Traditional technologies such as bamboo drip method of irrigation of terrace rice which makes rice crops free from contaminated water borne diseases as seen in Arunachal Pradesh, bench terracing in higher elevation for soil conservation *vis-à-vis* nutrient loss to avoid diseases of crops due to nutrient deficiencies, well adapted techniques of growing potato in the higher elevation compared to paddy at lower elevation to match the soil fertility gradient, emphasis of farmers to grow tuberous crops in shorter jhum cycles as compared to cereals under longer jhum cycles are some of the need based techniques adopted by traditional farmers to avoid nutrient deficiency diseases in crops under traditional land use system in north east.	Chhetry, 2008

18	Traditional ecological knowledge for the management of pest and diseases of crops grown in rotational jhum and sedentary agricultural system.	In a traditional jhum fallow ecosystem, one often finds socially valued early successional nepalese alder (*Alnus nepales*) left *in situ* during slash and burn operation. Sometimes even plants this trees for multipurpose such as: (i) Fixing atmospheric nitrogen because its nodulating roots by frankia -a mycorrhiza which fixes nitrogen efficiently to the tune of 125kg/yr/h (Ramakrishnan, 2004). As such diseases of jhum fallow crops due to nitrogen deficiencies is either not recorded or not found. (ii) Provide healthy shades to shade loving crops like turmeric, ginger and others. The allelopathic effect of this plant species inhibits growth of pathogens over crops and decomposition of leaves inhibits the soil pathogens. (iii) Ramification of roots enriches the soil through nitrofixation and shoots with profuse branches provide shelter to hundreds of birds which prey on various crop pests and pathogens. Further excreta of these wild birds act as pesticides and fungicides of crops. Similarly, there are other nitrogen fixing legumes such as *Flemingia veslita*, pigeon pea intercropped with potato and maize are often found in sedentary agriculture system for similar purpose.	Chhetry, 2008

Comparative effects of water deficit on *Medicago laciniata* and *Medicago truncatula* lines sampled from sympatric populations

Mounawer Badri[1]*, Soumaya Arraouadi[1], Thierry Huguet[2] and Mohamed Elarbi Aouani[3]

[1]Laboratory of Legumes, Centre of Biotechnology of Borj Cedria, B.P. 901, 2050 Hammam-Lif, Tunisia.
[2]Laboratoire de Symbiose et Pathologie des Plantes, INP-ENSAT, B.P. 107, 31326 Castanet Tolosan Cedex, France.
[3]NEPAD/North Africa Biosciences Network, National Research Centre, El Buhouth St, Cairo, 12311, Egypt.

We evaluated the responses to water deficit of twelve lines of *Medicago laciniata* and *Medicago truncatula* including eight lines from four Tunisian sympatric populations of both species, and four references lines of *M. truncatula*. They were exposed to two water treatments, well irrigated and drought-stress (33% of field capacity) for a period of 45 days. At harvest, we measured five quantitative traits including the length of stems (LS), aerial dry weight (ADW), root dry weight (RDW), RDW/ADW ratio and leaf area (LA). Analysis of variance showed that the variation of measured traits among studied lines was significantly explained by the effects of species, line, treatment, and their interactions. Treatment and species had the largest effects. We also analyzed the broad-sense heritability of the drought response index (DRI), defined as the ratio between the observed values with and without water deficit treatment. DRI of most measured traits had high broad-sense heritability (H^2). The length of stems (LS) was the trait most affected by drought stress, while RDW/ADW ratio was not affected by drought. Generally, few significant differences were observed between DRI values of measured traits between lines within species as well as between lines of *M. laciniata* and *M. truncatula* collected from sympatric populations. Most of the correlations established between DRI values were positive. The environmental factors that most influenced variation of DRI values among populations were available phosphorus (P) and mean annual rainfall.

Key words: *Medicago laciniata*, *Medicago truncatula*, sympatric populations, lines, drought response index, environmental factors.

INTRODUCTION

Abiotic stresses such as drought, high salinity and extreme temperatures, together, represent the primary cause of crop loss worldwide, reducing average yields for most major crop plants by more than 50%. In contrast, the estimated yield loss caused by pathogens is typically about 10 - 20% (Kreps et al., 2002). Water deficit is a common environmental stress experienced by plants. It affects both development and growth, and has a negative effect on productivity (Altinkut et al., 2003; Slama et al., 2006; Verslues et al., 2006; Tuberosa and Salvi, 2006). Approximately one third of the world's arable land suffers from chronically inadequate supplies of water for agriculture, and in virtually all agricultural regions, yields of rain-fed crops are periodically reduced by drought (Boyer, 1982). Drought stress is common not only in arid and semi-arid regions, but also in places where total precipitation is high but is not evenly distributed over the growing season (Altinkut et al., 2003). Plant responses to drought occur through changes in their morphological, biochemical and metabolic processes (Li et al., 2004; Pérez-Pérez et al., 2009). Tolerance or susceptibility to drought stress is a very complex phenomenon, in part, because stress may occur at multiple stages of plant development and often more than one stress simultaneously affects the plant (Tuberosa and Salvi, 2006). Improving drought tolerance is probably one of the most difficult tasks for plant breeders. The difficulty comes

*Corresponding author. E-mail: mounawer_badri@yahoo.fr.

comes from the diversity and unpredictability of drought conditions in the field, and from the diversity of drought tolerance strategies developed by the plants that are targeted and subjected to selection criteria (Teulat et al., 2001). Water deficit is therefore the most important abiotic stress and strategies to sustainable use of water (Kang et al., 2003) and improve plant drought resistance are urgent and should integrate both conventional breeding and biotechnological approaches (Chaves et al., 2009). Plants have evolved different adaptive strategies to alleviate the adverse effects of these abiotic stresses (Verslues et al., 2006). Legumes are the third largest family of flowering plants and contain several major agricultural and economical species (Gepts et al., 2005). They have generated much interest in the plant scientific community, not only because of the important crop plants, but also because of their interactions with microbial symbionts (Nunes et al., 2008). In Tunisia, as in the world, the limiting factors for cultivation of legumes are their low adaptation to the diverse local eco-geographical factors and the slowness of genetic improvements due to their genomic complexity. Drought is one of the major factors that limit legumes crop productivity in Tunisia with 95% of her surface being located in semi-arid, arid and Saharan stages. Since the genus *Medicago* is endemic in Tunisia and represents an important proportion of the native flora in all bioclimatic stages (Seklani et al., 1996), most of agronomical traits of interest such as the tolerance to drought stress can be found within its annual species. *M. truncatula*, an annual forage plant and a close relative of alfalfa, was identified as being a suitable model legume because it is a diploid (2n = 16), self-pollinating species with a small genome and the resources to create transgenic plants (Cook, 1999). Genetic and genomic tools are rapidly evolving and the scope of work performed in *M. truncatula* is expanding and diversifying (Young and Udvardi, 2009). *M. truncatula* and *M. laciniata* have two different geographical distributions both globally and in Tunisia. The first one is omni-Mediterranean and ubiquitous in Tunisia. The second is the only species of the genus *Medicago* that is limited to southern Mediterranean regions (Heyn, 1963); it is restricted in Tunisia to the inferior semi-arid, arid and Saharan stages (Badri et al., 2007).

In the last two decades, molecular mechanisms of drought stress response and tolerance in plants have become an active area of investigation where many genes, which are regulated by drought stress, have been reported in a variety of plants (Yamaguchi-Shinozaki et al., 2002; Leung, 2008). Numerous studies have focused on the physiological and molecular mechanisms by which the model legume *M. truncatula* responds to abiotic stresses such as flooding stress (Limami et al., 2007), drought (González-Andres et al., 2007; Badri, 2008), salt stress (Merchan et al., 2007; Mhadhbi and Aouani, 2008), and cold stress (Avia and Lejeune-Hénaut, 2007). Nevertheless, few investigations have compared the responses of different *Medicago* annual species to water deficit by taking into account their geographical origins and the genetic variation within the studied species (Hamidi, 1991; Mefti et al., 2001; Turk, 2006).

Drought response index (DRI) is the ratio between the observed values under water-stressed and well-watered conditions; it is a measurement of change in plant traits caused by drought stress (Chen et al., 2007). There is a dearth of information on the use DRI as an indicator of drought tolerance in *M. laciniata* and *M. truncatula*. The DRI value was considered as the indicator for water deficit tolerance. The aims of this study were to (i) assess and compare the effects of drought stress on *M. laciniata* and *M. truncatula* lines sampled from sympatric populations; (ii) compare the effects of water deficit on the local and references lines of *M. truncatula*; and (iii) estimate the associations of water deficit responses in local lines of both species with site-of-origin environmental factors.

MATERIALS AND METHODS

Plant material and growth parameters

Eight lines from 4 sympatric populations of *M. laciniata* and *M. truncatula* and 4 references lines of *M. truncatula* were used. The sympatric populations from which studied lines were used are El Ghouilet, Jelma, Deguache and Medenine (Figure 1). The eco-geographical factors of sampling sites of these populations are summarized in the Table 1. Soil samples were collected from three locations at these sites, and analyzed by the Laboratory of Soils at the Ministry of Agriculture, Tunisia. To distinguish between natural populations of both species, we used two symbols; TN for *M. truncatula* and TNL for *M. laciniata*. The studied lines are TNL1.9 and TN1.21, TNL2.3 and TN2.12, TNL4.7 and TN4.20, and TNL9.2 and TN13.11 for *M. laciniata* and *M. truncatula*, respectively. Natural populations of both species from which studied lines have been chosen were characterized using quantitative traits and microsatellite markers (Badri et al., 2007). From each population, a single-seed per pod was used to initiate lines of progeny from self-pollination. Inbred lines were created by two generations of spontaneous self-pollination in isolation in the greenhouse, even if we know that maternal effects could persist after two generations of self-pollination. The offspring in each presumed line was considered genetically identical. Consequently, the within-line variance can be considered as environmental, while the among-lines variance component is assumed to be solely genetic (Arraouadi et al., 2009). We used four references lines of *M. truncatula*: Jemalong A17 (JA17) from an Australian collection, two Algerian lines DZA315.16 and DZA45.5, and the French line F83005.5 from the Var region. These lines are the parents of various crosses: JA17 x DZA315.16, JA17 x F83005.5, DZA315.16 x DZA45.5, DZA45.5 x DZA315.16, and DZA45.5 x F83005.5. A number of recombinant inbred lines (RILs) populations derived from these crosses were obtained and a set of framework genetic maps were developed and others are in progress (Huguet et al., 2005).

The present work was conducted in the Centre of Biotechnology of Borj Cedria (CBBC), Tunisia. Seeds were surface-sterilized and scarified by immersion in concentrated H_2SO_4 for 7 min and rinsed ten times with sterile distilled water. The soaked seeds were sown in Petri dishes on agar 0.9% medium before being vernalized at 4°C for 96 h in darkness. After germination, the seedlings were transferred to 400 ml plastic pots (9.3 cm diameter and 8.5 cm deep) filled with a mixture of compost and the soil of the CBBC at a

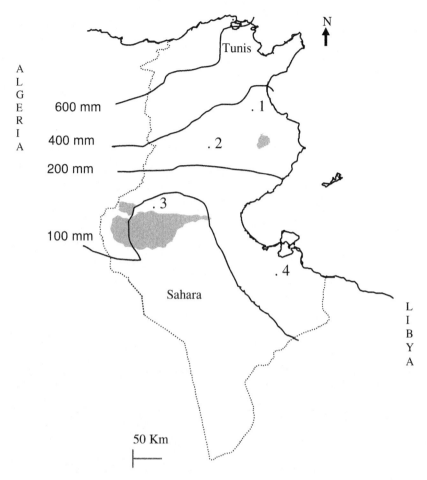

Figure 1. Map of Tunisia with the location of sympatric populations of *M. laciniata* and *M. truncatula* from which studied lines were collected. 1: El Ghouilet; 2: Jelma, 3: Deguache; and 4: Medenine.

Table 1. List of the four eco-geographical factors of sampling sites of natural populations of *M. laciniata* and *M. truncatula* from which studied lines were collected.

	El Ghouilet	Jelma	Deguache	Medenine
Organic matter (%)	1.6	1.8	0.9	0.7
Phosphorus (mg/kg)	26	18	22	18
Mean annual rainfall (mm)	350	250	50	175
Altitude (m)	30	300	25	119

ratio of 3:1 having 22.5% of organic matter, pH 8, available phosphorus (P) = 224 mg/kg, K_2O = 0.36 g/kg, total carbon = 13%, active lime = 4%, electrical conductivity (ECe) = 3 mmho/cm and saturation = 43 ml/100 g. All plants were grown in individual pots in a growth chamber at temperature of 25°C, 80% of relative humidity and a photoperiod of 16/8 h. The experimental design was completely randomized. The lines under study underwent two water treatments, well irrigated and kept under drought (33% of field capacity) as described by Badri et al. (2005). Five replicates for each line per treatment were used. Plants were harvested after a period of 45 days. At harvest, five quantitative traits of the aerial and root growth were measured including length of stems (LS), aerial dry weight (ADW), root dry weight (RDW), the RDW/ADW ratio and the sixth leaf area (LA). Plant organs were harvested and dried at 70°C for 48 h in order to determine the dry matter.

Statistical analyses

Data were subjected to an analysis of variance (ANOVA) using the Statistical Analysis System (SAS 7.02 Institute, Inc., 1998) and the means of measured traits were compared between lines with Duncan's multiple range tests. Drought response index (DRI) represents the relative change for each trait caused by drought. It was calculated using the following formula (Chen et al., 2007):

DRI = (Value from drought treatment/ Value from control)*100%

Table 2. Proportions and significance levels of line, treatment, line x treatment interaction, species and species x treatment interaction effects on measured traits for *M. laciniata* and *M. truncatula*.

Traits		*M. laciniata*					*M. truncatula*				*M. laciniata-M. truncatula*			
	Effects	df	F	%		Effects	df	F	%	Effects	df	F	%
LS	Line	3	10.25***	5.63			3	3.55*	1.40	Species	1	51.22***	16.10
	Treatment	1	163.86***	89.92			1	251.19***	98.60	Treatment	1	255.96***	80.45
	Line x Treatment	3	8.12***	4.46			3	0.03ns	0.01	Species x Treatment	1	10.97***	3.45
	Line	3	7.06***	4.19			3	10.60***	3.87	Species	1	131.66***	34.16
ADW	Treatment	1	155.24***	92.24			1	259.54***	94.85	Treatment	1	206.41***	53.55
	Line x Treatment	3	6.01**	3.57			3	3.50*	1.28	Species x Treatment	1	47.36***	12.29
	Line	3	11.28***	19.76			3	5.35**	6.35	Species	1	104.00***	53.26
RDW	Treatment	1	39.25***	68.74			1	75.25***	89.26	Treatment	1	62.35***	31.93
	Line x Treatment	3	6.57**	11.50			3	3.70*	4.39	Species x Treatment	1	28.93***	14.81
	Line	3	1.06ns	10.60			3	22.71***	33.63	Species	1	21.36***	52.10
RDW/ADW	Treatment	1	8.74**	87.57			1	36.80***	54.49	Treatment	1	19.56***	47.72
	Line x Treatment	3	0.18ns	1.83			3	8.03***	11.88	Species x Treatment	1	0.07ns	0.18
	Line	3	7.23***	10.38			3	148.10***	61.34	Species	1	78.79***	85.04
LA	Treatment	1	61.89***	88.89			1	80.41***	33.30	Treatment	1	10.98***	11.85
	Line x Treatment	3	0.51ns	0.73			3	12.93***	5.36	Species x Treatment	1	2.88ns	3.11

Significance levels; ns: not significant (p > 0.05), * p ≤ 0.05, ** p ≤ 0.01, *** p ≤ 0.001, F: Snedecor-Fisher coefficient. Length of stems (LS), aerial dry weight (ADW), root dry weight (RDW), the RDW/ADW ratio and leaf area (LA).

As described by Badri et al. (2007), total phenotypic variance (σ_T^2) was the sum of two components of variance (i) arising between lines genotypes (σ_g^2), and (ii) due to residual error arising between individuals within original genotype (σ_e^2). Broad-sense heritability (H^2) was estimated for each trait as the ratio of the variance arising between lines genotypes (σ_g^2) divided by the sum of (σ_g^2) and (σ_e^2) (Bonnin et al., 1997). Phenotypic correlations among measured traits were estimated by the SAS CORR procedure (SAS 7.02 Institute, Inc.). Correlations between DRI of measured traits for *M. laciniata* and *M. truncatula* lines and environmental factors were estimated by computing Pearson's correlation coefficient (r). Significance level was set to 0.05 and adjusted for multiple comparisons by Bonferroni correction (Badri et al., 2007).

RESULTS AND DISCUSSION

Analysis of variance showed that variation of measured traits among the studied lines was significantly explained by the effects of line, treatment and their interaction (Table 2). Treatment had the largest effect on trait values, with the exception of leaf area (LA), with effects ranging from 68.74 - 92.24% and from 54.49 - 98.60%, respectively, for *M. laciniata* and *M. truncatula*. Furthermore, analysis of variance of species and treatment effects showed that while the highest effect was observed for the treatment factor for length of stems (LS) and aerial dry weight (ADW), the highest level was found for the species effect for root dry weight (RDW), RDW/ADW ratio and leaf area (LA). Overall, species effect is largely higher than that of lines for all measured traits indicating that the variation between *M. laciniata* and *M. truncatula* is higher

Table 3. Means of DRI values and broad-sense heritability (H^2) of traits measured for studied lines of *M. laciniata* (TNL) and *M. truncatula* (TN).

	TNL1.9	TNL2.3	TNL4.7	TNL9.2	TNL1.21	TNL2.12	TN4.20	TN13.11	JA17	DZA315.16	F83005.5	DZA45.5	H^2 (TNL)	H^2 (TN)
LS	24.38abc	16.22bc	11.74c	12.57c	18.28bc	25.56abc	19.44bc	31.03ab	33.64ab	36.90a	17.61bc	24.42abc	0.81	0.60
ADW	32.74ab	24.49abc	17.83bc	13.46c	23.60abc	21.64abc	22.95abc	27.10abc	25.36abc	28.75abc	21.20bc	37.29a	0.69	0.60
RDW	123.60a	24.64cd	25.91cd	18.41d	27.88cd	24.11cd	43.78bc	36.25cd	38.52bcd	25.03cd	35.53cd	59.96b	0.96	0.82
RDW/ADW	445.52a	177.49b	155.06b	212.57b	120.74b	115.48b	202.02b	140.79b	149.46b	106.35b	171.01b	105.16b	0.72	0.86
LA	60.38cd	66.96bc	59.45cd	66.57bc	65.25bcd	65.54bcd	43.48e	103.11a	79.56b	63.75bcd	65.46bcd	50.27de	0.26	0.95

In a row, DRI means followed by the same letter(s) are not significantly different at $P=0.05$ based on Duncan's multiple-range test. Length of stems (LS), aerial dry weight (ADW), root dry weight (RDW), the RDW/ADW ratio and leaf area (LA).

than that within either species.

Means comparison of DRI values and broad-sense heritability

Generally, few significant differences were observed between DRI values of measured traits in *M. laciniata*. The lowest DRI values were for TNL4.7 and TN9.2 (Table 3). Furthermore, no significant difference in DRI was observed among *M. truncatula* lines, with the exception of leaf area (LA). TN13.11 from Medenine has the largest change in LA trait between treatments. Few significant differences were observed between lines of both species collected from sympatric populations. While the highest values for RDW and LA were respectively registered for TNL1.9 and TNL4.7 lines of *M. laciniata*, the highest levels for LS and LA were for TN13.11 of *M. truncatula*. In response to water stress, the plant faces physiological changes including loss of cell turgor, closing of stomata, reduction in cell enlargement, and reduced leaf surface area (Aslam and Tahir, 2003). Altinkut et al. (2001) reported that leaf size and relative water content (RWC) tolerance are associated with enhanced tolerance to water stress in barley and wheat. On the other hand, there was no significant difference for DRI values between local and references lines of *M. truncatula* except for DZA315.16 and TN13.11 showing the lowest and highest DRI values, respectively, for length of stems (LS) and LA traits. In contrast to the RDW/ADW ratio, for most of both species lines, the length of stems (LS) is the most seriously affected trait. The root/shoot ratio increase was due to greater inhibition of the shoot as compared to root growth (Gomes et al., 1996). Numerous studies reported that water deficit reduced the growth and productivity of plants and increased the root/shoot ratio (Slama et al., 2006). The response of *M. laciniata* and *M. truncatula* to water stress deficit would be partially related to the preferential allocation of dry matter towards roots. This parameter is considered as a criterion of adaptation to drought (Ben Naceur et al., 1999), although some findings showed no clear correlation between root trait and water extraction ability (Petrie and Hall, 1992). It has been suggested that abscisic acid is involved in the control of the root/top ratio, mainly when the plant is under water stress (Munns and Sharp, 1993). A rise in abscisic acid in drought-exposed seedlings of *Qualea grandiflora* has been reported by Sassaki et al. (1997). Understanding the physiological mechanisms that make some species tolerant and others sensitive is fundamental in identifying clearly the recognizable traits for use in breeding programs aimed at developing cultivars that would adapt to the environmental stress (Abdel-Nasser and Abdel-Aal, 2002). To our knowledge, this is the first report about the analysis of the effects of water deficit on pure lines coming from sympatric populations of two annual *Medicago* species. Previous studies comparing responses under water deficit of medics had not taken in consideration the variation within species as well as the site-of-origin eco-geographical factors of plant material (Hamidi, 1991). Broad-sense heritability (H^2) of DRI values of measured traits ranged from 0.26 - 0.96 and from 0.60 - 0.95 for *M. laciniata* and *M. truncatula*, respectively, indicating that these traits are genetically controlled.

Correlation among DRI values

Among the 20 possible correlations between DRI of measured traits for both species, six and five were respectively significant for *M. laciniata* and *M. truncatula* (Table 4). Five and three of these correlations are positive, respectively, for *M. laciniata* and *M. truncatula*. These positive correlations suggest that the measured traits vary in similar ways under well-watered and water-stressed conditions in both species. Negative

Table 4. Estimated correlations between DRI values of measured traits for *M. laciniata* (above) and *M. truncatula* (below).

	LS	ADW	RDW	RDW/ADW	LA
LS	1.00	0.78***	0.67**	0.03^{ns}	0.16^{ns}
ADW	0.75***	1.00	0.48*	-0.32^{ns}	0.50*
RDW	0.34*	0.69***	1.00	0.58*	-0.16^{ns}
RDW/ADW	-0.41**	-0.50**	0.12^{ns}	1.00	-0.58*
LA	0.29^{ns}	0.09^{ns}	-0.16^{ns}	-0.19^{ns}	1.00

Significance levels; ns: not significant (p > 0.05), *p ≤ 0.05; **p ≤ 0.01, ***p ≤ 0.001. Length of stems (LS), aerial dry weight (ADW), root dry weight (RDW), the RDW/ADW ratio and leaf area (LA).

Table 5. Estimated correlations between DRI values of measured traits for *M. laciniata* and *M. truncatula* and eco-geographical factors.

	M. laciniata				*M. truncatula*			
	O.M	P	An. rain	Altitude	O.M	P	An. rain	Altitude
LS	0.47	0.51	0.63	-0.13	-0.18	-0.43	-0.06	0.26
ADW	0.47	0.40	0.47	-0.04	-0.21	-0.06	-0.03	-0.09
RDW	0.39	0.82*	0.68*	-0.45	-0.60	0.01	-0.60	-0.44
RDW/ADW	0.21	0.49	0.48	-0.28	-0.56	0.07	-0.76*	-0.47
LA	0.03	-0.23	0.03	0.23	-0.30	-0.44	0.22	0.20

*Significant after using Bonferroni correction at α = (0.05/20=0.0025). Organic matter (O.M), available phosphorus (P), and mean annual rainfall (An rain). Length of stems (LS), aerial dry weight (ADW), root dry weight (RDW), the RDW/ADW ratio and leaf area (LA).

correlation between DRI of RDW/ADW ratio and leaf area (LA) for *M. laciniata*, indicates that plants showing the highest reduction percentage for LA were the most tolerant ones. Nevertheless, no consistent pattern of association was found for *M. truncatula* between RDW/ADW ratio and LA.

Associations of DRI values with environmental factors

Among the 40 possible correlations between DRI of measured traits and environmental factors, only 2 and 1 out of them were significant, respectively for *M. laciniata* and *M. truncatula* (Table 5). Root dry weight (RDW) is positively correlated with assimilated P_2O_5 and mean annual rainfall in *M. laciniata*. A negative correlation was detected between RDW/ADW ratio and mean annual rainfall for *M. truncatula* lines. Correlations between DRI of measured traits and environmental factors suggest that these particular characters have adapted in response to the regional differences in eco-geographical factors. This is similar to Kloss and McBride (2002), who reported significant correlations between variables of growth measured for California's blue oak (*Quercus douglasii*), cultivated in greenhouse under three different water regimes, and mean annual rainfall and altitude factors.

DRI of most measured traits had high broad heritability indicating that these characters can be used as good descriptors in the genetic analysis of tolerance to drought in both species. Most of the correlations between DRI of the measured traits were positive, suggesting that behaviors of lines in drought follow the same general pattern. Correlations between DRI of scored traits and environmental factors suggest that the response to drought is partially dependent upon local adaptation of these lines in their original sites.

Conclusions

The significant variation in drought tolerance – as measured by the DRI - found between the studied lines from *M. laciniata* and *M. truncatula* suggests that a rich genetic resource for drought tolerance exists in different genotypes, and that it is feasible to select water deficit tolerant lines in both species. Furthermore, breeding schemes should be based upon heritability of traits. It will be relatively easier to select lines for these traits and pass them onto the offspring. The correlation analyses performed in this study indicate that different drought tolerance traits are inter-correlated; it is thus more appropriate to evaluate a breeding strategy comprehensively based on multiple traits.

Further study is needed to analyze the water deficit responses in sympatric populations of *M. laciniata* and *M. truncatula*. Numerous traits related to water deficit tolerance should be taken into account including leaves, stems and roots ions content of Na^+, Cl^- and K^+, accumulation of proline, chlorophyll quantity and relative water content (RWC). In spite of the fact that a large number of drought-induced genes have been identified in a wide range of plant species (Verslues et al., 2006; Shinozaki and Yamaguchi-Shinozaki, 2007), a molecular basis for plant tolerance to water stress still remains far from being completely understood. Thus, lines from *M. laciniata* and *M. truncatula* with opposite behaviors under water deficit will be inter-crossed to obtain recombinant inbred lines (RILs) populations. The long term purpose of this study is to identify quantitative trait loci (QTLs) and/or genes of the tolerance to drought stress. Indeed, Identifying genomic regions (QTLs) that contribute to drought resistance will help to develop *M. truncatula* lines suitable for water-limiting environments through marker-assisted breeding. Detailed characterization of these genomic regions through the development and evaluation of near-isogenic lines will lead to an improved understanding of drought tolerance and might set the stage for the positional cloning of drought tolerance genes.

ACKNOWLEDGMENTS

We thank Houcine Ilahi (IPEIB, Tunisia), and an anonymous reviewer for their helpful comments on the manuscript. We are also grateful to Adel Zitoun for technical assistance. This research work was supported in part by a Tunisian-French collaborative project (PICS 712).

REFERENCES

Abdel-Nasser LE, Abdel-Aal AE (2002). Effect of elevated CO_2 and drought on proline metabolism and growth of safflower (*Carthamus mareoticus* L.) seedlings without improving water status. Pakistan J. Biol. Sci., 5(5): 523-528.

Altinkut A, Kazan K, Gozukirmizi N (2003). AFLP marker linked to water-stress-tolerant bulks in barely (*Hordeum vulgare* L.). Genet. Mol. Biol., 26(1): 77-82.

Altinkut A, Kazan K, Ipekei Z, Gozukirmizi N (2001). Tolerance to paraquat is correlated with the traits associated water stress tolerance in segregating F_2 populations of barley and wheat. Euphytica, 121: 81-86.

Arraouadi S, Badri M, Abdul Jaleel C, Djébali N, Ilahi H, Huguet T, Aouani ME (2009). Analysis of genetic variation in natural populations of *Medicago truncatula* of southern Tunisian ecological areas, using morphological traits and SSR markers. Tropical Plant Biol., 2: 122-132

Aslam M, Nadeem TMH (2003). Morpho-physiological response of maize inbred lines under drought environment. Asian J. Plant Sci., 2(13): 952-954.

Avia K, Lejeune-Hénaut I (2007). Acclimation of *Medicago truncatula* to cold stress. In *Medicago truncatula* handbook, pp. 17-22.

Badri M, Arraouadi S, Huguet T, Aouani ME (2005). Genetic variation of the model legume *Medicago truncatula* (Fabaceae) lines under drought and salt stress. Proceedings of the 6th Tunisian-Japanese seminar on culture, science and technology, Sousse, Tunisia. pp. 18-21.

Badri M, Ilahi H, Huguet T, Aouani ME (2007). Quantitative and molecular genetic variation in sympatric populations of *Medicago laciniata* and *M. truncatula* (Fabaceae): relationships with eco-geographical factors. Genet. Res., 89: 107-122.

Badri M (2008). Analyse comparative de la variabilité génétique chez *Medicago laciniata* et *M. truncatula* et identification des QTLs liés au stress hydrique. PhD Thesis, Faculty of Sciences of Tunis, Tunisia. pp. 1-172.

Ben NM, Naily M, Selmi M (1999). Effet d'un déficit hydrique survenant à différents stades de développement du blé sur l'humidité du sol, la physiologie de la plante est sur la composante du rendement. Méditerranée, 2: 53-60.

Bonnin I, Prosperi JM, Olivieri I (1997). Comparison of quantitative genetic parameters between two natural populations of a selfing plant species, *Medicago truncatula* Gaertn. Theor. App. Genet. 94: 641-651.

Boyer JS (1982). Plant productivity and environment. Science 218: 443-448.

Chaves MM, Flexas J, Pinheiro C (2009). Photosynthesis under drought and salt stress: regulation mechanisms from whole plant to cell. Ann. Bot., 103: 551-560.

Chen C, Tao C, Peng H, Ding Y (2007). Genetic analysis of salt stress responses in Asparagus Bean (*Vigna unguiculata* (L) ssp Sesquipedalis Verdc). J. Hered., 98(7): 655-665.

Cook D (1999). *Medicago truncatula*: a model in the making! Curr. Opin. Plant Bio., 2: 301-304.

Gepts P, Beavis WD, Brummer EC, Shoemaker RC, Stalker HT, Weeden NF, Young ND (2005). Legumes as a model plant family. Genomics for food and feed report of the cross-legume advances through genomics conference. Plant Physiol., 137: 1228-1235.

Gomes FE, Enéas FJ, Prisco JT (1996). Effects of osmotic stress on growth and ribonuclease activity in *Vigna unguiculata* (L.) Walp. seedlings differing in stress tolerance. R. Bras. Fisiol. Veg., 8(1): 51-57.

Gonzalez-Andres F, Casquero PA, San-Pedro C, Hernandez-Sanchez E (2007). Diversity in white lupin (*Lupinus albus* L.) landraces from northwest Iberian plateau. Genet. Resour. Crop Ev., 54: 27-44.

Hamidi A (1991). Production de matière sèche, transpiration et efficience de l'utilisation de l'eau chez onze espèces de luzernes annuelles. Engineer's Thesis, INA, Algeria. pp. 1-88.

Heyn CC (1963). The annual species of *Medicago*. Scr. Hierosolymitana, 12: 1-154.

Huguet T, Gherardi M, Prosperi JM, Chardon F (2005). Construction of framework genetic maps based on a set of *Medicago truncatula* recombinant inbred lines (RILs): a powerful tool for mapping genes and detection of quantitative trait loci (QTLs). Model legume congress, Asilomar conference grounds, Pacific Grove, California, p. 149.

Kang S, Davies WJ, Shan L, Cai H (2003). Water-saving agriculture and sustainable use of water and land resources, vols. 1 and 2. Shaanxi Science and Technology Press.

Kloss S, McBride JR (2002). Geographic patterns of variation in biomass production of California blue oak seedlings as a response to water availability. USDA Forest Service Gen. Tech. Rep. PSW-GTR-184.

Kreps JA, Wu Y, Chang HS, Zhu T, Wang X, Harper JF (2002). Transcriptome changes for *Arabidopsis* in response to salt, osmotic, and cold stress. Plant Physiol., 130: 2129-2141.

Leung H, (2008). Stressed genomics - bringing relief to rice fields. Curr. Opin. Plant Biol., 11: 201-208.

Li Z, Zhao L, Kai G, Yu S, Cao Y, Pang Y, Sun X, Tang K (2004). Cloning and expression analysis of a water stress-induced gene from *Brassica oleracea*. Plant Physiol. Biochem., 42: 789-794.

Limami AM, Ricoult C, Planchet E (2007). Response of *Medicago truncatula* to flooding stress. *Medicago truncatula* handbook, pp. 3-6.

Mefti M, Abdelguerfi A, Chebouti A (2001). Etude de la tolérance à la sécheresse chez quelques populations de *Medicago truncatula*

(L.) Gaertn. Cah. Options Méditerr., 45(14): 173-176.

Merchan F, de Lorenzo L, Gonzalez RS, Niebel A, Manyani H, Frugier F, Sousa C, Crespi M (2007). Identification of regulatory pathways involved in the reacquisition of root growth after salt stress in *Medicago truncatula*. Plant J., 51: 1-17.

Mhadhbi H, Aouani ME (2008). Growth and nitrogen-fixing performances of *Medicago truncatula-Sinorhizobium meliloti* symbioses under salt (NaCl) stress: Micro- and macrosymbiont contribution into symbiosis tolerance. Biosaline Agriculture and High Salinity Tolerance, Abdelly C, Öztürk M, Ashraf M, Grignon C (eds.), Birkhäuser Verlag, Switzerland, pp. 91-98.

Munns R, Sharp RE (1993). Involvement of abscisic acid in controlling plant growth in soils of low water potential. Aust. J. Plant Physiol., 20: 425-437.

Nunes CM, Araujo SS, Silva JM, Fevereiro PS, Silva AB (2008). Physiological responses of the legume model *Medicago truncatula* cv. Jemalong to water deficit. Environ. Exp. Bot., 63: 289-296.

Pérez-Pérez JG, Robles JM, Tovar JC, Botia P (2009). Response to drought and salt stress of lemon 'Fino 49' under field conditions: Water relations, osmotic adjustment and gas exchange. Sci. Hortic., 122: 83-90.

Petrie CL, Hall AE (1992). Water relations in cowpea and pearl millet under water deficits. II. Water use and root distribution. Aust. J. Plant Physiol., 19: 591-600.

SAS Institute (1998). SAS/STAT User' Guide. Version 7. SAS Institute Inc., Cary, NC, USA.

Sassaki RM, Machado EC, Lagôa AM, Felippe GM (1997). Effect of water deficiency on photosynthesis of *Dalbergia miscolobium* Benth., a Cerrado tree species. R. Bras. Fisiol. Veg., 9(2): 83-87.

Seklani H, Zoghlami A, Mezni M, Hassen H (1996). Synthèse des travaux de recherche réalisés sur les *Medicago* à l'Institut National de la Recherche Agronomique de Tunisie. Cah. Options Méditerr., 18: 31-37.

Shinozaki K, Yamaguchi-Shinozaki K (2007). Gene networks involved in drought stress response and tolerance. J. Exp. Bot., 58(2): 221-227.

Slama I, Messedi D, Ghnaya T, Savouré A, Abdelly C (2006). Effects of water deficit on growth and proline metabolism in *Sesuvium portulacastrum*. Environ. Exp. Bot., 56: 231-238.

Teulat B, Souyris I, This D (2001). QTLs for agronomic traits from Mediterranean barely progeny grown in several environments. Theor. Appl. Genet., 103: 774-787.

Tuberosa R, Salvi S (2006). Genomics-based approaches to improve drought tolerance of crops. Trends Plant Sci., 11(8): 405-412.

Turk MA (2006). The effect of drought stress on germination and emergence development of annual medics. EUCARPIA, fodder crops and amenity grasses section *Medicago* spp. group, Perugia, 3-7 September, Faculty of Agriculture.

Verslues PE, Agarwal M, Katiyar-Agarwal S, Zhu J, Zhu JK (2006). Methods and concepts in quantifying resistance to drought, salt and freezing, abiotic stresses that affect plant water status. Plant J., 45: 523-539.

Yamaguchi-Shinozaki K, Kasuga M, Liu Q, Nakashima K, Sakuma Y, Abe H, Shinwari ZK, Seki M, Shinozaki K (2002). Biological mechanisms of drought stress response. JIRCAS Working Report 1-8.

Young ND, Udvardi M (2009). Translating *Medicago truncatula* genomics to crop legumes. Curr. Opin. Plant Biol., 12: 193-201.

Permissions

All chapters in this book were first published in JPBCS, by Academic Journals; hereby published with permission under the Creative Commons Attribution License or equivalent. Every chapter published in this book has been scrutinized by our experts. Their significance has been extensively debated. The topics covered herein carry significant findings which will fuel the growth of the discipline. They may even be implemented as practical applications or may be referred to as a beginning point for another development.

The contributors of this book come from diverse backgrounds, making this book a truly international effort. This book will bring forth new frontiers with its revolutionizing research information and detailed analysis of the nascent developments around the world.

We would like to thank all the contributing authors for lending their expertise to make the book truly unique. They have played a crucial role in the development of this book. Without their invaluable contributions this book wouldn't have been possible. They have made vital efforts to compile up to date information on the varied aspects of this subject to make this book a valuable addition to the collection of many professionals and students.

This book was conceptualized with the vision of imparting up-to-date information and advanced data in this field. To ensure the same, a matchless editorial board was set up. Every individual on the board went through rigorous rounds of assessment to prove their worth. After which they invested a large part of their time researching and compiling the most relevant data for our readers.

The editorial board has been involved in producing this book since its inception. They have spent rigorous hours researching and exploring the diverse topics which have resulted in the successful publishing of this book. They have passed on their knowledge of decades through this book. To expedite this challenging task, the publisher supported the team at every step. A small team of assistant editors was also appointed to further simplify the editing procedure and attain best results for the readers.

Apart from the editorial board, the designing team has also invested a significant amount of their time in understanding the subject and creating the most relevant covers. They scrutinized every image to scout for the most suitable representation of the subject and create an appropriate cover for the book.

The publishing team has been an ardent support to the editorial, designing and production team. Their endless efforts to recruit the best for this project, has resulted in the accomplishment of this book. They are a veteran in the field of academics and their pool of knowledge is as vast as their experience in printing. Their expertise and guidance has proved useful at every step. Their uncompromising quality standards have made this book an exceptional effort. Their encouragement from time to time has been an inspiration for everyone.

The publisher and the editorial board hope that this book will prove to be a valuable piece of knowledge for researchers, students, practitioners and scholars across the globe.

List of Contributors

Owino Charles Onyango
Kenya Plant Health Inspectorate Service. P.O Box 249, Kitale, Kenya

Maarouf I. Mohammed
Shambat Research Station, ARC, P.O. Box 30, Khartoum North

Pitipong Thobunluepop
Faculty of Technology, Department of Agricultural Technology, Mahasarakham University, Talard Sub-District, Muang District, Maha Sarakham 44000, Thailand, 44000

V. V. Singh
Directorate of Rapeseed-Mustard Research, Bharatpur, Rajasthan 321 303, India

Vandana Verma
Directorate of Rapeseed-Mustard Research, Bharatpur, Rajasthan 321 303, India

Aniruddh K. Pareek
Directorate of Rapeseed-Mustard Research, Bharatpur, Rajasthan 321 303, India

Monika Mathur
Directorate of Rapeseed-Mustard Research, Bharatpur, Rajasthan 321 303, India

Rajbir Yadav
Directorate of Rapeseed-Mustard Research, Bharatpur, Rajasthan 321 303, India

Poonam Goyal
Directorate of Rapeseed-Mustard Research, Bharatpur, Rajasthan 321 303, India

Ajay Kumar Thakur
Directorate of Rapeseed-Mustard Research, Bharatpur, Rajasthan 321 303, India

Y. P. Singh
Directorate of Rapeseed-Mustard Research, Bharatpur, Rajasthan 321 303, India

K. R. Koundal
NRCPB, IARI, New Delhi, India

K. C. Bansal
NRCPB, IARI, New Delhi, India

A. K. Mishra
Directorate of Rapeseed-Mustard Research, Bharatpur, Rajasthan 321 303, India

Arvind Kumar
Directorate of Rapeseed-Mustard Research, Bharatpur, Rajasthan 321 303, India

Sandeep Kumar
Directorate of Rapeseed-Mustard Research, Bharatpur, Rajasthan 321 303, India

R. E. Oliver
USDA-ARS, Small Grains and Potato Germplasm Research Unit, Aberdeen, ID 83210, USA

C. Yang
Hebei Institute of Food and Oil Crops, Shijiazhuang, Hebei 050031, People's Republic of China

G. Hu
USDA-ARS, Small Grains and Potato Germplasm Research Unit, Aberdeen, ID 83210, USA

V. Raboy
USDA-ARS, Small Grains and Potato Germplasm Research Unit, Aberdeen, ID 83210, USA

M. Zhang
Hebei Institute of Food and Oil Crops, Shijiazhuang, Hebei 050031, People's Republic of China

George Ouma
Department of Botany and Horticulture, Maseno University, P. O. BOX 333, Maseno, Kenya

R. H. Luo
Crop Genetic Improvement and Biotechnology Laboratory, Guangxi Academy of Agricultural Sciences, Nanning, P.R. China 530007

V. A. Dalvi
Crop Genetic Improvement and Biotechnology Laboratory, Guangxi Academy of Agricultural Sciences, Nanning, P.R. China 530007

Y. R. Li
Crop Genetic Improvement and Biotechnology Laboratory, Guangxi Academy of Agricultural Sciences, Nanning, P.R. China 530007

K. B. Saxena
International Crops Research Institute for the Semi-Arid Tropics (ICRISAT), Patancheru, A.P. India

List of Contributors

Guohua Mi
The Key Lab of Plant Nutrition, MOA, College of Resources and Environmental Sciences. China Agricultural University, Beijing, 100193, China

Fanjun Chen
The Key Lab of Plant Nutrition, MOA, College of Resources and Environmental Sciences. China Agricultural University, Beijing, 100193, China

Fusuo Zhang
The Key Lab of Plant Nutrition, MOA, College of Resources and Environmental Sciences. China Agricultural University, Beijing, 100193, China

Ephraim Nuwamanya
Department of Crop Science, Makerere University Kampala, P. O. Box 7062, Kampala, Uganda

Yona Baguma
National Crops Resources Research Institute (NaCRRI), P. O. Box 7084, Kampala, Uganda

Naushad Emmambux
Department of Food Science, Faculty of Agriculture, University of Pretoria, South Africa

John Taylor
Department of Food Science, Faculty of Agriculture, University of Pretoria, South Africa

Rubaihayo Patrick
Department of Crop Science, Makerere University Kampala, P. O. Box 7062, Kampala, Uganda

G. Olaoye
Department of Agronomy, University of Ilorin, Ilorin, Nigeria

O. B. Bello
Department of Agronomy, University of Ilorin, Ilorin, Nigeria

J. L. Okuto
Department of Botany, Maseno University, P. O. Box 333, Maseno, Kenya

G. Ouma
Department of Botany, Maseno University, P. O. Box 333, Maseno, Kenya

F. Ardila
Ex Aequo asbl, Rue Locquenghien, 41 B-1000 Brussels, Belgium
Instituto de Genética CICV y A, CNIA, INTA. C.C. 25 (1712) Castelar, Buenos Aires, Argentina

M. M. Echeverría
Ex Aequo asbl, Rue Locquenghien, 41 B-1000 Brussels, Belgium
Unidad Integrada: Facultad de Ciencias Agrarias, Universidad Nacional de Mar del Plata - Estación Experimental INTA Balcarce. C.C. 276 (7620) Balcarce, Buenos Aires, Argentina

R. Rios
Instituto de Genética CICV y A, CNIA, INTA. C.C. 25 (1712) Castelar, Buenos Aires, Argentina

R. H. Rodríguez
Unidad Integrada: Facultad de Ciencias Agrarias, Universidad Nacional de Mar del Plata - Estación Experimental INTA Balcarce. C.C. 276 (7620) Balcarce, Buenos Aires, Argentina

Christopher O. Akinbile
Department of Agricultural Engineering, Federal University of Technology, Akure, Nigeria

S. Sarduie-Nasab
Graduate student of Agronomy and Plant Breeding Department, Shahid Bahonar University of Kerman, Kerman, Iran

G. R Sharifi-Sirchi
Horticulture and Dates Research Institute, College of Agriculture, Shahid Bahonar University of Kerman, Kerman, Iran

M. H. Torabi-Sirchi
Department of Agriculture Technology, Faculty of Agriculture, University Putra Malaysia, 43400 Serdang, Selangor Darul Ehsan, Malaysia

O. M. Agbogidi
Faculty of Agriculture, Delta State University, Asaba Campus, Asaba, Nigeria

G. Olaoye
Department of Agronomy, University of Ilorin, P.M.B. 1515, Ilorin, Nigeria

O. B. Bello
Department of Agronomy, University of Ilorin, P.M.B. 1515, Ilorin, Nigeria

A. K. Ajani
Department of Agronomy, University of Ilorin, P.M.B. 1515, Ilorin, Nigeria

T. K. Ademuwagun
Department of Agronomy, University of Ilorin, P.M.B. 1515, Ilorin, Nigeria

Vandana Vinayak
CCS Haryana Agricultural University, Regional Research Station, Uchani, Karnal

A. K. Dhawan
CCS Haryana Agricultural University, Regional Research Station, Uchani, Karnal

V. K. Gupta
CCS Haryana Agricultural University, Regional Research Station, Uchani, Karnal

O. T. Adeniji
AVRDC – The World Vegetable Center, Regional Center for Africa, PMB 25, Duluti, Arusha, Tanzania

I. Swai
AVRDC – The World Vegetable Center, Regional Center for Africa, PMB 25, Duluti, Arusha, Tanzania

M. O. Oluoch
AVRDC – The World Vegetable Center, Regional Center for Africa, PMB 25, Duluti, Arusha, Tanzania

R. Tanyongana
AVRDC – The World Vegetable Center, Regional Center for Africa, PMB 25, Duluti, Arusha, Tanzania

A. Aloyce
Horticultural Research and Training Institute, Tengeru, Arusha, Tanzania

Nadeem Akbar
Department of Agronomy, University of Agriculture, Faisalabad, Pakistan

Asif Iqbal
Department of Agronomy, University of Agriculture, Faisalabad, Pakistan

Haroon Zaman Khan
Department of Agronomy, University of Agriculture, Faisalabad, Pakistan

Muhammad Kashif Hanif
Department of Agronomy, University of Agriculture, Faisalabad, Pakistan

Muhammad Usman Bashir
Department of Agronomy, University of Agriculture, Faisalabad, Pakistan

Joseph Kamau
Kenya Agricultural Research Institute- Katumani, Nairobi, Kenya

Rob Melis
University of KwazuluNatal – Petermaritzburg, South Africa

Mark Laing
University of KwazuluNatal – Petermaritzburg, South Africa

John Derera
University of KwazuluNatal – Petermaritzburg, South Africa

Paul Shanahan
University of KwazuluNatal – Petermaritzburg, South Africa

Eliud Ngugi
University of Nairobi, Kabete Campus, Nairobi, Kenya

M. Sié
Africa Rice Center (WARDA), 01 B.P. 2031, Cotonou, Benin

S. A. Ogunbayo
Africa Rice Center (WARDA), 01 B.P. 2031, Cotonou, Benin

D. Dakouo
Institut de l'Environnement et de Recherches Agricoles (INERA), Programme Riz et Riziculture, Centre Régional de Recherches Environnementales et Agricoles de l'Ouest, BP 910 Bobo-Dioulasso, Burkina Faso

I. Sanou
Institut de l'Environnement et de Recherches Agricoles (INERA), Programme Riz et Riziculture, Centre Régional de Recherches Environnementales et Agricoles de l'Ouest, BP 910 Bobo-Dioulasso, Burkina Faso

Y. Dembélé
Institut de l'Environnement et de Recherches Agricoles (INERA), Programme Riz et Riziculture, Centre Régional de Recherches Environnementales et Agricoles de l'Ouest, BP 910 Bobo-Dioulasso, Burkina Faso

B. N'dri
Africa Rice Center (WARDA), 01 B.P. 2031, Cotonou, Benin

K. N. Dramé
Africa Rice Center (WARDA), 01 B.P. 2031, Cotonou, Benin

K. A. Sanni
Africa Rice Center (WARDA), 01 B.P. 2031, Cotonou, Benin

B. Toulou
Africa Rice Center (WARDA), 01 B.P. 2031, Cotonou, Benin

List of Contributors

R. K. Glele
Universite d'Abomey Calavi, 01 BP 526, Cotonou, Benin

Bishun D. Prasad
Department of Agricultural Biotechnology, Assam Agricultural University, Jorhat, Assam, India
Depatment of Biology, the University of Western Ontario, ON, Canada N6A 5B7, Canada

Ganesh Thapa
Department of Biochemistry, Assam Agricultural University, Jorhat, Assam, India
Department of Biotechnology, Indian Institute of Technology Guwahati, Assam, India

Samindra Baishya
Department of Biochemistry, Assam Agricultural University, Jorhat, Assam, India

Sangita Sahni
Depatment of Biology, the University of Western Ontario, ON, Canada N6A 5B7, Canada

K. Mbuya
University of Kinshasa, B.P 117 Kinshasa 11, RD–Congo
Institut National pour l' Etude et la Recherche Agronomiques (INERA), B.P. 2037, Kinshasa 1, RD-, Congo

K. K. Nkongolo
Department of Biological Sciences, Laurentian University, Sudbury, Ontario, Canada

A. Kalonji-Mbuyi
University of Kinshasa, B.P 117 Kinshasa 11, RD-Congo
Nuclear Research Center, Kinshasa, RD-Congo

R. Kizungu
University of Kinshasa, B.P 117 Kinshasa 11, RD–Congo
Institut National pour l' Etude et la Recherche Agronomiques (INERA), B.P. 2037, Kinshasa 1, RD-, Congo

Dandena Gelmesa
Haramaya University P.O. Box 138, Dire Dawa, Ethiopia

Bekele Abebie
Adama University, P.O. Box., 1888, Adama Ethiopia

Lemma Desalegn
Ethiopian Institute of Agricultural Research-Melkassa Agricultural Research Center, P. O. Box 436 Nazreth Ethiopia

Mudasir Hafiz Khan
Department of Agriculture, Srinagar (J&K), India

Sunil Dutt Tyagi
Department of Plant Breeding and Genetics, Kisan (P. G.), College, Simbhaoli, Ghaziabad, (U. P.), India

Gopal Kumar Niroula Chhetry
Manipur University, Department of Life Sciences, Imphal 795003, India

Lassaad Belbahri
Agronomy Department, University of Applied Sciences Western Switzerland technology, architecture and landscape, 150 Route de Presinge, 1254 Jussy, Switzerland

Mounawer Badri
Laboratory of Legumes, Centre of Biotechnology of Borj Cedria, B.P. 901, 2050 Hammam-Lif, Tunisia

Soumaya Arraouadi
Laboratory of Legumes, Centre of Biotechnology of Borj Cedria, B.P. 901, 2050 Hammam-Lif, Tunisia

Thierry Huguet
Laboratoire de Symbiose et Pathologie des Plantes, INP-ENSAT, B.P. 107, 31326 Castanet Tolosan Cedex, France

Mohamed Elarbi Aouani
NEPAD/North Africa Biosciences Network, National Research Centre, El Buhouth St, Cairo, 12311, Egypt

CPSIA information can be obtained
at www.ICGtesting.com
Printed in the USA
BVOW10*1640161016
465179BV00005B/66/P